THE DOMESDAY GEOGRAPHY OF
SOUTH-WEST ENGLAND

TERRA WILLELMI DE OV. INSVMERSETA.

Willm̃ hr̃ .i. manẽ q̃ uocat̃ Watelega. qua̅ tenuit alestan' die qua rex
e .f. ĩi. &̃ .ŏ. Et reddit Gildũ p .i. hida. hanc por arare .i. carr̃. Ibi hr̃
Willm̃ .ij. uill' 7 vi. quadrag̃ nemorĩs i long̃. 7 uij. i lat̃. &̃ ual &̃ p anñ .x. sol.
&̃ quando Willm̃ recepit ualebat tantu̅de.

Witt hr̃ .i. manẽ q̃ uocat̃ hautona q̃ ten̅ Alestan̅ die q̃ rex E.f.
ii. &̃ .ŏ. &̃ reddt gildũ p .xiii. bidis. has poss arare .xii. carr̃. Inde hr̃
W .v. hidas. &̃ .iiii. carr̃ i dnĩo. &̃ uill .vii. hidas. &̃ .x. carr̃. Ibi
hr̃ W. xvi. uill' &̃ xxiiii. bord̃ &̃ v. seruos. &̃ xxxvi. aialia. &̃
xl iiii. porc̃ &̃ cc. oues. x. min̅. &̃ ii. mol q̃ reddũ. vii. sol. &̃ vi.
den̅. &̃ i. leuga̅ nemoris i loug̃. &̃ dim i lat̃. &̃ lx ag̃ p̃ti.
&̃ ual &̃ xv. lib. &̃ q̃ñ. w. recep ualebat .x ii. lib.

Willm̃ hr̃ .i. manẽ que uocat̃ Geuelcona. q̃ tenuit alestan de bosco ma.
de quarex E f. ii. 7 .ŏ. Et reddit Gildũ p. viii. hidis. has posse arare viii.
carr̃. q̃ tene Radulfus bloutiet. de. W. De his hr̃ Radulf̃ .iiii. hid. in dnĩo.
7 .iii. carr̃. 7 uill hñt .iiii. hid 7 v. carr̃. Ibi hr̃ .R. vi. uill' 7 .iiii. bord̃.
7 i de seruos. 7 ii. runand. 7 .ii. eq̃s in do mi tal. 7 x ii. annual. 7 xvi. porc̃
7 .C. oues. 7 ii. molend̃ños q̃ reddũ pannũ .xx. sol. 7 lxxx. p̃ti. 7
xt. agros pascue. &̃ ual &̃ pannũ .viii. libras. 7 q̃do Willm̃ recepit ualebat
tantu̅de huic manfioni fu addite .ii. hidg̃ t̃re quas tenuere .v. tagni par t̃.

·H·

THE
DOMESDAY GEOGRAPHY
OF
SOUTH-WEST ENGLAND

EDITED BY

H.C.DARBY

Emeritus Professor of Geography in the University of Cambridge

AND

R.WELLDON FINN

CAMBRIDGE UNIVERSITY PRESS

CAMBRIDGE

LONDON · NEW YORK · MELBOURNE

CAMBRIDGE UNIVERSITY PRESS
Cambridge, New York, Melbourne, Madrid, Cape Town, Singapore, São Paulo, Delhi

Cambridge University Press
The Edinburgh Building, Cambridge CB2 8RU, UK

Published in the United States of America by Cambridge University Press, New York

www.cambridge.org
Information on this title: www.cambridge.org/9780521118033

First published 1967
Reprinted with corrections 1979
This digitally printed version 2009

A catalogue record for this publication is available from the British Library

Library of Congress Catalogue Card Number: 67–11519

ISBN 978-0-521-04771-5 hardback
ISBN 978-0-521-11803-3 paperback

A century hence the student's materials will not be in the shape in which he finds them now. In the first place, the substance of Domesday Book will have been rearranged. Those villages and hundreds which the Norman clerks tore into shreds will have been reconstituted and pictured in maps, for many men from over all England will have come within King William's spell, will have bowed themselves to him and become that man's men.

From the concluding paragraph of F. W. MAITLAND'S
Domesday Book and Beyond (Cambridge, 1897)

CONTENTS

PREFACE

This book is the fifth of a number covering the whole of Domesday England, and it is built upon the same plan as the *The Domesday Geography of Eastern England*, first published in 1952. The greater part of the preface to that volume is equally relevant to this one, and its argument must be repeated here. The Domesday Book has long been regarded as a unique source of information about legal and economic matters, but its bearing upon the reconstruction of the geography of England during the early Middle Ages has remained comparatively neglected. The extraction of this geographical information is not always as simple as it might appear to be from a casual inspection of the Domesday folios. Not only are there general problems of interpretation, but almost every county has its own peculiarities. There is, moreover, the sheer difficulty of handling the vast mass of material, and of getting a general view of the whole. The original survey was made in terms of manors, villages and hundreds, but the clerks reassembled the information under the headings of the different landholders of each county. Their work must therefore be undone, and the survey set out upon a geographical basis.

The information that such an analysis makes available is of two kinds. In the first place, the details about plough-teams and about population enable a general picture of the relative prosperity of different areas to be obtained. In the second place, the details about such things as meadow, pasture, wood and salt-pans serve to illustrate further the local variations both in the face of the countryside and in its economic life. An attempt has been made to set out this variety of information as objectively as possible in the form of maps and tables. When all the maps have been drawn and all the tables compiled, we may begin to have a clearer idea of both the value and the limitations of the survey that has so captured the imagination of later generations.

But great though the bulk of the Domesday Book is, it is only a summary. The making of it not only omitted much, but has, too often, resulted in obscurity. No one works for long on the text before discovering how fascinating and tantalising that obscurity is. In reflecting over many Domesday entries we have been reminded, time and again, of some remarks in Professor Trevelyan's inaugural lecture at Cambridge in 1927: 'On the shore where Time casts up its stray wreckage, we gather corks and broken planks, whence much indeed may be argued and more guessed; but what the great ship was that has gone down into the deep, that we shall never see.' The scene that King William's clerks looked upon has gone, and the most we can do is to try to obtain some rough outline of its lineaments; this chapter in the history of the English landscape can only be a very imperfect one.

The Domesday Geography of Eastern England contained an introductory chapter which is not included here. Amongst other things that chapter explained that the counties are considered separately and that the treatment of each follows a more or less standard pattern. This method inevitably involves some repetition, but, after experiment, it was chosen because of its convenience. It enables the account of each county to be read or consulted apart from the rest, and it also has the advantage of bringing out the peculiar features that characterise the text of each county. For although the Domesday Book is arranged on a more or less uniform plan, there are many differences between the counties, both in the nature of their information and in the way it is presented. The relevance of each of the items to a reconstruction of Domesday geography is examined, and any peculiar features that occur in the phrasing of the Domesday text are also noted. All the standard maps have been reproduced on the same scale to facilitate comparison between one county and another. A final chapter sums up some of the salient features of the Domesday geography of the south-western counties as a whole. The treatment of the statistics for the boroughs is different from that in volumes one and two (see p. 364), but this does not appreciably affect the maps.

The maps in this volume have been drawn by Mr G. R. Versey, but this is only a part of our debt to him. At all stages of the work he has helped to check the material and has given much general assistance. For help from time to time we are grateful to Sir Frank Stenton and Professor V. H. Galbraith. We are also indebted to Professor R. R. Darlington for the loan of photographs of the *Liber Exoniensis*, and to Professor H. P. R. Finberg and Mr S. C. Morland for help with the identification of place-names. To our fellow contributors we owe warm thanks for their courtesy and patience. Our debt to the officials of the Cambridge University Press is also great.

H. C. DARBY

R. WELLDON FINN

KING'S COLLEGE
CAMBRIDGE
St Petrock's Day, 1967

LIST OF MAPS

Fig. 1. South-western Counties.

References to the *Liber Exoniensis* are in italics.
Those to the Exchequer folios are in roman type.

CHAPTER I
WILTSHIRE

BY R. WELLDON FINN, M.A.

Wiltshire and the four south-western counties seem to have comprised one circuit, examined by one body of Commissioners. Unfortunately, only a single Wiltshire entry—that relating to a holding in Sutton Veny—has survived in the Exeter Domesday, and the two versions are set out on p. 3 below. A comparison of them shows not only the usual differences of phrasing and content between Exchequer and Exeter entries, but also a difference in the amount of pasture recorded: half a league by one in the Exchequer version, and half a league by one furlong in that of the Exeter text. We are at once prepared for the fact that the Wiltshire folios have their full share of the doubts and difficulties common to the Domesday Book as a whole.

The Exeter Domesday is only a part of the *Liber Exoniensis*, which also includes Summaries of the Wiltshire possessions of Glastonbury Abbey and five lay magnates.[1] A comparison of the information in these Summaries with the corresponding details in the Exchequer Domesday Book is indicated in the table on p. 2. The large discrepancy in the figures for the hidage of the abbey's holdings 'seems inexplicable'.[2] The other differences are of the order commonly encountered when the Exchequer entries are compared with those of associated documents.

The *Liber Exoniensis* also includes three differing copies of the record of the payment of a geld levied at a time near that of the Inquest. These accounts give information about the fiefs within each hundred, but they rarely mention place-names. In spite of the latter fact, however, the information is of great service in allocating the place-names of the Domesday Book to their correct settlements, all the more so because the Domesday text does not contain hundredal rubrics.[3] The differences between the

---

[1] For the contents of the *Liber Exoniensis*, see Appendix I, p. 393 below.

[2] R. R. Darlington in *V.C.H. Wilts.* II (London, 1955), pp. 218–21, presents a full discussion of the details of the Summaries.

[3] Three hundreds are mentioned incidentally in the Wiltshire text. The *firma* of Malmesbury (64b) is said to include the profits of the courts of the hundreds of *Cicemetone* (later called Chedglow) and *Sutelesberg* (later called Startley). The third hundred is that of *Wrderusteselle*, which seems to imply the hundred of *Wrde* (Highworth) in which the manor of *Rusteselle* (Lus Hill) lay (69).

Some Wiltshire Fiefs

Comparison of the Exchequer Domesday
with the Summaries of the Liber Exoniensis

Note: (1) The information for Glastonbury Abbey and its knights is more detailed than that for the other fiefs.

(2) The value of the land of the knights cannot be calculated because sub-tenancies of manors are involved.

(3) For a detailed discussion, see R. R. Darlington in *V.C.H. Wiltshire*, II (London, 1955), pp. 218–21.

(4) Some of the coscets and cottars of the Domesday entries are called bordars in the Summary for Glastonbury Abbey.

	Glastonbury Abbey (*527b*, 66b)		Knights of Glastonbury Abbey (*527b*, 66b)	
	Summary	D.B.	Summary	D.B.
Hides	337	231	54	86 h. 3 v. 8 ac.
Teams	111½	112½	28	33½
Villeins	112	126	34	33
Bordars, etc.	142	139	18	22
Coliberts	38	38	—	—
Serfs	35	35	9	9
Burgesses	2	2	—	—
Value	£169. 10s.	£173	£55. 5s.	?

The combined plough-land figures for the Abbey and its knights are as follows: Summary, 156½; Domesday, 161½.

	Ralph de Mortemer (*530b*, 72–72b)		Miles Crispin (*530b*, 71)	
	Summary	D.B.	Summary	D.B.
Hides	46½	46½	70⅜	71⅞
Plough-lands	40	40	57½	56¾
Value	£38. 9s.	£40. 5s.	£59. 10s.	£59. 9s.

	Durand of Gloucester (*531*, 71b)		Gilbert de Bretteville (*531*, 71–71b)	
	Summary	D.B.	Summary	D.B.
Hides	32¾	32½	50½	49¼
Plough-lands	24	24	31½	32
Value	£26. 17s.	£26. 15s.	£34. 17s.	£41. 10s.

William de Mohun's holding in Sutton Veny

Exeter Domesday (47)

W[illelmus] de Moine habet i mansionem quae vocatur Sutona, quam tenuit Colo ea die qua rex E[dwardus] fuit vivus et mortuus et modo tenet eam W[alterus] Hosatus de W[illelmo]. Haec reddidit geldum tempore regis E[dwardi] pro v hidis. Has possunt arare iiii carucae. De his habet W[illelmus] iii hidas et i virgatam et ii carucas in dominio, et villani ii hidas i virgata minus et ii carrucas. Ibi habet W[illelmus] iii villanos et vi bordarios et iii servos et i roncinum et ccc oves et i molendinum qui reddit iiii solidos per annum et iiii agros prati et dimidiam leugam pascuae in longitudine et i quadragenariam in latitudine et ii agros nemoris. Haec villa valet c solidos et quando W[illelmus] recepit valebat iiii libras.

Exchequer Domesday (72)

Willelmus de Moiun tenet de rege Sutone et Walterus de eo. Colo tenuit tempore Regis Edwardi et geldabat pro v hidis. Terra est iiii carucis. De ea sunt in dominio iii hidae et una virgata terrae et ibi ii carucae et iii servi et iii villani et vi bordarii cum ii carucis. Ibi molinus reddens iiii solidos et iiii acrae prati et ii acrae silvae. Pastura dimidia leuua longa et una lata. Valuit iiii libras. Modo c solidos.

---

details of the Geld Accounts and those of the Domesday folios are very many, and may testify to the imperfection of the Domesday record as well as to our lack of knowledge. They confirm, for example, the suspicion that there must be many virtual duplications in the text.

The Geld Accounts also reinforce the view that the Inquest officials usually presented their material in terms not of a vill but of a manor. The Domesday totals for men, teams, mills and so on in some entries are suspiciously large, and the existence of sub-tenancies is sometimes indicated. We are, however, not only unable to apportion these totals, but also unable even to tell whether or not the unnamed holdings were at places described elsewhere in the Wiltshire folios. Thus Brokenborough (67) was a manor of 50 hides, 60 plough-lands, 64 teams, 102 recorded people and 8 mills. Separate details are given for one of its components, Corston; reference is also made to three sub-tenancies, but where they were, or what each contained, we cannot say. The same was true of the four large manors of the bishop of Salisbury, described on fo. 66: Potterne, Bishop's Cannings, Ramsbury and Salisbury. The entries for some large manors do not even mention the existence of sub-tenancies, although the large quantities involved lead us to suppose them to have existed.[1] Thirty entries tell us in each case of 75 or more recorded inhabitants; it is highly unlikely that any of these entries referred to a single village. The entries for some manors refer to *appendicii*, but without telling us where they were.[2]

The Domesday folios themselves often provide evidence of virtual duplication in their treatment of holdings in dispute.[3] This especially affects hidage totals, but occasionally other information as well. Thus the description of Calne ends with a reference to the 5 hides held by Alfred d'Epaignes but claimed by Nigel the Physician (64b); this may refer to Alfred's holding at Yatesbury described in detail on a later folio (73). Or, again, the account of Chalke (68) records a sub-tenancy of 7½ hides (with 5 teams and worth £7) which was claimed by the abbess of Wilton and which is described in detail under the name of Trow on fo. 73; but the 5 teams of fo. 68 appear on fo. 73 as only 4.

The present-day county of Wiltshire in terms of which this study is written is not exactly the same as the Domesday county. Kilmington (*91*, *453b*, *520*; 86b, 98) and Yarnfield (*447*, 95), the latter in Maiden Bradley, formed part of Somerset and were not transferred to Wiltshire until 1895–6. There has also been gain from Berkshire. The folios of both counties contain entries for the Wiltshire village of Shalbourne (57b, 73, 74 *bis*, 74b) and the Berkshire portion was not transferred to Wiltshire until 1895. In the same year the nearby Bagshot, described only in the Berkshire folios (60b), also became part of Wiltshire. There have been losses as well as gains. In the north, the following four areas were transferred between 1897 and 1930 to Gloucestershire: (1) Kemble (67), Poole Keynes (69b) and Somerford Keynes (66) together with its hamlet Shorncote (73); (2) Ashley (71b) and Long Newnton (67); (3) Poulton (68b), hitherto a detached portion of Wiltshire; (4) Minety, not named in Domes-

---

[1] Such, for example, were Aldbourne (65) with 150 recorded inhabitants, Coombe Bissett (65) with 85, and Heddington (69) with 39.

[2] 'Appendages' (*appendicii*) are mentioned in entries relating to the following: Amesbury (64b), Bradford on Avon (67b), Chippenham (64b), Corsham (65), Melksham (65), Netheravon (65), Rushall (65) and Stert (70b). One 'appendage' of Bradford on Avon and one of Corsham are named, but to the others, and to those of other manors, we are given no clue.

[3] For these parallel entries see: (1) R. Welldon Finn, 'The assessment of Wiltshire in 1083 and 1086', *Wilts. Archaeol. and Nat. Hist. Mag.* L (Devizes, 1946), pp. 382–401; (2) R. R. Darlington in *V.C.H. Wilts.* II, pp. 45–7.

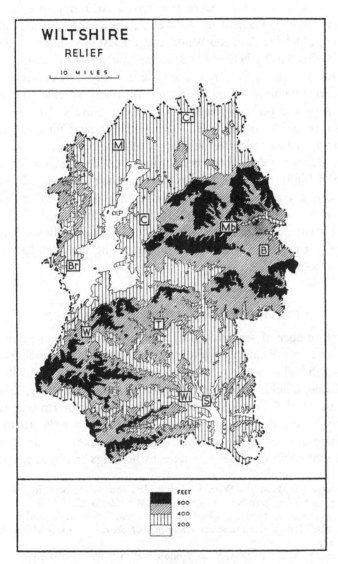

Fig. 2. Wiltshire: Relief.

Domesday boroughs are indicated by initials: B, Bedwyn; Br, Bradford on Avon; C, Calne; Cr, Cricklade; M, Malmesbury; Mb, Marlborough; S, Salisbury; T, Tils-head; W, Warminster; Wi, Wilton.

day Book. The following were transferred to Hampshire in 1895:
Bramshaw (74 *bis*), Damerham (66b, 67b), and Plaitford (74), together
with Martin, Melchet Park and Whitsbury, the last three not named in the
Domesday Book.[1] Tytherley was described under both counties (42, 48,
48b, 50 *bis*, 74), although it lay in Hampshire. There have also been losses
to Berkshire. Standen (72) and Charlton (71b) became part of Hungerford
in 1895. A postscriptal entry at the foot of fo. 72b in the Wiltshire folios
contains a reference to one hide in Coleshill, the remaining 24 hides of
which are described under Berkshire (59b, 61, 63), to which the village still
belongs.[2] There are also other complications. A Wiltshire folio (66b) says
that one of Idmiston's 10 hides 'lies in Hampshire' (*iacet in Hantescire*).
Conversely, a Hampshire folio (50) says that 1½ virgates at Wellow
were 'put outside the county and put in Wiltshire' (*misit foras comi-
tatum et misit in Wiltesire*).[3] Finally, the Wiltshire folios mistakenly
include Gussage St Michael (69) and one holding in Gillingham (73b),
both in Dorset.[4]

SETTLEMENTS AND THEIR DISTRIBUTION

The total number of separate places mentioned in the Domesday Book
for the area now forming Wiltshire is approximately 335, including the ten
places which had or seem to have had burgesses—Bedwyn, Bradford on
Avon, Calne, Cricklade, Malmesbury, Marlborough, Salisbury, Tilshead,
Warminster and Wilton.[5] This figure, for a variety of reasons, cannot
accurately reflect the total number of settlements in 1086. In the first
place, there are ten entries in which no place-names appear, and it is
possible that some of these may refer to holdings at places not named

[1] *Witeberge* (72b), in the Wiltshire folios, has sometimes been identified with
Whitsbury, but it has here been identified with Woodborough.
[2] For references to Standen, Charlton and Coleshill, see H. C. Darby and Eila
M. J. Campbell (eds.), *The Domesday Geography of South-east England* (Cambridge,
1962), p. 240.
[3] *Ibid.* p. 295 n. Before 1895, West Wellow was in Wiltshire and East Wellow in
Hampshire. They have been counted as one Hampshire village in this analysis.
[4] See p. 71 below.
[5] The total of 335 does not include the two unidentified places of *Quintone* (64b)
and *Scepeleia* (66) which are named incidentally in entries relating to other places.
They are omitted because we do not even know whether they were in Wiltshire. The
Dorset folios also mention a *Scipeleia* (77) which appears to refer to the latter place, but
we cannot tell whether it was in either county or neither.

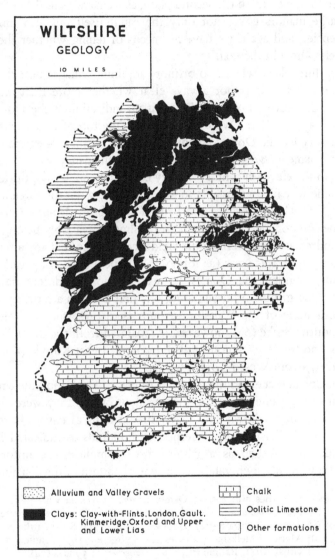

Fig. 3. Wiltshire: Surface geology.

Based on Geological Survey One-Inch Sheets (New Series)
265–7, 281–3, 298–9, 313–15, and Quarter-Inch Sheet 19.

elsewhere in the text.[1] In the second place, as we have seen, the entries for some large manors cover not only the caput itself but also unnamed dependencies, and again we have no means of telling whether these are named elsewhere in the text.[2]

In the third place, when two or more adjoining villages bear the same basic name today, it is not always clear whether more than one unit existed in the eleventh century. There is no indication in the Domesday text that, say, the East Overton and West Overton of today existed as separate villages; the Domesday information about them is entered under only one name (*Ovretone*, 65b, 68), though they may well have been separate in the eleventh century as they certainly were by the thirteenth.[3] In the same way it is impossible to distinguish between the three units of Manningford Abbots, Bohun and Bruce (*Maneforde*, *Maniford*, 67, 73b, 74).[4] The components of each such group almost invariably belong to the same hundred as deduced from the Geld Accounts, so that any attempt to separate them on this basis (as for some counties) does not greatly help. For some counties, the Domesday text occasionally differentiates the units of these groups by designating one unit by such a term as *alia* or *altera*, but none of the Wiltshire groups is distinguished in this way. Winterbourne Stoke (69), however, is separated from the other Winter-bournes, and the Geld Accounts also distinguish East Knoyle (*Cheneuuel Regis*, 1, 65) from West Knoyle (*Chenvel*, 68).

The problem is acute in three groups of places which took their respective Domesday names from the streams along which they were aligned. The table on p. 10 shows how these three groups of entries have been allocated on the basis of their respective hundreds as indicated by the Geld Accounts. Taken as a whole, these complications, unfortunate though they are, hardly invalidate the general pattern of the distributions.

---

[1] The relevant entries are: Bishop Osmund, *dimidia ecclesia* with ¼ hide (65b); Arnulf of Hesdin, *unum manerium* of 1 hide (70); Durand of Gloucester, ¼ virgate (70b); Durand of Gloucester, ¼ virgate (71); Godescal, ¼ hide (73); Stephen the carpenter, 3 hides (73b); Alvric of Melksham, 2 hides less 1 virgate (73b); Edmund, 1 virgate (74); Saieva, 1 virgate (74); Rainburgis, *unum manerium* of 5 hides (74).

[2] See p. 3 above.

[3] For the history of these, and of all other names mentioned in this chapter, see J. E. B. Gover, A. Mawer and F. M. Stenton, *The Place-Names of Wiltshire* (Cambridge, 1939).

[4] The adjacent parishes of Kington Langley, Kington St Michael and Langley Burrell have been treated as three because: (1) the first appears as *Langhelei* (66b) and the second as *Chintone* (72b), both in Thorngrove hundred; (2) the third appears as *Langefel* (69b) in Chippenham hundred.

Quite different are those groups of places bearing the same Domesday name but not lying adjacent. Here are some examples: Barford in Downton (*Bereford*, 72) and Barford St Martin (*Bereford*, 72b, 74, 74b); Charlton now in Hungerford in Berkshire (*Cerletone*, 71b), Charlton near Malmesbury (*Cerletone*, 67), and another parish of Charlton (near Pewsey) not described in the Domesday Book; Draycot Cerne (*Draicote*, 74b), Draycot Fitz Payne in Wilcot (*Draicote*, 66), and Draycot Foliat in Chisledon (*Dracote*, 71); Easton Grey (*Estone*, 72b) and Easton Piercy in Kington St Michael (*Estone*, 70, 73); Littleton Drew (*Liteltone*, 66, 66b) and Littleton Pannell in West Lavington (*Liteltone*, 71b). The absence of hundredal rubrication increases the difficulty of identifying each holding, but the Geld Accounts and later manorial history help to show which holding belongs to which village.

The total of 335 includes half a dozen or so places about which very little information is given. We are told nothing, for example, about Poulshot (65) except that it had a church which lay in (*adjacet*) the manor of Corsham and which was worth 5s.; the details of the holding have apparently been included in those of Corsham itself. Or, again, all that we hear of *Gategram* (74) is that it was rated at one hide, that it had half a plough-land and that its annual value was 10s. Hurdcott (in Winterbourne Earls), with half a hide and half a plough-land, was worth 20s. (72); it also had half a mill, but we hear nothing of the other half, or of the men who worked the mill, or of those who produced grain for it. All we hear of Avebury (65b) is that it had a church to which belonged 2 hides, and that it was worth 40s. Teams and population are likewise not entered for any of the three holdings that comprised Buttermere (70, 72, 74b), yet together they were rated at nearly 1½ hides and had 2¼ plough-lands, and their combined annual value was 24s. 3d. Here might also be mentioned those other villages with no teams or no population or neither entered for them.[1]

Not all the Domesday names appear on the present-day map of Wiltshire parishes. Over 110 out of a total of 335 Domesday place-names are represented today by the names of hamlets, or of individual houses and farms, or of topographical features. Thus *Rochelie* (69b, 70b) is now the hamlet of Rockley in Ogbourne; *Fistesberie* (72b) is Fosbury House in Tidcombe; *Trole* (73b) is Trowle farm, common and hill in Bradford on Avon. The parish of Hilmarton to the north of Calne has at least four

[1] See pp. 22–3 below.

Domesday names represented within its limits apart from its own.[1] The name of Aldred's holding of *Ferstesfeld* (73b) has changed to that of himself, and appears from the twelfth century onwards as variants of *Alderstone*, now represented by a farm of that name in Whiteparish. Some names

Three Groups of Wiltshire Settlements

Deverill (now Upper Wylye)

Brixton Deverill, Hill Deverill and Longbridge Deverill	*Devrel*, 66 b, 68 b, 69 b, 70, 72 b, 74, 74 b	Heytesbury hundred
Kingston Deverill and Monkton Deverill	*Devrel*, 66 b, 68 b	Mere hundred

Winterbourne (now R. Bourne)

Winterbourne Dauntsey, Winterbourne Earls, Winterbourne Gunner, and Gomeldon (in Idmiston)	*Wintreburne*, 66, 66 b, 69 b, 70, 73, 73 b, 74	Alderbury hundred

Winterbourne (now R. Till)

Addeston (in Maddington), Berwick St James, Elston (in Orcheston St George), Maddington, Rolleston and Shrewton	*Wintreburne*, 67 b, 68 b, 69 *bis*, 69 b, 70, 72 b *bis*, 73, 73 b	Dole hundred
Winterbourne Stoke	*Wintreburne*, *Wintreburnestoch*, 65, 69	Dole hundred

Note: (1) The holding on fo. 65 comprised 2 hides and 1 virgate, which appear in the Geld Accounts as *Winterburnestoca* (*V.C.H. Wilts.* II, pp. 118 and 179).

(2) Quite separate from these two groups of Winterbournes are Winterbourne Bassett and Winterbourne Monkton to the north-west of Marlborough.

[1] Hilmarton (70, 71 b, 73 b), Beversbrook (71 b, 73), Highway (66 b, 72), Littlecott (71) and Witcomb (70). There may also be two other Domesday names in the parish. W. H. Jones thought that *Bichenehilde* (70) is represented by Beacon hill (*Domesday for Wiltshire*, London and Bath, 1865, p. 198). The editors of *The Place-Names of Wiltshire* thought that *Cowic* (68 b) is represented by Cowage farm (p. 269). In this analysis, *Bichenehilde* has been left unidentified, and *Cowic* has been identified with Conock.

SETTLEMENTS AND THEIR DISTRIBUTION 11

have disappeared entirely from the map. *Withenham* (66) is the lost Wittenham (in Wingfield parish) and the last record of it comes from the seventeenth century. *Berrelege* (73 b) appeared in the fifteenth century as Barley's Court and was probably near South Wraxall, and has been plotted there.[1] There are some names which remain unidentified or which cannot be identified with certainty.[2] It is unlikely that any two scholars will agree about some of them. Whether they will yet be located, or whether the places they represent have completely disappeared, leaving no trace or record behind, we cannot say.

On the other hand, some 60–70 parishes on the modern map are not mentioned in the Domesday text. Their names do not appear until the twelfth or thirteenth century or even later; and, presumably, if they existed in 1086, they are accounted for under the statistics relating to other place-names. Thus, so far as record goes, Roundway was first mentioned in 1149, Chittoe in 1167, Worton in 1173, Marston in 1198; they may have formed part of the large manor of Potterne (66) with 40 teams and 89 recorded people; so may have Devizes, first mentioned in 1135. Other groups of villages likewise may have been included in such large manors as those of Chalke (68), Downton (65 b) and Ramsbury (66).[3] Several modern parish names (Chute, Clarendon, Savernake) are found in infertile forested areas. Some of the places not mentioned in the Domesday Book must have existed, or at any rate borne names in Domesday times, because they are named in pre-Domesday documents and again in the thirteenth century. Such are Everleigh, Patney, Semley and Woodford.[4] Occasionally, although a modern parish is not named in the Domesday Book, it contains a name or names that are mentioned. Thus the name of Box was not recorded until 1144, but its parish includes the Domesday names of Ditteridge (*Digeric*, 71b) and Hazelbury (*Haseberie*, 65 b, 71, 73, 73b). From this account it is clear that there have been many changes in the village geography of the county, and that a list of Domesday names differs appreciably from a list of present-day parishes.

[1] *V.C.H. Wilts.* VII (London, 1953), p. 18.
[2] The full list of unidentified names in this analysis is as follows: *Alvestone* (67b), *Bechenehilde/Bichenehilde* (70 *bis*), *Celdewelle* (72), *Chenebuild* (70), *Cuvlestone* (73), *Gategram* (74), *Getone* (71b), *Sclive* (70). It is possible that *Quintone* (64b) and *Scepeleia* (66) must be added to these, but we do not know whether they were even in Wiltshire (see p. 6 above).
[3] *V.C.H. Wilts.* II, pp. 83–4 and 91.
[4] It must be noted, however, that some of these pre-Domesday charters appear only in manuscripts of the twelfth, thirteenth, or fourteenth century.

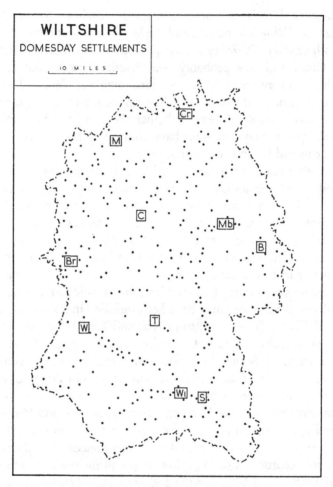

Fig. 4. Wiltshire: Domesday place-names.

Domesday boroughs are indicated by initials: B, Bedwyn; Br, Bradford on Avon; C, Calne; Cr, Cricklade; M, Malmesbury; Mb, Marlborough; S, Salisbury; T, Tils-head; W, Warminster; Wi, Wilton.

Even so, the distribution of Domesday names (and also of Domesday settlements, could we but reconstruct the composition of many manors) was very similar, in a general way, to that of present-day villages (Fig. 4). In the chalk area, the river valleys were marked by lines of villages (along the Ebble, the Nadder, the Wylye, the Salisbury Avon, the Bourne and the Kennet), and these were separated one from another by expanses of open chalk upland. Below the chalk escarpment, on the Lower Greensand outcrop, were located spring-line settlements. To the north and north-west, in the Clay Vale itself, villages were more generally distributed, e.g. along By Brook valley in the Cotswolds, along the Bristol Avon, on the Thames gravels, or in association with the porous outcrops of the Clay Vale. There was, however, one strikingly empty area in what was later known as the Forest of Braydon—on the cold Oxford Clay to the south-west of Cricklade; villages were also absent from the forest areas of the south-east and south-west.

THE DISTRIBUTION OF PROSPERITY AND POPULATION

Some idea of the nature of the information in the Domesday folios for Wiltshire, and of the form in which it is presented, may be obtained from the account of the village of Sutton Veny in the Wylye valley. It was owned by three landholders and the entries relating to their respective estates are set out below:

Fo. 72. William de Mohun holds Sutton of the king, and Walter of him. Colo held it in the time of King Edward, and it paid geld for 5 hides. There is land for 4 plough-teams. Of this, there are in demesne 3 hides and 1 virgate of land, and there [are] 2 plough-teams and 3 serfs, and 3 villeins and 6 bordars with 2 plough-teams. There is a mill paying 4s., and 4 acres of meadow and 2 acres of wood. Pasture half a league long and 1 broad. It was worth £4. Now 100s.

Fo. 72. William fitzGuy holds Sutton of the King. Alwold and his sister held it in the time of King Edward, and it paid geld for 8 hides. There is land for 6 plough-teams. Of this, there are in demesne 4 hides, and there [are] 2 plough-teams and 4 serfs, and 6 villeins and 8 bordars with 4 plough-teams. There [are] 2 parts of a mill paying 13s. 4d., and 6 acres of meadow. Pasture 1 league long and 2 furlongs broad, and as much wood. It was worth £8. Now £10.

Fo. 73. The same Nigel [the Physician] holds 4 hides in Sutton, and it paid geld for as much in the time of King Edward. There is land for 3 plough-teams. [The abbey of] St Mary of Montebourg [in northern France] holds it of Nigel. Of this land, there are in demesne 2 hides, and there [are] 1 plough-team and 3 serfs, and 5 villeins and 5 bordars with 2 plough-teams. There [is] a third part of a mill paying 6s. 8d., and 3 acres of meadow. Pasture half a league long and 1 furlong broad. Wood 1 league long and 1 furlong broad. It was worth £4. Now 100s.

The first of these is the only entry relating to Wiltshire that has survived in the *Liber Exoniensis*, and the two Latin versions are given on p. 3 for comparison.

These entries are fairly representative and straightforward, and each sets out the recurring standard items that are found for most villages. These are five in number: (1) assessment, (2) plough-lands, (3) plough-teams, (4) population, and (5) values. The bearing of these five items of information upon regional variations in the prosperity of the county must now be considered.

(1) *Assessment*

The Wiltshire assessment is stated in terms of hides and virgates, and, very occasionally, of what appear to be geld acres. Fractions of both hides and virgates are common. The usual formula is *geldabat pro n hidis*, but there are variations, as the following examples show:

> Maiden Bradley (71 b): *Comes Tosti tenuit T.R.E. et pro x hidis se defendit.*
> Blunsdon (74): *Lanch tenuit de rege E. Blontesdone et defendebat se pro ii hidis.*
> Hannington (66 b): *Ibi sunt xv hidae.*
> Ditchampton (68): *Ipsa ecclesia habet dimidiam hidam terrae in Dicehantone.*
> Bincknoll (71): *Idem Gislebertus tenet in Bechenhalle v hidas. Hacun tenuit T.R.E.*

It is difficult to see what significance, if any, lies behind these variations.[1] The last two formulae are especially frequent on and after fo. 73 b where the lands of the king's serjeants, thegns and officers are described; some of these entries have the word *geldantem* interlined. The assessment of the demesne portion of a holding is usually stated up to folio 68 b, and afterwards only spasmodically, e.g. in the account of the lands of Edward of Salisbury (69–69 b).

Some people have thought the hides of the south-west were small, un-like the 120-acre hides which Round and Maitland believed to be common throughout England.[2] Eyton thought that the Dorset hide contained 48 acres.[3] Tait thought that the Wiltshire hide was also a small one, and he based his view on the details given for the assessment of the hundred of Calne in the Geld Accounts; but his case was weakened by the fact that

---

[1] For a discussion see R. Welldon Finn, 'The assessment of Wiltshire in 1083 and 1086', *Wilts. Arch. and Nat. Hist. Mag.* L (1946), pp. 382–401.

[2] J. H. Round in P. E. Dove (ed.), *Domesday Studies*, I (London, 1888), p. 213; F. W. Maitland, *Domesday Book and Beyond* (Cambridge, 1897), pp. 478–86.

[3] See p. 80 below.

his equations of acres with fractions of virgates were not exact but only
to 'the nearest round number in each case', e.g. 7 acres as ⅔ virgate and
3 acres as ⅓ virgate.[1] Darlington has also pointed to details in the Geld
Accounts that indicate a hide of either 40 acres or 48 acres.[2] There is no
entry for more than 24 acres in the Wiltshire folios and the numbers that
frequently occur (2, 4 and 6) seem to favour a 48-acre hide. On the other
hand, there are also assessments involving amounts of 5 acres.[3] The fact
is that we cannot be sure how many acres were reckoned to the hide in
Wiltshire.

The account of the Ancient Demesne of the Crown on fos. 64b–65
includes six entries, in each of which appears one of the following statements:

Nunquam geldavit, ideo nescitur quot hidae sint ibi.

Nunquam (or Non) geldavit nec hidata fuit.

Each of the six large manors concerned was responsible for the ancient
render of one night's *firma*.[4] There were also other holdings that had not
paid geld, at Highworth and at Bagshot, the latter described in the
Berkshire folios. Corsham did not pay the full amount of its geld, nor did
Calcutt, nor Kilmington, the last described in the Somerset folios. The
relevant entries for these five places are as follows:

Highworth (65 b): *Radulfus presbyter tenet ecclesiam de Wrde et ad eam
pertinent iii hidae quae non geldabant T.R.E.*

Bagshot (60b): *Duae hidae non geldabant quia de firma regis erant et ad opus
regis calumniatae sunt.*

Corsham (65): *Ibi sunt xxxiiii hidae sed pro xviii hidis reddit geldum.*

Calcutt (73 b): *Ibi sunt v hidae. T.R.E. geldabat pro dimidia hida.*

Kilmington (453b, 98): *Ibi sunt v hidae sed pro una hida geldat.*

Here, too, must be mentioned the hide at Alderbury (68b) which had
never paid geld. There were also two half-hides at Upton Scudamore for
which geld had not been paid since King William came to England (70,
71 b). The discovery of the missing hide is mentioned in the Geld Accounts

---

[1] J. Tait, 'Large hides and small hides', *Eng. Hist. Rev.* XVII (1902), pp. 280–2.

[2] R. R. Darlington in *V.C.H. Wilts.* II, pp. 182–3.

[3] E.g. 2 acres at Whaddon (72), 4 acres at Shrewton (*Wintreburne* 69b), 6 acres at
Hilperton (73b, 74) and 24 acres at Somerford (70b *bis*). Amounts of 5 acres were to
be found, for example, at Allington near All Cannings (70) and Buttermere (70).

[4] The manors were those of Calne, Bedwyn, Amesbury, Warminster, Chippenham
and Tilshead. See pp. 30–1 below.

(*1 b*) under the hundred of Warminster.[1] An entry for Pewsey (65 b) speaks not of hides but presumably of carucates: *Rainboldus presbyter tenet ecclesiam de Pevesie cum una car' terrae. Valet xx solidos*. The description of Cawden hundred in the Geld Accounts also speaks of carucates (59 hides and 3 carucates).[2] More interesting are the five references in the Geld Accounts to carucates that did not pay geld (*nunquam geldantes*).[3] The total amounts to 41 carucates, and they are not mentioned in the Domesday text.[4] A reasonable assumption seems to be that these represent land brought into cultivation, or added to a manor, since the time when its assessment was made. An alternative explanation is that the owners were trying to account for land the hidage of which did not tally with that of the charters.[5]

A glance through the Wiltshire folios reveals many examples of the 5-hide unit. Thus Sopworth (71b) was assessed at 5 hides, Trowbridge (73b) at 10 hides and Bromham (65) at 20 hides. Two Wiltshire manors are called *Fifhide*. One is represented today by Fifield in Enford, and it was assessed at 5 hides (65 b). The other is represented by Fifield Bavant in Ebbesborne Wake; Alfred of Marlborough's holding here was assessed at 5 hides (70b); later on the folio we read that Ulmar held here 1 hide of Alfred, but we cannot say whether this was included within Alfred's 5 hides or was in addition to them. When a village was divided amongst a number of owners, the same feature can often be demonstrated. Thus Uffcott (71 b, 74) with two holdings was assessed at 5 hides (1½ + 3½); and Burbage (65 b, 70b, 74 b *bis*) with four holdings was assessed at 7½ hides (¼ + 2½ + 2½ + 2¼). Eleven places in the modern county were assessed at 2½ hides each; 30 places at 5 hides each; 5 places at 7½ hides each; 43 places at 10 hides each; and 4 places at 15 hides each. There were also a number of places assessed at larger multiples of five, e.g. 20, 25 and 40 hides. We cannot be far wrong in saying that the 5-hide principle is readily apparent in well over one third of the Wiltshire villages. It is also possible that some villages were grouped in blocks for the purpose of assessment. Thus Codford (71 b, 72, 72 b) with 1½ + 6 + 1½ hides and Ashton Gifford (70 b) with 6 hides formed a 15-hide block. In the absence of information about such possible

[1] See V. H. Galbraith, *The Making of Domesday Book* (Oxford, 1961), p. 226.
[2] *V.C.H. Wilts.* II, p. 201.
[3] For the hundreds of Dole, Elstub, Melksham, Westbury and Collingbourne, see *V.C.H. Wilts.* II, pp. 179, 184, 193, 212 and 213.
[4] For the Dorset and Somerset references to non-gelding carucates in the Domesday text, see pp. 83, 153 below. [5] See p. 82 below.

groups we cannot be certain about the full extent of the 5-hide unit in Wiltshire.

It is clear, however, that the assessment was largely artificial in character, and bore no constant relation to the agricultural resources of a vill. The variation among a representative selection of five-hide vills speaks for itself:

	Plough-lands	Plough-teams	Popula-tion	*Valuit*	*Valet*
Beechingstoke (67b)	5	5	14	60s.	100s.
Chisbury (71)	9	9	36	£8	£12
Smithcot (70b)	4	5	10	40s.	60s.
Surrendell (72b)	6	6	19	£7	£7
Westlecott (73)	2	2	10	40s.	40s.

An extreme example of low rating was the king's holding of 1 hide at Britford (64) where there were 48 recorded people with 19 teams at work. At Winterbourne Stoke (65) there was another royal holding of 2¼ hides with 46 recorded people and 11 teams at work. Taken as a whole, however, Wiltshire was a highly rated county.

The total assessment, including non-gelding hides, amounted to 3,891 hides, 3½ virgates and 74 acres.[1] This is for the area included within the modern county, and so is not strictly comparable with Maitland's total for the Domesday county.[2] The nature of some entries, and the possibility of unrecognised duplicate entries, make exact calculation very difficult.[3] All that any estimate can do is to indicate the order of magnitude involved. We must furthermore remember the six royal manors of which it was said, *nescitur quot hidae sint ibi* or *nec hidata fuit*, and also those 41 carucates of land which the Geld Accounts say had never been assessed.

[1] In view of the uncertainty about the number of acres to the hide, the totals for hides and virgates on the one hand, and for acres on the other, have been kept separate. Two acres have been taken away from the total of acres to allow for the Whaddon entry (72), which speaks of 3½ virgates less 2 acres. One Cholderton entry also speaks of 'less 4 acres' (71b) but this is cancelled by another Cholderton entry which speaks of 'and 4 acres' (70).

[2] F. W. Maitland, *op. cit.* p. 400, reckoned 4,050 hides. An earlier estimate by R. Welldon Finn came to 4,084⅞ hides and 74 acres (*art. cit.* p. 383), but this made no allowance for duplicate entries.

[3] It is very difficult, for example, to determine whether the hidage attributed to churches should be counted or not. The Geld Accounts show that some of the assessments given for churches were not included in those of the manors to which they were attached.

(2) Plough-lands

Plough-lands are systematically entered for the Wiltshire holdings, and the normal formula runs: 'There is land for *n* teams' (*Terra est n carucis*). Unusual entries include the following:

North Newnton (67 b): *Totum manerium possunt arare x carucae* (instead of the usual formula).
Tollard Royal (69 b): *geldabant pro ii hidis et dimidia. Terra est totidem carucis* (there were 2 teams in demesne and half a team with the tenants).

A reckoning in half-teams is not unusual, but there are only very occasional references to oxen, such as that for Chedglow below and that in an entry for Buttermere (74 b). In two entries the phrase *Terra est* is followed by a blank space in which a figure was never inserted: Biddestone (71) and Tockenham (71 b). There are, moreover, four entries (for three places) which make no reference to plough-lands although they state the number of teams at work, e.g. an entry for Swindon (66) and another for Stanton St Quinton (72 b); these account for a total of 16 teams. There are also 25 entries which mention neither plough-lands nor teams, e.g. one for Beversbrook (73), another for Blunsdon (74), and some of the entries relating to ecclesiastical holdings in the left-hand column of fo. 65 b.

Some plough-land figures look suspiciously artificial, e.g. Chippenham (64 b) with 100 (for 82 teams) and Corsham (65) with 50 (for 45 teams). Other figures, however, coincide with the totals of teams, e.g. for Ramsbury (66), where 54 plough-lands equal the sum of 8 + 29 + 11 + 6 teams, or for Bedwyn (64 b), where 79 plough-lands equal the sum of 12 + 67 teams.

The relation between plough-lands and plough-teams varies a great deal in individual entries. They are equal in number in about 57 per cent of the entries which record both. Occasionally, this is implied in various ways, e.g. by the phrase *Terra est n carucis quae ibi sunt* which occurs in several entries in the left-hand column of fo. 72 b.[1] Other phrases that indicate equality of plough-lands and teams include the following:

Choulston (68 b): *Terra est i caruca et dimidia et tantum est ibi.*
Highworth (65 b): *Terra est ii carucis. Has habent ibi presbyter.*
Chedglow (71): *Terra est vi bobus qui ibi sunt arantes.*
Winterslow and Tytherley (74): *Terra est i caruca. Hanc habent ibi iiii rustici.*[2]

[1] A formula used in the Summaries of the *Liber Exoniensis* (*527 b*) is *Haec terra sufficit n carrucis*, where the plough-lands and teams are equal or nearly equal in number.   [2] Tytherley is now in Hampshire—see p. 6 above.

A deficiency of teams is also quite common, and is encountered in 37 per cent or so of the entries which record both plough-lands and teams. The values of some deficient holdings had fallen in the years prior to 1086, but on others they had remained constant, and on yet others they had even increased. No general correlation between plough-team deficiency and decrease in value is therefore possible, as the following table shows:

		Plough-lands	Teams	*Valuit*	*Valet*
Decrease in value	Castle Eaton (68 b)	12	8½	£15	£12
	Norton (67)	8	5	£6	£4
Same value	Alderton (73)	3	1	60s.	60s.
	Hullavington (72)	14	10	£12	£12
Increase in value	Badbury (66 b)	10	6	£8	£10
	Chisledon (67 b)	22	15	£18	£24

These figures are merely indications of unexplained changes on individual holdings.

An excess of teams is a good deal less frequently encountered, being found in only 6 per cent of the entries which record both plough-lands and teams. The values on some of these holdings had risen, e.g. at Somerford (67) and Smithcot (70 b), but frequently this was not so. Other holdings with excess teams were valued the same in 1086 as at an earlier date, e.g. at Fifield in Enford (65 b) and Milston (68 b). The values of yet other holdings with excess teams had even dropped, e.g. at Broughton Gifford (70 b) and Stratton St Margaret (73).

The total number of plough-lands recorded for the area within the modern county is 3,343¾. If, however, we assume that the plough-lands equalled the teams on those holdings where a figure for the former is not given, the total becomes 3,359¾. For various reasons the figure given by Maitland is not strictly comparable with the present total.[1] In view of the nature of some entries and of the possibility of unrecognised duplicate entries, no estimate can be anything other than an approximation; and none of these totals takes account of the non-gelding carucates mentioned in the Geld Accounts.

[1] F. W. Maitland, *op. cit.* pp. 401 and 410. Maitland's total of 3,457 was for the Domesday county, but, as in the present estimate, allowance was made for the missing plough-land figures.

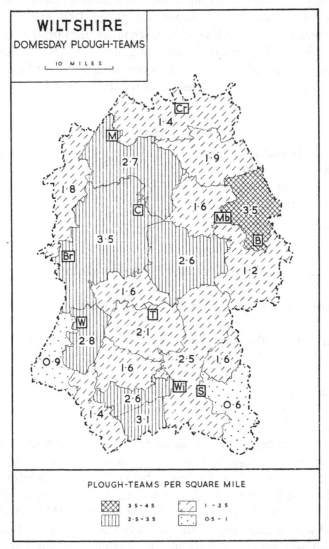

Fig. 5. Wiltshire: Domesday plough-teams in 1086 (by densities).

Domesday boroughs are indicated by initials: B, Bedwyn; Br, Bradford on Avon; C, Calne; Cr, Cricklade; M, Malmesbury; Mb, Marlborough; S, Salisbury; T, Tils-head; W, Warminster; Wi, Wilton.

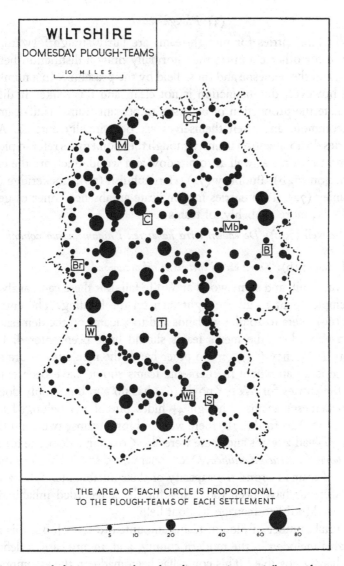

Fig. 6. Wiltshire: Domesday plough-teams in 1086 (by settlements).

Domesday boroughs are indicated by initials: B, Bedwyn; Br, Bradford on Avon; C, Calne; Cr, Cricklade; M, Malmesbury; Mb, Marlborough; S, Salisbury; T, Tilshead; W, Warminster; Wi, Wilton.

(3) *Plough-teams*

The Wiltshire entries for plough-teams are fairly straightforward, and, like those for other counties, they normally draw a distinction between the teams on the demesne and those held by the peasantry. In a number of entries, however, the distinction is not clear, and this makes it difficult to calculate the proportion of demesne to peasant teams. Half-teams are not uncommon, and oxen themselves are occasionally entered. About 100 entries do not mention teams, although they sometimes refer to plough-lands and to people as well as to meadow and wood. Such are the entries for Alderton (72b), Budbury (74), Hardenhuish (70), Lockeridge (71b) and Stanley (72). Two entries may indicate an absence either of general livestock or, more probably, of teams:

> Cumberwell (70b): *De eadem terra habet rex i hidam in suo dominio et nil ibi est.*
>
> Wolf Hall (74b): *Terra est iii carucis et nil pecuniae.*

There was a mill and some wood at Wolf Hall, but these can hardly have given employment to all the eight men on the holding. The entry for Wilcot (69) refers to 10 plough-lands and to 3 teams on the demesne, but leaves a gap where the men's teams should have been entered. Under Rodbourne Cheney (71), on the other hand, the men's teams are stated but there is a gap where the demesne teams should have been entered; one of the entries for West Knoyle (68) has no gap but simply does not refer to demesne teams although 4½ hides out of ten belonged to the demesne. At Durnford (67b) there was land for 3 teams; two Englishmen (*duo Angli*) had 2 teams and there were also 6 oxen in a demesne team: *Ibi sunt vi boves in caruca dominica.* Occasional vills seem to have possessed a number of teams beyond the capacity of their inhabitants. Thus 3 teams would seem to have been too many for the 4 recorded inhabitants of Groundwell (70b) to manage without help.

The total number of plough-teams amounted to 2,925¾, but this refers to the area included in the modern county, and, in any case, a definitive total is hardly possible.[1] This count has been made on the assumption of an eight-ox team. We cannot say what allowance should be made for the teams of the non-gelding carucates that are mentioned only in the Geld Accounts.[2]

[1] F. W. Maitland's total for the Domesday county was 2,997 teams (*op. cit.* p. 401).
[2] See p. 16 above.

(4) *Population*

The bulk of the population was comprised in the six main categories of villeins, bordars, serfs, coscets, cottars and coliberts. In addition to these were the burgesses and a miscellaneous group that included swineherds, bee-keepers and others, together with unspecified numbers of potters and 'men'. The details of the groups are summarised on p. 28. This estimate is not comparable with that of Sir Henry Ellis, which was in terms of the Domesday county and which included tenants-in-chief and sub-tenants.[1] In any case, an estimate of Domesday population can rarely be definitive, and all that can be claimed for the present figures is that they indicate the order of magnitude involved. These figures are those of recorded population, and must be multiplied by some factor, say 4 or 5, in order to obtain the actual population; but this does not affect the relative density as between one area and another.[2] That is all that a map such as Fig. 7 can roughly indicate.

It is impossible to say how complete were the Domesday figures. We should have expected to hear more, for example, of priests. Then there are some 70 or so entries which do not mention inhabitants although they sometimes refer to plough-lands and even teams and to such other resources as wood or meadow. Thus there were two holdings at Uffcott (71b, 74) reckoned at a total of 5 hides. These comprised 2½ plough-lands and one team, and the annual value of the whole amounted to 45*s*., yet no other resources and no inhabitants are entered. Who, then, we may ask, worked the team? Or who worked the team at Harding (74b) unless it was men from Richard Sturmy's other holdings at the nearby villages of Burbage, Grafton or Shalbourne (74b)? Like Harding, Buttermere was a vill along the thinly populated eastern border of the county. Two of its holdings (72, 74b) were rated at nearly 1½ hides, and they had 2¼ plough-lands but no population or teams or other resources. We can hardly suppose the holdings were waste, because they were worth 23*s*. 4*d*. a year; but when we see that a third holding (70) was rated at only 5 acres, we may perhaps envisage them as fields away from human habitation and worked by men from neighbouring vills. Then there are those other

---

[1] Sir Henry Ellis, *A General Introduction to Domesday Book*, II (London, 1833), pp. 501–3. His total for the Domesday county (excluding tenants and sub-tenants) came to 9,708.

[2] But see p. 388 below for the complication of serfs.

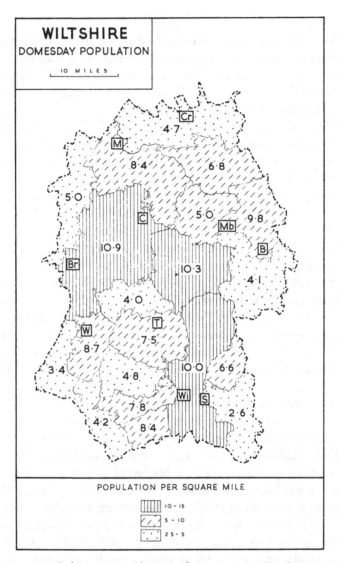

Fig. 7. Wiltshire: Domesday population in 1086 (by densities).

Domesday boroughs are indicated by initials: B, Bedwyn; Br, Bradford on Avon; C, Calne; Cr, Cricklade; M, Malmesbury; Mb, Marlborough; S, Salisbury, T, Tilshead; W, Warminster; Wi, Wilton.

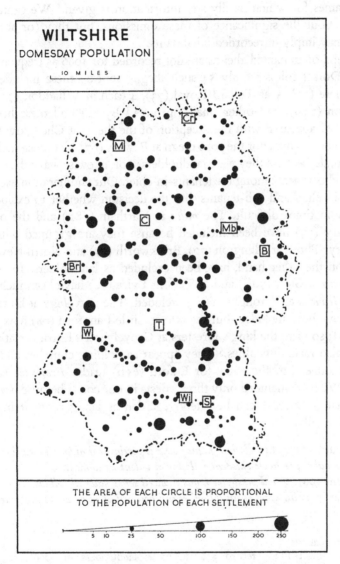

Fig. 8. Wiltshire: Domesday population in 1086 (by settlements).

Domesday boroughs are indicated by initials: B, Bedwyn; Br, Bradford on Avon; C, Calne; Cr, Cricklade; M, Malmesbury; Mb, Marlborough; S, Salisbury; T, Tilshead; W, Warminster; Wi, Wilton.

place-names for which hardly any information is given.[1] We cannot be certain about the significance of these omissions, but they (or some of them) may imply unrecorded inhabitants.

Groups of unnamed thegns are not recorded for 1066 as frequently as in the Dorset folios.[2] Only 54 such thegns appear; these include 8 at Buttermere (72), 5 at Tollard Royal (73), 4 each at Whaddon (72) and Hilperton (73b), and other smaller groups.[3] By 1086 all these had apparently disappeared with the exception of the thegn at Chedglow (70).[4] But neither this man nor the 1086 thegn at Pewsey (67b) has been included in our totals, because they were landholders; nor, likewise, have the named thegns who appear among the tenants-in-chief. Our enumeration excludes tenants-in-chief and sub-tenants, but the decision whether to exclude or not is sometimes difficult. The *miles* at Hartham (68b) and the one at Yatesbury (73) have been included because they are grouped with the peasantry. Those at Allington (70), Brinkworth (67) and North Newnton (67b), on the other hand, have been excluded as landholders. Likewise, the *francus* at Stert (70b) and at Teffont Evias (70) have been included, but the *francus* at Wroughton (70) excluded. The two *Angli* at Durnford (67b) have been included, but the two at Little Langford (68) have been excluded; so have the king's foresters at Grovely (74). Entries relating to priests also raise difficulties.[5] They appear as holding churches and land or land alone, e.g. the entry for Enford (65b) reads: *Presbyter habet i hidam.* We have included only three priests in our count because it would seem that only these can be safely regarded as being resident in their respective vills:

Aldbourne (65): *Ad ecclesiam hujus manerii pertinent ii hidae. Terra ii carucis. Has habet presbyter ejusdem ecclesiae et valent xl solidos.*
Potterne (66): *Quod presbyter hujus manerii tenet valet xl solidos.*
Rowde (70): *iiii servi et iiii villani et viii bordarii et xi cosceƷ et presbyter cum iiii carucis.*

---

[1] See p. 9 above.
[2] The Dorset folios record 340 thegns as landholders in 1066, see p. 91 below.
[3] Groups of named individuals occasionally appear as holders in 1066, e.g. the 4 each at Clyffe Pypard (70b) and Alderton (72b), but we cannot be sure whether or not these were thegns.
[4] A thegn held 2 hides at Pewsey (67b) in 1086, but we are not told whether he had been there in 1066.
[5] See p. 50 below.

Only the priest at Rowde is enumerated amongst the peasantry. We also hear that Gerald the priest of Wilton (*Giraldus presbyter de Wiltone*) held the tithe of the church at Collingbourne Ducis (65), but he may well have been born at Wilton or resident there rather than at Collingbourne. Villeins constituted the most important element numerically in the population, and amounted to 35 per cent of the total. There was an unspecified number of villeins at Swallowfield (73b), the entry for which merely says that the villeins hold (*villani tenent*) $1\frac{1}{2}$ teams. There was also an unspecified number on a sub-tenancy at North Newnton (67b). An entry for Overton (65b) leaves a space for the number of villeins, and the letters *rq* (*require*) are to be found in the margin in red ink. Bordars amounted to 28 per cent of the total, and on a few estates they equalled or exceeded the number of villeins. Serfs, who amounted to 16 per cent, are usually enumerated along with the demesne teams. They are most numerous on the 22 royal manors, where they constitute 25 per cent of the recorded population. There were, on the other hand, some quite large manors that seem to have been without serfs, e.g. the abbess of Shaftesbury's manor of Tisbury (67b) with 80 inhabitants and 40 teams. The fourth main group (14 per cent) was that of coscets—*cosceʒ, cosceti* or (on most royal manors) *coʒets*; and they amounted to nearly 80 per cent of the total number recorded in the Domesday Book as a whole. Their status is not easy to define;[1] we occasionally find them as the sole class on a holding, e.g. the 6 coscets at Thickwood (69b). There is a blank where the number of coscets should have been inserted in one of the entries for Tollard Royal (71b). Cottars accounted for nearly 3 per cent of the total, and coliberts for nearly 2·5 per cent; the latter were to be found only on royal manors and on those of major monastic houses. They amounted to over one quarter of the total number of coliberts recorded in the Domesday Book as a whole.

The miscellaneous group included 87 swineherds (*porcarii*); they were entered in substantial groups for only four manors: 22 for the abbess of Shaftesbury's manor of Bradford on Avon (67b), and the rest for three royal manors, 23 for Chippenham (64b), 13 for Warminster (64b) and 29 for Westbury (65). There were also 9 bee-keepers (*mellitarii*) at Westbury (65), and perhaps to these should be added the *serviens* who paid a honey rent of 7 sesters of honey at Bradford on Avon (67b). On

---

[1] Some Wiltshire coscets and cottars are called bordars in the Summary for Glastonbury Abbey—see p. 2 above. For the status of the Somerset coscets, see p. 166 below.

Recorded Population of Wiltshire in 1086
A. Rural Population

Villeins	3,418
Bordars	2,766
Serfs	1,550
Coscets	1,385
Cottars	276
Coliberts	232
Miscellaneous	108
Total	**9,735**

Details of Miscellaneous Rural Population

Swineherds	87		Knights (*milites*)		2
Bee-keepers	9		*Serviens*		1
Rustici	2		Potters	unspecified	
Priests	3		Men (*homines*)	unspecified	
Angli	2		Total		108
Franci	2				

B. Urban Population

BEDWYN	25 burgesses (see p. 54 below).
BRADFORD ON AVON	33 burgesses (see p. 54 below).
CALNE	73 burgesses, 1 *domus* (see p. 54 below).
CRICKLADE	'many burgesses', 33 burgesses, 2 *domus* (see p. 55 below).
MALMESBURY	75¼ *masurae*, 11 burgesses, 9 coscets, 5½ *domus*, 8¼ *vastae masurae* (see pp. 51–4 below).
MARLBOROUGH	No enumeration (see p. 55 below).
SALISBURY	No enumeration (see p. 57 below).
TILSHEAD	66 burgesses (see p. 57 below).
WARMINSTER	30 burgesses (see p. 58 below).
WILTON	25 burgesses, 5 *domus* (see p. 58 below).

Note: Coscets at Malmesbury are also included in Table A.

the manor of Westbury (65) there was also an unspecified number of potters (*potarii*) who rendered 20*s*. a year,[1] and an unspecified number of *homines* at Lavington (73). The term *rustici* is used in the joint entry for Winterslow and Tytherley (74), which tells of a team with 4 *rustici* but with no other inhabitants.[2] As we have seen, a few priests, *Angli*, *franci* and *milites* who were not landholders have also been counted.

(5) *Values*

The value of an estate is generally given for two dates—for 1086 and for some earlier year.[3] A number of entries give a figure only for 1086, e.g. on fo. 74; whether such amounts also refer to an earlier date, we cannot tell. We are, however, often specifically told that an amount had remained constant by means of the formula *valuit et valet*. The Domesday use of the word *valuit* leaves the exact date of the earlier valuation in doubt, but some entries state that it refers to the time when an existing holder received an estate. Certainly, for the one Wiltshire estate surveyed in both the Exchequer version and the Exeter text, the latter uses *quando recepit* as the equivalent of *valuit*.[4] It would seem as if this might be true generally.[5] On the other hand, there are a few estates where the earlier valuation is specifically said to be *T.R.E.*, e.g. at Castle Eaton (68b) and Coombe Bissett (65). Another holding with a *T.R.E.* valuation was at Damerham (66b), now in Hampshire. The entry for Brixton Deverill (68b) states the earlier value to be that during the lifetime of the late Queen Matilda. The word *reddit* is occasionally used instead of *valet*. The values of land in demesne and of land held by sub-tenants are sometimes given separately. The following examples illustrate some of the variations to be found. The two entries for Hilperton are not only on the same folio but belong to the same fief and are next to one another:

Bradfield (72): *Valuit et valet xxx solidos.*
Collingbourne Ducis (65): *Valuit xl libras. Modo lx libras.*
Donhead (67b): *Dominium abbatissae valet xxii libras. Hominum vero x libras, et prius tantundem valuit.*

---

[1] This might be a scribal error for *porcarii*.
[2] Only two *rustici* have been counted here because Tytherley is in Hampshire. For the equation of *rustici* with *villani* in Dorset, see p. 94 below.
[3] One of the holdings at Shalbourne (57b), surveyed in the Berkshire folios, is valued at three dates.   [4] Sutton Veny. See p. 3 above.
[5] The Dorset evidence is more abundant on this point, and indicates an equivalence. See pp. 95–6 below.

WILTSHIRE

Lackham (71 b): *Valuit v libras. Modo similiter.*
Ogbourne (65 b): *Valet xxv libras.*
Bremhill (67): *Valuit xiiii libras quando abbas recepit. Modo xvi libras.*
Hilperton (73 b): *Valuit iiii libras. Modo x solidos minus.*
Hilperton (73 b): *Reddit xl solidos.*
Compton Chamberlayne (65): *Hoc manerium reddit xii libras ad pensum.*
Castle Eaton (68 b): *Valuit xv libras T.R.E. Modo xii libras.*
Coombe Bissett (65): *Hoc manerium reddit xxiiii libras ad pensum. T.R.E. tantundem ad numerum.*
Brixton Deverill (68 b): *Valebat xv libras vivente regina Mathilde.*
Ramsbury (66): *Dominium episcopi valet lii libras et xv solidos. Quod alii tenent xvii libras.*

The amounts are usually stated in round numbers of pounds and shillings, e.g. £6 or 100s. Occasional precise figures suggest careful appraisal, e.g. the £6. 13s. at Steeple Ashton (68), the 13s. at Hilmarton (73 b) and the 7s. at Swindon (74). For the royal manors, described on fo. 65, the amounts are sometimes said to be by weight (*ad pensum*) or by tale (*ad numerum*): for Britford and Compton Chamberlayne, the reckoning was by weight; for Westbury it was by tale; for Coombe Bissett, as we have just seen above, both methods were used. So were they in the three successive entries on fo. 65 for Aldbourne, Corsham and Melksham, where there were disagreements about the amounts:

Aldbourne: *Hoc manerium reddit lxx libras ad pensum, sed ab Anglis non appreciatur nisi lx libras ad numerum.*
Corsham: *Hoc manerium cum appendiciis reddit xxx libras ad pensum. Angli vero appreciantur ad xxxi libras ad numerum.*
Melksham: *Hoc manerium reddit c et xi libras et xi solidos ad pensum. Angli vero appreciantur ad totidem libras ad numerum.*

There was also disagreement about the values of Damerham (66 b), now in Hampshire, and of Newton Tony (70). The entry for the latter reads: *Valuit x libras. Modo xviii libras. Ab Anglis appreciantur xii libras.*

A few valuations give us a glimpse of an archaic system that had almost disappeared. This was the arrangement by which the king had received a rent in kind from certain royal manors—the *firma unius noctis cum omnibus consuetudinibus suis.* This phrase, or something like it, is found in the accounts of six manors on fos. 64b–5: Calne, Bedwyn, Amesbury,

Warminster, Chippenham and Tilshead. None of the manors paid geld or seems to have been assessed in hides.[1] We are not given the names of the constituents of each manor as we are for similar manors in Dorset,[2] but the entries for Amesbury and Chippenham speak of *cum appendiciis suis*.[3] The accounts of two manors state the value of the *firma*, £110 by tale for Chippenham, and £100 by tale for Tilshead. No reference is made to the *firma unius noctis* in the entries for the royal manors of Melksham and Westbury; but the former is said to pay £111. 11s. by weight and the latter £100 by tale. It seems likely that they also rendered or had recently rendered a night's *firma*. Unlike the other six manors, they were assessed in hides.

Values remained the same or declined or, more usually, increased. Occasionally the variations were considerable, e.g. at Urchfont (68) from £15 to £30, at Trowbridge (73b) from £4 to £8, and at Huish (74b) from 30s. to 60s. The rise at Upton Scudamore (71b) from 15s. to 60s. is striking; so is the fall at Wilsford from 100s. to 8s., but the latter figure may be an error for £8.[4]

Generally speaking, the greater the number of teams and men on an estate, the higher its value; but there is much variation, and it is impossible to discern any consistent relationship, as the following figures for five holdings, each rendering £3 in 1086, show:

	Teams	Population	Other resources
Fisherton Anger (69)	1	9	Mill, meadow, pasture
Somerford (70b)	2	23	Meadow, pasture
Luckington (72b)	4	6	Meadow, wood
West Dean (72)	3	13	Mill, meadow, pasture
Wilsford (72b)	1	4	Mill, meadow, pasture

There are, moreover, some perplexities. The hide at Cumberwell (70b), where we are told there was nothing (*nil ibi est*), was nevertheless worth 8s.; part of this may well be the half-hide of waste land (*vastata terra*) which the Geld Accounts say the king held in Bradford hundred.[5]

---

[1] See p. 15 above.   [2] See p. 97 below.
[3] There is one other reference to *appendicii*—in the entry for Bradford on Avon (67b), which is unusual in that it names one.   [4] *V.C.H. Wilts.* II, p. 160.
[5] *Ibid.* p. 199. The Geld Accounts also record half a virgate of waste land belonging to Bishop Walkelin of Winchester (9b, 16). Nothing is said about this in the Domesday account of the bishop's manor of Ham (65b), but Ham is in the bare country by Inkpen Beacon, and near the empty settlement at Buttermere (see p. 23 above).

Conclusion

For the purpose of calculating densities, Wiltshire has been divided into twenty units, based largely upon the main physical and geological features. The boundaries of the density units have been drawn to coincide with those of civil parishes, and this gives the units an artificial character because many parishes extend across more than one kind of soil and country. Occasional adjustments have been made to meet this artificiality. There is, for example, a marked difference between the Grovely–Great Ridge upland to the south of the Wylye and the downland to the north. Four long narrow parishes cross the river—Heytesbury, Fisherton de la Mere, Wylye and Steeple Langford. These parishes have therefore been divided between the units to the north and south of the river. The western tongue of Warminster has likewise been detached from the parish and included within another unit. Another difficulty arises from the fact that the totals for some manors cover a number of anonymous settlements that may not be contained within the same unit. The division into twenty units is therefore not a perfect one, but it may provide a basis for distinguishing broad variations over the face of the countryside.

Of the five standard formulae, those relating to plough-teams and to population are most likely to reflect something of the distribution of wealth and prosperity throughout the county. Taken together, certain common features stand out on Figs. 5 and 7. As might be expected, the most densely occupied areas lay in the lower valley of the Bristol Avon, the Vale of Pewsey and the valley of the Salisbury Avon; here there were 10 or more people per square mile and 2·5 or more teams. The areas with densities below the average for the county were the forest areas of the east (Savernake, Chute, Clarendon and Melchet) and the south-west (Selwood), the Grovely–Great Ridge area, the north-west of Salisbury Plain, the Braydon Forest district in the north of the county and the Cotswolds in the north-west; all these had about 5 recorded people or under per square mile and 1·5 teams or under. Of these poor districts, the Clarendon and Melchet areas in the south-east had particularly low densities—2·6 for population and 0·6 for teams.

Figures 6 and 8 are supplementary to the density maps, but it is necessary to make reservations about them. As we have seen on p. 8, some Domesday names covered two or more settlements, and several of the symbols should thus appear as two or more smaller symbols. This limitation, although serious locally, does not affect the main impression conveyed by the maps. Generally speaking, they confirm and amplify the information on the density maps.

WOODLAND AND FOREST

Types of entries

The Wiltshire folios give information both for wood itself and for underwood. The amount of each is normally expressed in one of two ways —in terms of linear dimensions or in terms of acres. There are also a few variant entries.

Woodland is called *silva* in the Exchequer folios, and its amount is usually recorded by stating its length and breadth in terms of leagues and furlongs.[1] The following examples illustrate the kind of entry that is encountered:

> Amesbury (64b): *Silva vi leuuis longa et iiii leuuis lata.*
> Bishop's Cannings (66): *Silva i leuua longa et x quarentenis lata.*
> Ramsbury (66): *Silva xvi quarentenis longa et iiii quarentenis lata.*
> Dauntsey (66b): *Silva dimidia leuua longa et tantundem lata.*

Where the quantities are in furlongs, the number is always less than the twelve usually supposed to make a Domesday league, with the exception of the amount for Ramsbury quoted above. The largest amount entered under a single name is the 6 leagues by 4 leagues for Amesbury (64b). The exact significance of these linear dimensions is far from clear, and we cannot hope to convert them into modern acreages.[2] All we can do is to plot them diagrammatically as on Fig. 9, and we have assumed that a league comprised 12 furlongs. The measurements in an entry for a large manor, such as Ramsbury, may well cover a number of widely separate unnamed places, and this seems to imply, though not necessarily, some process of addition whereby the dimensions of separate tracts have been consolidated into one sum. The entry for Bedwyn (64b) is unusual in that it speaks specifically of two woods but gives only one set of measurements: *Duae silvae habentes ii leuuas longitudine et unam leuuam latitudine.* The account of Collingbourne Ducis (65) also mentions two woods; the

---

[1] The single Wiltshire entry in the Exeter folios uses *nemus* instead of *silva*. See p. 3 above.
[2] See p. 372 below.

first measured one league by one league, and the other was a third part of the wood called Chute (*tercia pars silvae quae vocatur Cetum*).[1]

In six entries, the amount of wood is stated to be 'between length and breadth' (*inter longitudinem et latitudinem*) or 'in length and breadth' (*in longitudine et latitudine*).[2] When drawing Fig. 9, we assumed that all such entries imply that the length and breadth were the same.[3] Two entries give only one dimension: 1 furlong of wood (*una quarentena silvae*) at Tollard Royal (71b) and 3 furlongs at Witherington (74). It may be that these entries also imply that the length and breadth were the same, but on Fig. 9 the entries have been plotted as single lines.[4]

Very frequently the amount of wood is stated in acres. The figures range from a single acre at, for example, Middleton (69b) to 200 acres at South Newton (68), but the great majority do not rise above 15 acres. The figures are often in round numbers that suggest estimates, e.g. the 50 acres at Stitchcombe (74); but occasional entries seem to be precise, e.g. the 7 acres at Kennett (70b) and the 9 acres at Smallbrook (74b). No attempt has been made to convert figures into modern acres, and they have merely been plotted as conventional units on Fig. 9. Occasionally, both acres and linear dimensions appear in entries for one place-name, e.g. Westbury with 4 acres (74b) and 3 leagues by half a league of wood (65).

There are a number of entries in which the acre seems to be used as a linear measure:

> Calne (64b): *Silva ii quarentenis longa et una quarentena et xxiiii acris lata.*
> Chedglow (70b): *i acra silvae in longitudine et latitudine.*
> Woodhill (66): *Silva una quarentena longa et iii acris lata.*

On Fig. 9 these have been indicated under the category of 'other mention'.[5] At Grafton (74b) yet another unit was in use, for we hear of arpents of wood: *ii arpenz silvae.*[6]

Two other unusual entries must be mentioned. South Newton and Washern each had in Melchet Wood rights that involved pasture for

[1] See p. 38 below.
[2] The entries for Chippenham (64b), Corsham (65), Grimstead (72), Ebbesborne (72b), Lacock (69b) and Melksham (65).
[3] See p. 100 below for the Dorset evidence that points to this.
[4] See p. 101 below for the Dorset evidence.   [5] See pp. 101–2 below.
[6] For a discussion of the arpent see Sir Henry Ellis, *A General Introduction to Domesday Book*, I, p. 117.

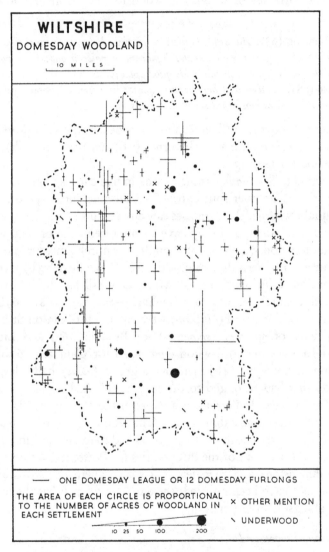

WILTSHIRE
DOMESDAY WOODLAND

10 MILES

—— ONE DOMESDAY LEAGUE OR 12 DOMESDAY FURLONGS

THE AREA OF EACH CIRCLE IS PROPORTIONAL
TO THE NUMBER OF ACRES OF WOODLAND IN   × OTHER MENTION
EACH SETTLEMENT
                                          ＼ UNDERWOOD
10 25 50   100   200

Fig. 9. Wiltshire: Domesday woodland in 1086.

When the wood of a settlement is entered partly in linear dimensions and partly in some other way, both are shown. When wood and underwood are entered for the same settlement, both are shown.

80 swine, 80 cartloads of wood, and wood for the repair of houses and fences:

> South Newton (68): *Ad istum manerium pertinet habere per consuetudinem in silva Milcheti quater xx caretedes lignorum et paisson quater xx porcorum et ad domos et sepes reemendandas quod opus fuerit.*
>
> Washern (68): *In silva Milchete pastura quater xx porcis et quater xx caretedes lignorum et ad domos et ad sepes quod opus est.*

Two holdings, now in Wiltshire, but described in the Berkshire folios, refer to *silva ad clausuram* at Shalbourne (57b) and Bagshot (60b), which implies wood for fencing.

Underwood (*silva minuta*) is mentioned in five entries,[1] and *silva parva* or *parva silva* in another four entries.[2] As all nine entries give the size, we may doubt whether the witnesses saw any real difference between *silva minuta* and *silva parva*, and they have been plotted alike on Fig. 9; both types of underwood were usually, though not entirely, measured in acres. Of the entries with linear dimensions, that for Cowesfield (73b) gives only one dimension, 2 furlongs of underwood. One vill had both wood and underwood entered for separate holdings—Hartham (68b, 69b, 71). Three acres of alder-wood (*alnetum*) were entered for Stanton St Bernard (67b), 4 acres of spinney (*spinetum*) for Bremhill (67) and Littlecott (71), and 10 acres of thorn-wood (*runcetum*) for Chilmark (67b). The description of Bedwyn (64b) mentions a grove (*lucus*) half a league by 3 furlongs that had belonged to the manor *T.R.E.* but had fallen into the hands of Henry de Ferrers; it appears in the account of his fief (72) as *silva*, but without any dimensions. The wood on the royal manor of Britford (65) is said to be in the king's hand and to render 40s. in his *firma*. Finally, each of two out of the three entries for Potterne (69b, 73b) gives a set of dimensions and then leaves a space where the word *silva* or *pastura* should have been inserted; the entry on fo. 73b has the letter 'r' (*require*) alongside in the margin.[3]

Distribution of woodland

As Fig. 9 shows, the main recorded wood was upon the clay areas of the northern and western parts of the county. Large quantities were

---

[1] At Cowesfield (73b), Ditteridge (71b), Sherston (71), Wilcot (69) and Woodborough (72b).
[2] At Colerne (71), Coombe Bissett (65), Hartham (69b) and Wylye (68).
[3] The entries have not been plotted on either the wood or the pasture map.

entered for some manors, and there were smaller amounts for the villages
in and around the vales of Pewsey, Warminster and Wardour. Beyond,
along the extreme north-west margin of the county, wood was relatively
scarce on the Oolitic outcrops of 'Cotswold' Wiltshire.

To the south and east, wood was relatively infrequent on the chalk
outcrops, but moderate amounts appear to have been associated with the
Grovely Forest area and with the forests of the east, Savernake, Chute,
Clarendon and Melchet. The exceptionally large amount of wood (6 by
4 leagues) entered for Amesbury may reflect the proximity of the two
latter forests.[1]

Forest

Forest (*foresta*) is specifically mentioned in five entries:

Grovely (74): *Forestarii regis tenent i hidam et dimidiam in foresta de Gravelin-
ges. Valet xxx solidos.*
Laverstock (68): *Hujus terrae quarta pars in foresta regis est posita.*
Milford (71): *Medietas hujus terrae est in foresta regis.*
Milford (74): *Dimidia haec terra* [sic] *est in foresta.*
Downton (65b): *rex habet in sua foresta iiii hidas.*

The first entry refers to Grovely Forest on the high ground between the
Wylye and the Nadder where the Chalk is covered by Clay-with-flints.
The entries for Laverstock and Milford probably refer to Clarendon
Forest (Fig. 10). Nearby are the Winterbourne villages; the Domesday
entries for them make no references to forest but there are two references
in the Geld Accounts under Alderbury Hundred; one of these specifically
mentions forest and the other probably refers to the same Winterbourne
holding.[2]

(*a*) *Edwardus vicecomes iii hidas, de his sunt ii in foresta regis* (*8b*).
(*b*) *Edwardus ii hidas quas rex accepit de eo* (*2b*).

The entry under Downton may refer, as W. H. Jones believed, to the New
Forest or to Melchet Forest.[3] Two of the entries relating to Downton

[1] But note that Amesbury was a complex and unusual manor. Bowcombe (51, 64b)
in the Isle of Wight and Lyndhurst (39) in the New Forest had contributed to the *firma*
of Amesbury at some period before 1086.   [2] *V.C.H. Wilts.* II, p. 203.
[3] W. H. Jones, *Domesday for Wiltshire* (London and Bath, 1856), p. xliii. See
H. C. Darby and Eila M. J. Campbell (eds.), *The Domesday Geography of South-east
England* (Cambridge, 1962), p. 337.

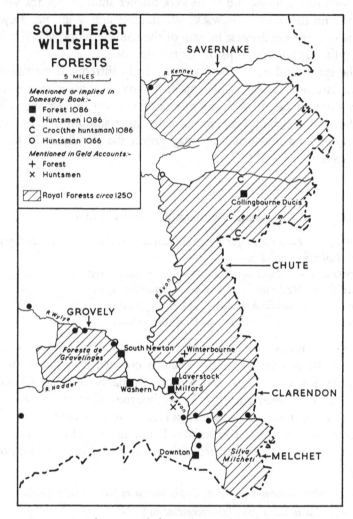

Fig. 10. South-east Wiltshire: Domesday and later Forests.

The extent of the royal forests about A.D. 1250 is taken from
V.C.H. Wiltshire, IV (London, 1959), pp. 391–460.

hundred in the Geld Accounts add further details which do not appear in the Domesday entry:[1]

(a) *Inde habet Walchelinus episcopus xxx hidas, et ii hidas de quibus homines ibi manentes fugati sunt propter forestam regis (2b).*

(b) *Duae hidae de terra Walchelini episcopi sunt ibi wastae propter forestam regis (9, 15b).*

The Domesday text itself makes reference not to Melchet Forest but to Melchet Wood (*silva Milcheti*) in two entries on fo. 68 for South Newton and Washern.[2] In the same way the entry for Collingbourne Ducis (65) refers not to Chute Forest but to the wood called Chute (*tercia pars silvae quae vocatur Cetum*).[3] The other Wiltshire forests go unnoticed in the Domesday text unless we take the great woodlands (each 4 leagues by 4) of the manors of Chippenham and Melksham to refer to the forests that bore those names in later records. Other Wiltshire forests that appear in later records are those of Braydon, Savernake and part of Selwood. Their presence in the eleventh century is suggested only by the sparsity of vills in their neighbourhood, and by the low densities of population and teams in the areas they occupied. Some landholders are described as huntsmen in the Domesday entries, and still others in the Geld Accounts. One of the Geld Account huntsmen held one hide at Britford,[4] where, on another holding, the Domesday Book records: *Silva est in manu regis* (65).

MEADOW

Types of entries

The entries for meadow in the Wiltshire folios are comparatively straightforward. For holding after holding, the same phrase is repeated— '*n* acres of meadow' (*n acrae prati*). The amount of meadow entered under each place-name ranges from only half an acre at North Wraxall (69b), for example, to 200 acres each at Ashton Keynes (67b) and Bedwyn (64b); there is also an amount of 200 acres recorded in the joint entry for Latton and Eisey (68b). Such large quantities are exceptional. Not many entries record more than 100 acres, and some of these clearly cover more than one settlement, e.g. the 130 acres for the manor of Melksham (65) and the 100 acres for that of Chippenham (64b). By far the great majority of vills

---

[1] *V.C.H. Wilts.* II, p. 208.   [2] See pp 34–6 above.
[3] See pp. 33–4 above.   [4] *V.C.H. Wilts.* II, p. 202.

had under 30 acres of meadow apiece. Round figures, such as 10 or 40 acres, suggest that they are estimates, but there are also detailed figures that suggest precision, e.g. the 1½ acres at Chedglow (70), the 9 acres at Little Langford (69 b), and the 108 acres at All Cannings (68). As for other counties, no attempt has been made to translate these figures into modern acreages, and they have been plotted as conventional units on Fig. 11.

There are some unusual entries. The meadow of seven places is measured in terms of linear dimensions, e.g. the entry for Aldbourne (65) says: *pratum i leuua longum et v quarentenis latum*. With the exception of Tilshead (65), all these places lie in the north-east corner of the county. Four entries use the arpent, a measure usually reserved for estimating vineyards: those for Burbage (74 b) with 2 *arpenʒ* of meadow, for Marten (74 b) with 2, and for Shalbourne (73, 74 b) with one and then three. One can only note that three of these entries occur close together on the upper part of the left-hand column of fo. 74 b, and that among them is an entry for Grafton which does not mention meadow but which measures woodland by the arpent.[1] Each of two entries gives a combined figure for both meadow and pasture:

Edington (74 b): *tantum prati et pasturae quantum convenit i hidae.*[2]
Marten (74): *Ibi xxx acrae inter pratum et pasturam.*[3]

At Edington, where the meadow and pasture were 'appropriate' for 1 hide, the holding itself answered for 1 hide.

Distribution of meadowland

As might be expected, the northern clayland, well-watered by the Thames, the Bristol Avon, and a network of tributaries, had a great deal of meadow. Almost every village here had some registered for it. There were large amounts for a number of villages near the Thames itself— 200 acres for Ashton Keynes (67 b), 200 jointly for Latton and Eisey (68 b), 100 for Castle Eaton (68 b) and 60 for Calcutt (73 b). To the south, quantities of meadow marked the lines of villages along the alluvial valleys that

---

[1] See p. 34 above.

[2] The Edington entry has been plotted as 'other mention'; another Edington entry separately mentions meadow and pasture (68). For Marten, 15 acres of meadow and 15 acres of pasture have been plotted.

[3] R. R. Darlington has suggested that this may imply 'that the extent of the meadow of this settlement on high ground varied in accordance with the rainfall' (*V.C.H. Wilts.* II, p. 56).

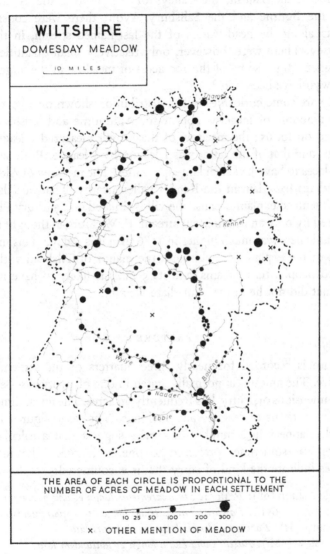

WILTSHIRE
DOMESDAY MEADOW
⌐ 10 MILES ⌐

THE AREA OF EACH CIRCLE IS PROPORTIONAL TO THE
NUMBER OF ACRES OF MEADOW IN EACH SETTLEMENT

10 25 50 100 200 300

× OTHER MENTION OF MEADOW

Fig. 11. Wiltshire: Domesday meadow in 1086.

When the meadow of a settlement is entered partly in acres
and partly in some other way, both are shown.

break the chalk plateau, the valleys of the Wylye, the Nadder, the Ebble, the Bourne and the Salisbury Avon; there were considerable amounts along the head-waters of the last-named stream, in the Vale of Pewsey. There were, however, only relatively small quantities along the Kennet. Might some of the 200 acres of meadow in the large manor of Bedwyn have been here?

There are some curiosities in the distribution shown on Fig. 11. Substantial amounts of meadow appear for Aldbourne and Tilshead, both described on fo. 65; the meadow of the former measured 1 league by 5 furlongs, and that of the latter was 1 league by ¼ league. Both seem unlikely places to have had such large quantities. The meadow of Aldbourne may perhaps have lain in the Kennet valley nearby. That of Tilshead is more difficult to explain because the vill is situated on Salisbury Plain and is watered by only an intermittent stream. F. W. Morgan thought that the water-table here was much higher in 1086 than it is today.[1] This may well be part of the explanation, although the manor of Tilshead with its 98 recorded people, its 27 teams, and its 9 mills is likely to have included much that did not lie in the immediate vicinity.

Types of entries   PASTURE

Pasture is recorded for nearly three quarters of the settlements of Wiltshire. The amount is normally stated in one of two ways—either in linear dimensions or, a little less frequently, in acres. The linear dimensions are given in terms of leagues and furlongs. The large figures for furlongs that appear in a number of entries suggest that a calculation in furlongs was sometimes preferred to one in leagues. The following examples indicate the kind of entry that is encountered:

Steeple Ashton (68): *Pasturae xix quarentenis longa et una quarentena lata.*
Wanborough (65 b): *Pastura dimidia leuua longa et xv quarentenis lata.*
Westbury (65): *Pastura iii leuuis longa et iii leuuis lata.*
Winterslow (68 b): *Pastura una leuua longa et tantundem lata.*

The exact significance of these linear dimensions is far from clear,[2] and we cannot hope to convert them into modern acreages. All we can do is

---

[1] F. W. Morgan, 'The Domesday geography of Wiltshire', *Wilts. Arch. and Nat. Hist. Mag.* XLVIII (1936), p. 76.
[2] See p. 372 below.

to plot them diagrammatically as on Fig. 12, and we have assumed that a league comprised 12 furlongs.[1] The measurements (3 by 3 leagues) in an entry for a large manor, such as Westbury (65), may well cover a number of widely separated places, and this may imply, though not necessarily, some process of addition whereby the dimensions of separate tracts have been consolidated into one sum.

Three entries state the amount of pasture to be 'in length and breadth'; thus for Melksham (65) we hear of *vii leuuae pasturae in longitudine et latitudine*.[2] We have assumed that this implies the same facts as the more usual formula.[3] Nine entries for eight places give only one measurement,[4] e.g.:

Ansty (73 b): *ii quarentenae pasturae.*
Winterbourne (73 b): *dimidia leuua pasturae.*

Whatever be the precise significance of such entries, they have been plotted as single lines on Fig. 12.[5]

Quite frequently the amount of pasture is stated in terms of acres. Amounts are generally below 30 acres, but they range from 3 acres at Thickwood (69b) to 150 acres at South Newton (68). The figures are often in round numbers that suggest estimates, but some seem to be precise, e.g. the 9 acres at Wishford (68) and the 13 at Winterbourne (69). Like the acreages for meadow, they have all been treated as conventional units. Occasionally, both acres and linear dimensions appear in entries for one place-name, e.g. Burcombe with 20 acres (68b) and with 8 furlongs by 1 furlong of pasture (68). The acre seems to be used as a linear measure in two entries:[6]

Crofton (70): *pastura vi acris longa et totidem lata.*
Orcheston (72b): *Pastura dimidia leuua et xl acris longa et lata.*

---

[1] The pasture (1 league by ¼ league) at Latton and Eisey has been plotted at the former, and a symbol for 'other mention' plotted for the latter.
[2] The other two entries are for Manningford (74) and Whaddon (73b).
[3] The Exeter text for the single holding common to both it and the Exchequer version does not provide conclusive proof because the dimensions for breadth happen to be different, although this may be due to scribal error—see p. 1 above.
[4] The entries for Ansty (73b), East Coulston (73b), Horningsham (70b), Lus Hill (69, 70b), Porton, (74), Tilshead (73b), Grimstead (74) and Winterbourne (73b).
[5] See p. 101 below.
[6] Indicated under the category of 'other mention' on Fig. 12.

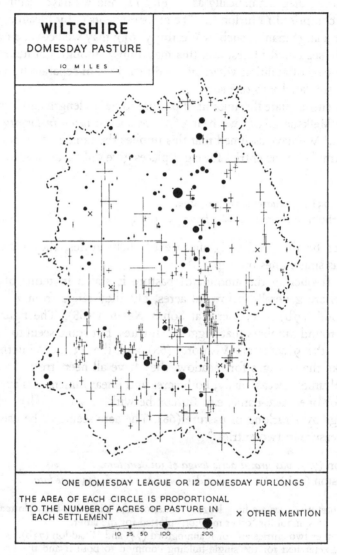

WILTSHIRE

DOMESDAY PASTURE

10 MILES

—— ONE DOMESDAY LEAGUE OR 12 DOMESDAY FURLONGS

THE AREA OF EACH CIRCLE IS PROPORTIONAL
TO THE NUMBER OF ACRES OF PASTURE IN
EACH SETTLEMENT   × OTHER MENTION

10 25 50   100   200

Fig. 12. Wiltshire: Domesday pasture in 1086.

When the pasture of a settlement is entered partly in linear
dimensions and partly in some other way, both are shown.

Three holdings each comprised two separate quantities of pasture:

Odstock (73 b): *Pastura i leuua longa et ii quarentenis lata et in alia parte v acrea pasturae.*

Salisbury (66): *Pastura xx quarentenis longa et x quarentenis lata et alibi v quarentenae pasturae longitudine et una latitudine.*

Stratford Tony (69): *ii acrae pasturae juxta flumen et alia pastura i leuua longa et vi quarentenis lata.*[1]

There are a number of unusual entries. In that for Corsham (65) we hear of *una hida pasturae* instead of the usual formula. That for Chedglow (71) speaks of 1½ virgates but omits what should be, it would seem, the word *pasturae*. The entry for Porton (69 b) is unique in the Wiltshire folios because it mentions sheep: *Pastura l ovibus.* Each of two entries gives a combined figure for both meadow and pasture: for Edington (74 b) and Marten (74).[2] Each of two out of the three entries for Potterne (69 b, 73 b) gives a set of dimensions and then leaves a space where the word *silva* or *pastura* should have been inserted; the entry on fo. 73 b has the letter 'r' (*require*) alongside in the margin.[3]

Distribution of pasture

Generally speaking, the main feature of Fig. 12 is the absence of recorded pasture from many villages in the north-west of the county. Elsewhere, the distribution of villages with pasture is general. Holdings with pasture measured in acres seem to be confined to certain areas, but how significant geographically or tenurially this is, we cannot say.

MILLS

Mills are mentioned in connection with 197 of the 335 Domesday settlements in Wiltshire; a mill also appears in an entry for one anonymous holding.[4] The Latin form *molinum* is usually used but the form *molendinum* is occasionally to be found, e.g. in the entry for Alderton (72 b). The number of mills on a holding is stated, and normally their annual value, ranging from a mill at Lus Hill (70 b) that rendered only 12*d.* to the half-mill at

---

[1] R. R. Darlington suggests that the 2 acres probably comprised marshy land near the Ebble, while the 'other pasture' was on the downs above the village (*V.C.H. Wilts.* ii, p. 56).

[2] See p. 40 above.   [3] See p. 36 above.

[4] The anonymous holding was that of Bishop Osmund (65 b). See p. 8 above.

Salisbury (64b) worth 30s. and the 2 mills at Keevil (69b) worth 65s. No render was entered for the 2 mills at Latton and Eisey (68b). Such sums as 5s., 10s., 12s. 6d. and 25s. were frequent. Renders of uneven amounts were rare, e.g. the mill at Alderton (72b) returned 37d. and the four mills at Salisbury (66) returned 47s. 7d. No eels were recorded as part of any of the Wiltshire mill renders.

Fractions of mills are frequently encountered. Thus two holdings at Langford (68b, 72) each had half a mill rendering 30d. There were also two holdings at Sherrington (72b bis) each with half a mill that rendered 7s. 6d. At Compton Bassett (70b, 71b, 74) three holdings each had 'a third part' of two mills rendering 10s.'. At Sutton Veny 'two parts' of a mill rendered 13s. 4d. (72) and, on another holding, a 'third part' rendered 6s. 8d. Three Winterbourne entries presumably refer to the same place; one speaks of 'half a mill' rendering 3s. 9d. (73), and two others each speak of 'part of a mill' rendering 22½d. (73b, 74). Presumably there was one mill rendering 7s. 6d. on the River Bourne.[1] The fractions of mills and their renders do not always combine so neatly and we may wonder whether some fractions have been entered as whole mills. Thus on one holding at Manningford there was a 'third part' of a mill rendering 50d. (73b), and on another holding there were 'two parts' of a mill rendering 12s. 6d. (74). Could this latter figure refer to the whole mill, the respective renders being 50d. and 100d.? There was, incidentally, yet another mill entered for Manningford, and this did render 12s. 6d. (67). Or, again, at Alderton there was one mill rendering 37d. on one holding (72b), and a 'part of a mill' rendering 22d. on another holding (73). Might these two references really be to a single mill that rendered 5s., a common figure for mill payments? There were two holdings at Codford (71b, 72b), each with a quarter of a mill rendering 3s. and 3s. 1½d. respectively. We hear of another whole mill paying 10s. at Codford (72). Could the missing portion have been the half-mill rendering 6s. 3d. nearby at Ashton Gifford (70b)? At Somerford there were four separate parts of a mill, each yielding 15d., on four holdings (69b, 73b ter); there was also another mill rendering 5s. (70b); and there was a 'third part' of a mill rendering 8s. (70b), but we are told nothing of the other two parts. Could all six entries be referring to 1 mill yielding 18s.? The Domesday text certainly does not warrant such a conclusion; but it is interesting to note that there was yet another mill rendering 20s. on another holding at Somerford (67). In our

[1] See p. 10 above.

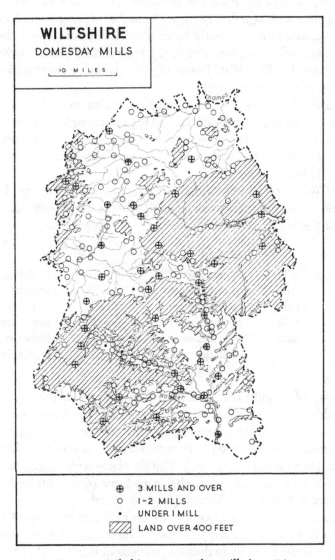

Fig. 13. Wiltshire: Domesday mills in 1086.
There was also a mill on an anonymous holding—see p. 45.

48   WILTSHIRE

reckoning, we have not made any assumptions but have counted Somerford as a place with 3⅓ mills. There are a number of other fractions which,
apparently, cannot conveniently be combined with others, e.g. the half-
mills at Cheverell (70), Ditteridge (70b), Hurdcott in Winterbourne Earls
(72), and the 1½ mills at West Dean (72). Clearly, many uncertainties enter
into a count of Domesday mills.

There were a number of anomalies, or apparent anomalies, about the
incidence of these mills. Wolf Hall (74b), for example, with no teams,
had a mill rendering 16s. Conversely, there were large and seemingly
prosperous manors for which no mills were recorded, e.g. Wootton
Rivers (65) with 16 teams.

Many Wiltshire villages with mills had one or two apiece, but the
table below cannot be very accurate in view of the uncertainties associated
with many entries:

Domesday Mills in Wiltshire in 1086

Under 1 mill	5 settlements	6 mills	2 settlements
1 mill	102 settlements	7 mills	4 settlements
2 mills	52 settlements	8 mills	6 settlements
3 mills	11 settlements	9 mills	2 settlements
4 mills	7 settlements	10 mills	1 settlement
5 mills	4 settlements	13 mills	1 settlement

The group of 13 mills (12¼ in fact) was at Chippenham (64b, 73), and
presumably this large manor with 177 recorded adults and 85¼ plough-
teams covered a number of settlements; the same was true of Ramsbury
with 10 mills (66). The two places with 9 mills apiece were the boroughs
of Calne (64b) and Tilshead (65). The six places with 8 mills each were
substantial manors: Bedwyn (64b), Amesbury (63b), Brokenborough
(67), Donhead (67b), Melksham (65) and Westbury (65, 74b). These
examples illustrate the unsatisfactory nature of the table above. We might
go so far as to say that a large number of mills entered under one name
creates a suspicion that more than one settlement is involved. This is not
to say that the table is without value, because, at any rate, it suggests that
there were some villages with more than one mill at work.

Fig. 13 shows how the mills were associated with the rivers, and
especially with the Bristol Avon, the Salisbury Avon, the Wylye and, to
a less extent, with the Kennet, the Nadder and the Ebble. Areas where

villages were without mills lay (*a*) in the northern plain between Malmesbury, Calne and Bradford on Avon and (*b*) on the downland to the north and south of the Kennet.

CHURCHES

Churches are not regularly enumerated in the folios for Wiltshire, and they are mentioned in connection with only 29 out of the 335 places recorded for the county. We are also told that Bishop Osmund of Salisbury held half a church (65 b), but there is no clue as to where. Fourteen of these places were among the 22 royal manors of the county (64 b–65 b). Another ten were named in the account of certain churches and their holders which immediately follows the description of the royal manors (65 b). Of the remaining five places, four belonged to ecclesiastical fiefs and one to that of Edward of Salisbury.[1] On the royal manor of Wootton Rivers (65) there were said to be two churches.

The endowments of most of these churches appear to be ancient because they were assessed for geld. Their values were often substantial. The church at Calne (64 b), for example, assessed at 6 hides, was worth £8; those at Corsham (65), Heytesbury (65 b) and Highworth (65 b), each assessed at 3 hides, were worth £7, £3 and £5 respectively. The church at Upavon (65 b), assessed at 2½ hides, was worth as much as £10. 15 s. The lands of a number of other churches were assessed at 2 hides or less. To the church at Alderbury (68 b) belonged 1 hide which had never paid geld; and at Pewsey (65 b) the church had one carucate of land (*una car' terrae*).[2] Three entries provide unusual details. The church at Netheravon was ruined, roofless and on the point of collapse; that at Collingbourne Ducis was ruined and decayed but its tithe was worth 10s.;[3] the church at Wilcot, on the other hand, was a new one:

Netheravon (65): *Ecclesiam hujus manerii cum una hida tenet Nigellus medicus. Haec cum omnibus appendiciis suis valet xxxii libras. Ipsa vero vasta est et ita discooperta ut pene corruat.*

Collingbourne Ducis (65): *Ad ecclesiam pertinet dimidia hida. Hujus ecclesiae decimam tenet Giraldus presbyter de Wiltone et valet x solidos. Ecclesia vasta est et dissipata.*[4]

Wilcot (69): *ecclesia nova.*

[1] The places are Alderbury (68 b), Brixton Deverill (68 b), Cricklade (67), Downton (65 b), and Wilcot (69). [2] See p. 16 above.
[3] This is the only reference to tithe in the Domesday entries for Wiltshire.
[4] For Gerald the priest of Wilton, see p. 27 above.

Eleven of these churches are said to be held by priests, who are usually named. Thus the entry for Heytesbury (65b) runs: *Alwardus presbyter tenet ecclesiam de Haseberie*. There is a variant in the entry for Westbury (65): *De eadem terrae hujus manerii habet ecclesia hidam et dimidiam et quidam clericolus tenet.*[1] We also hear of priests in other entries that do not mention churches, but they appear either as named or unnamed landholders. At Ramsbury (66) there were priests holding 4 hides. Only at Aldbourne (65), Potterne (66) and Rowde (70) can we be sure that the priests were resident, although no churches were specifically entered for Potterne and Rowde.[2]

Clearly the record of churches for Wiltshire is very incomplete. No church is recorded for Warminster (64b), but the name suggests an early church and we know that one was there. We know, also from non-Domesday evidence, that there were pre-Conquest churches at Bradford on Avon, Bremhill, Burcombe and Limpley Stoke.[3] Then, too, there were some churches not specifically mentioned but which were named as tenants-in-chief: Amesbury, Malmesbury and Wilton.

URBAN LIFE

There seem to have been ten boroughs in Domesday Wiltshire. The one true county borough was Malmesbury, described at the beginning of the record for the shire; more information is given about it than about any of the others. Four boroughs—Bedwyn, Calne, Tilshead and Warminster —formed parts of large royal manors. Another two also formed parts of substantial manors, Bradford on Avon held by the abbess of Shaftesbury, and Salisbury held by its bishop; the latter place is not described as a borough or said to have burgesses, but its burghal status is attested by a reference to the yield of its 'third penny'. Finally, Cricklade, Marlborough and Wilton are not even the subjects of regular entries, but, again, the burghal status of the first two is indicated by references to the 'third penny' of each; we also hear of appurtenant burgesses at Cricklade, and Wilton is specifically called a borough.

A market is recorded only for Bradford on Avon, and a mint only for

---

[1] *Clericolus* may imply a young clerk.

[2] For the count of priests see p. 26 above.

[3] *V.C.H. Wilts* II, pp. 32–3. Limpley Stoke is not named in the Domesday text, but it was at one time a detached part of Bradford on Avon.

Malmesbury, although we know from non-Domesday evidence that there were mints at Bedwyn, Cricklade, Marlborough, Salisbury, Wilton and, in pre-Conquest times, Warminster. The information about all ten boroughs is extremely meagre, especially for those other than Malmesbury.

Malmesbury

The description of Wiltshire opens with an account of Malmesbury (64b) and there are also incidental allusions to the borough on other folios. The main entry begins by saying that the king had 26 occupied messuages (*hospitatae masurae*) and 25 other *masurae* in which were houses that did not pay geld any more than waste land did. Each *masura* paid a rent of 10*d.*, and the total is said to be 43*s.* 6*d.*, that is one shilling less than fifty-one times 10*d.*, yet another example of the anomalies so often presented by Domesday arithmetic:

In burgo Malmesberie habet rex xxvi masuras hospitatas et xxv masuras in quibus sunt domus quae non reddunt geldum plus quam vasta terra. Unaquaque harum masurarum reddit x denarios de gablo; hoc est simul xliii solidos et vi denarios.

The entry also gives a list of 20¼ *masurae* and 2½ waste *masurae* held by different lords. To these must be added the 9 coscets belonging to the abbot of Malmesbury; they were outside the borough but they paid geld with the burgesses (*foris burgum ix coscez qui geldant cum burgensibus*). A later paragraph on the same folio says that there were 4 *masurae* and 6 waste *masurae* on an acre or field (*un' agr' terrae*) which Earl Harold had held. We also hear, on other folios, of 11 *burgenses* and 5½ *domus* in Malmesbury but attached to various nearby rural manors, most of which formed part of one or other of the fiefs of the lords named in the main entry.[1]

The two sets of information relating to the holdings of these lords are set out in the table on p. 53. It is difficult to see what relation one list bears to the other. There appear to be both coincidence and discrepancy between them. The main entry, for example, does not mention Miles Crispin, part of whose fief was Wootton Bassett (71). On the other hand, no Malmesbury holdings are mentioned for the rural manors of some of the lords named in the main entry. We may suspect that the manorial descriptions do not give us all the information they might have done. It

[1] The implications of the connection lie outside the scope of this chapter. For a discussion of the general problem, with references, see Carl Stephenson, *Borough and Town* (Cambridge, Mass., 1933), pp. 81 ff.

may be significant that those lords named in the main entry but without
record of rural connections all held estates in the neighbourhood of
Malmesbury, estates to which unrecorded appurtenant holdings may well
have been attached.[1]

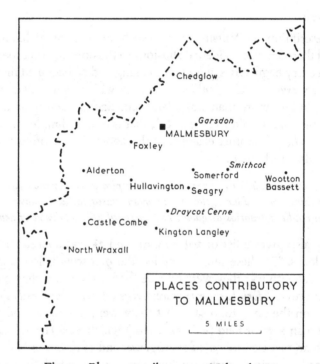

Fig. 14. Places contributory to Malmesbury.

Malmesbury is not specifically named in the entries for
Draycot Cerne, Garsdon and Smithcot.

On the assumption that the two lists are mutually exclusive, the total
of *masurae, domus, burgenses* and *cosceʒ* amounts to 100¾ together with 8½
waste *masurae*. We may therefore believe that Malmesbury was a borough
with at least 400–500 inhabitants, probably more when one remembers the

[1] Durand of Gloucester held estates in Ashley (now in Glos.), Luckington, Seagry,
and Tockenham (71 b); William of Eu in Beversbrook, Compton Bassett, Ditteridge,
Hilmarton, Lackham, Sevington, Sopworth and Yatton Keynell (71); Osbern Gifford
in Stanton St Quinton and Tytherton (72b); Tovi in Chelworth in Crudwell (67);
Edric's wife in Calstone Wellington (70).

presence of the abbey with its monks and servants, and also the possibility of unrecorded appurtenant holdings. If one assumes that the two lists are not mutually exclusive, the general estimate is not greatly affected. That Malmesbury was a centre of trade may be gathered from the mention of a mint (*moneta*) which rendered 100*s*. We hear also of a mill rendering 10*s*., but what else entered into the economic life of the borough, we cannot say. It seems to have been little affected by the Conquest; at any rate its annual

Places contributory to Malmesbury

Entered under Malmesbury (64b)		Entered under rural manors	
	Masurae	*Burgensis* (b) or *Domus* (d)	
Bishop of Bayeux	½ *vasta*	—	
Abbot of Malmesbury	4½	Somerford (67)	1 b of 12*d*.
		Garsdon (67)	1 b of 3*s*.
Abbot of Glastonbury	2	Kington Langley (66b)	1 b of 15*d*.
Edward the sheriff	3	Somerford (69b)	1 d of 15*d*.
		North Wraxall (69b)	2 b of 2*s*.
Ralph de Mortemer	1½	Hullavington (72)	1 d of 1*s*.
		Alderton (72b)	1 b of 7*d*.
Durand of Gloucester	1½	not mentioned	
William of Eu	1	not mentioned	
Humphrey de l'Isle	1	Somerford (70b)	1 b of 1*s*.
		Smithcot (70b)	1 b of 8*d*.
		Castle Combe (71)	2 b of 18*d*.
Osbern Gifford	1	not mentioned	
Alfred of Marlborough	½ *vasta*	Chedglow (70b)	½ d of 6*d*.
Geoffrey the marshal	½ *vasta*	Draycot Cerne (74b)	1 b of 1*s*.
Tovi	1¼	not mentioned	
Drew fitzPonz	½	Seagry (72b)	1 d of 9*d*.
Edric's wife	1	not mentioned	
Roger of Berkeley	1	Foxley (72b)	1 d
Arnulf of Hesdin	1	not mentioned	
The king	1 *vasta*	not mentioned	
Miles Crispin	not mentioned	Wootton Bassett (71)	1 d of 13*d*.

Note: 1. The names 'entered under Malmesbury' are set out in the order in which they occur on fo. 64b.

2. The entries for Garsdon, Smithcot and Draycot Cerne do not specifically name Malmesbury, but this is the borough probably meant.

3. The king's waste *masura* belonged to the land which Azor held *T.R.E.*

render was the same in 1086 as in 1066. The yield from the 'third penny' amounted to £6, the same as that of Salisbury, but more than that of Cricklade and Marlborough. With such fragments of information as these, we must be content.

When the king went on an expedition by land or sea, it was the duty of the borough to provide 20*s*. for feeding his sailors (*ad pascendos suos bu*ʒ*ecarl'*); as an alternative, the king might take with him (*ducebat secum*) one man from each honour (*pro honore*) of 5 hides.

Bedwyn

Bedwyn was the head of a large royal manor that in 1086 still rendered a night's *firma* (64 b). It had as many as 8 mills, 79 plough-teams, and a recorded population of 172 together with 25 burgesses. The form of the entry makes it impossible to say how this population and these resources were distributed between the borough and the rest of the manor, and we cannot say what was the relation of the 25 burgesses to the total number of inhabitants in the borough. Nor are we given any clue to the features that distinguished Bedwyn from the settlements around. We hear of a church (65 b), but there is no reference, for example, to a mint; yet we know from the evidence of coins that one was there.[1]

Bradford on Avon

Bradford was the head of a substantial manor belonging to the abbess of Shaftesbury (67 b). It had 40 plough-teams and a recorded population of 126 together with '33 burgesses paying 35*s*. 9*d*.'. The form of the entry makes it impossible to say how this population and these resources were distributed between the borough and the rest of the manor, and we cannot say what was the relation of the 33 burgesses to the total number of inhabitants in the borough. The mention of a market rendering 45*s*. provides an indication of the general life of the settlement.

Calne

Calne was the head of a large royal manor that in 1086 still rendered a night's *firma*. It had as many as 9 mills, 34 plough-teams, and a recorded population of 159 together with 45 burgesses on the king's demesne and another 25 on the land of the church (64 b). The form of the entry makes

[1] *V.C.H. Wilts.* II, pp. 16–20.

it impossible to say how this population and these resources were distributed between the borough and the rest of the manor. We hear of appurtenant holdings at two places nearby, to the south-east of the borough:

Bishop's Cannings (66): 1 *domus* rendering 20d. *per annum*.
Calstone Wellington (70): 1 *burgensis* rendering 11d.
Calstone Wellington (73): 2 *burgenses* rendering 20d.

Exactly what was the relation of this sum of 73 burgesses and 1 house to the total number of inhabitants in the borough, we cannot say. Calne was clearly a centre with at least 300–400 people, and probably substantially more than this.

Cricklade

There is no Domesday account of Cricklade, and we hear of it only incidentally in two entries connected with renders and in a number of entries which record holdings appurtenant to nearby manors. Of the first two entries, one says that the king received £5 from the 'third penny' of Cricklade (64b) and the other reads as follows: *Ecclesia S. Petri Westmonasteriensis tenet ecclesiam de Crichelade et habet ibi plures burgenses et tercium denarium ejusdem villae. Totum simul reddit ix libras* (67).

The contributory holdings are set out in the table on p. 56. They amount to a total of 33 burgesses, 2 houses and a garden. This can have been only a fraction of the 'many burgesses' mentioned on fo. 67. With its lower yield from the 'third penny' (£5 as compared with £6) it may have been somewhat smaller than Malmesbury. There is no reference to a mint, yet we know from the evidence of coins that one was there.[1] Cricklade, where the Roman road from Silchester to Cirencester crossed the Thames, must have been a centre of some importance.[2]

Marlborough

No burgesses are recorded for Marlborough, nor is it described as a borough (64b). It is not even the subject of a regular entry, but its burghal status is indicated by a reference to the 'third penny', which yielded £4 (*De tercio denario Merleberge, iiii librae*). The 'third penny'

[1] *V.C.H. Wilts.* II. pp. 16–20.
[2] For Cricklade as a crossing-place for armies, see, for example, the *Anglo-Saxon Chronicle* under the years 905 and 1016.

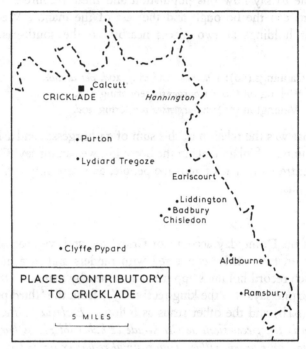

Fig. 15. Places contributory to Cricklade.

Note: The entry for Hannington does not specifically mention the name Cricklade, but this may well have been the borough that was meant.

Aldbourne (65)	6 *burgenses*	64*d.*
Badbury (66b)	1 *burgensis*	5*d.*
Calcutt (73b)	3 *burgenses*	21*d.*
Chisledon (67b)	6 *burgenses*	4*s.* 1*d.*
Clyffe Pypard (70b)	3 *burgenses*	3*d.*
Clyffe Pypard (74b)	1 *domus*	3*d.*
Earlscourt (73b)	1 *ortus*	2*d.*
Hannington (66b)	1 *domus*	5*d.*
Liddington (67b)	1 *burgensis*	6*d.*
Lydiard Tregoze (70)	7 *burgenses*	5*s.*
Purton (67)	1 *burgensis*	6*d.*
Ramsbury (66)	5 *burgenses*	5*s.*

of Salisbury and Malmesbury each yielded £6, and that of Cricklade yielded £5, but to what extent this variation reflects relative size, we cannot say. The only other mention of Marlborough is in connection with a church that was there, assessed at 1 hide and worth 30s. There is, for example, no reference to a mint, yet we know from the evidence of coins that there was one there.[1]

Salisbury

Salisbury was the head of a large 50-hide manor in the hands of its bishop (66). The manor had 32 plough-teams and a recorded population of 95, and it was worth £64. 10s. per annum. The entry that gives us this information does not mention burgesses, nor does it describe the settlement as a borough. We know that there must have been a borough because, on another folio, we are told that its 'third penny' amounted to £6, the same sum as that for Malmesbury (64b). This entry tells us also that the king had 20s. by weight from half a mill here. It is impossible to say what was the relation of the borough to the manor, and we are completely without information about its urban activities. Yet from non-Domesday sources we can assume it to have been a place of some consequence, what with its mint and its episcopal see.[2]

Tilshead

Tilshead was the head of a royal manor that in 1086 still rendered a night's *firma* said to be worth £100 by tale (65). It had as many as 9 mills, 27 plough-teams, and a recorded population of 98 together with '66 burgesses paying 50s.'. Four small estates at Tilshead are also separately listed on fo. 73b but without mention of teams or population; these were 'small portions of this large royal manor, probably given to royal servants'.[3] The form of the main entry makes it impossible to say how this population and these resources were distributed between the borough and the rest of the manor, and we cannot say what was the relation of the 66 burgesses to the total number of inhabitants of the borough. Clearly Tilshead was a borough with a population of at least about 300, and maybe considerably more. It is surprising to find such an urban centre set on the chalk expanse of Salisbury Plain, and the location of its meadow and

[1] *V.C.H. Wilts.* II, pp. 16–20.
[2] *V.C.H. Wilts.*, VI (London, 1962), p. 62. [3] *Ibid.* p. 193.

its 9 mills raises problems.[1] Some earlier writers identified *Theodulveside* not with Tilshead but with Devizes,[2] but place-name evidence points to the former place.[3]

Warminster

Warminster was the head of a substantial royal manor that in 1086 still rendered a night's *firma* (64b). It had as many as 7 mills, 42 plough-teams, and a recorded population of 74 together with 30 burgesses. The form of the entry makes it impossible to say how this population and these resources were distributed between the borough and the rest of the manor, and we cannot say what was the relation of the 30 burgesses to the total number of inhabitants in the borough. Nor are we given any hint of the economic activity that entered into the urban life of the settlement. There is, for example, no reference to a mint, yet we know from the evidence of coins that there was one here in pre-Conquest days.[4]

Places contributory to Wilton

Note: The entries for Dinton and Sutton Mandeville do not specifically name Wilton, but this may well have been the borough that was meant. Dinton is, however, tentatively assigned to Warminster in *V.C.H. Wilts.* II, p. 21.

Castle Combe (71)	1 *burgensis*	5s.
Dinton (67b)	2 *burgenses*	10d.
Durnford (71b)	4 *domus*	4s.
Fifield Bavant (70b)	2 *burgenses*	18d.
Marden (73)	1 *domus*	10d.
Netheravon (65)	5 *burgenses*	6s.
Odstock (73b)	1 *burgensis*	12d.
Salisbury (66)	7 *burgenses*	65d.
Sherrington (72b)	1 *burgensis*	3s.
Stratford Tony (69)	1 *burgensis*	20d.
Sutton Mandeville (72)	5 *burgenses*	40d.

Wilton

There is no Domesday account of Wilton, and we hear of it only incidentally in two entries connected with renders and in a number of entries which record holdings appurtenant to eleven rural manors. Of the

[1] *V.C.H. Wilts.* II, p. 56 n. See p. 48 above.
[2] E.g. A. Ballard, *The Domesday Boroughs* (Oxford, 1904), p. 9.
[3] J. E. B. Gover, Allen Mawer and F. M. Stenton, *op. cit.* pp. 236–7.
[4] *V.C.H. Wilts.* II, pp. 16–18.

59

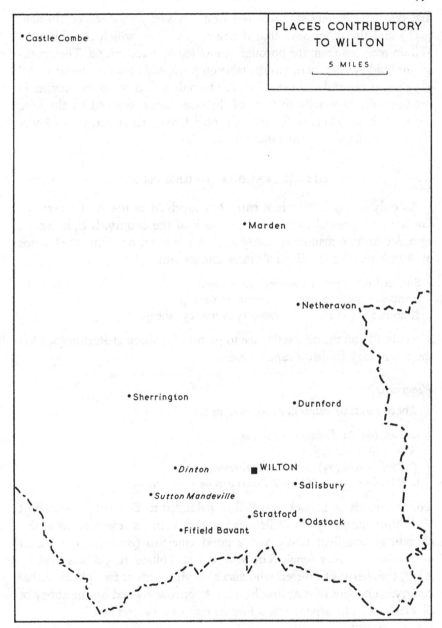

Fig. 16. Places contributory to Wilton.

Wilton is not specifically named in the entries for Dinton and Sutton Mandeville.

first two entries, one says that the king received £50 *de burgo Wiltunie* (64b) and the other says that the rents (*redditus*) which the abbess of Wilton received from the borough amounted to £10. 17s. 6d. The appurtenant holdings are set out in the table on p. 58, and they amount to a total of 25 burgesses and 5 houses. This can be only a fraction of the number in the borough, especially in view of the large sums received by the king and the abbess. There is, for example, no reference to a mint, yet we know from the evidence of coins that one was there.[1]

MISCELLANEOUS INFORMATION

Livestock

As only a single Wiltshire entry has survived in the Exeter text, we can have no general view of the livestock of the county. It is, however, recorded for two manors formerly in Somerset but now within the bounds of Wiltshire. The details of all three entries are:

Sutton Veny (*47*): 1 rouncey, 300 sheep.
Yarnfield (*447*): 2 cows, 25 swine, 134 sheep.
Kilmington (*453*): 14 *animalia*, 15 swine, 137 sheep.

There is also an unusual reference to pasture for sheep at Porton (69b) in the Domesday folios: *Pastura l ovibus.*

Vineyards

There are four references to vineyards:

Lacock (69b): *dimidia acra vineae.*
Wilcot (69): *vinea bona.*
Tollard Royal (73): *Ibi ii arpenni vineae.*
Bradford on Avon (67b): *Ibi unus arpennis vineae.*

The vineyards at Lacock and Wilcot belonged to Edward the sheriff of Wiltshire; the entry for Wilcot is unusual with its reference to a new church, an excellent house and a good vineyard (*ecclesia nova, domus obtima* [sic] *et vinea bona*). The vineyard at Tollard Royal was held by Aiulf, the sheriff of Dorset, who also held vineyards at two places in that county.[2] The vineyard at Bradford on Avon was owned by the abbey of Shaftesbury. The arpent was a French unit of measurement.[3]

[1] *V.C.H. Wilts.* II, p. 20.   [2] See p. 125 below.
[3] For a discussion of the arpent see Sir Henry Ellis, *A General Introduction to Domesday Book*, I, p. 117.

Other references

Waste is not mentioned, but at Cumberwell (70b) there was nothing (*nil ibi est*). The Geld Accounts occasionally refer to waste (see p. 31).

The entry for Earlscourt (73b) mentions a garden (*ortus*) at Cricklade.

A market is entered for the borough of Bradford on Avon (67b): *Mercatum reddit xlv solidos*; and a mint for that of Malmesbury (64b): *De moneta reddit ipsum burgum c solidos*.

At Fifield Bavant (70b) there was a forge: *Ibi i ferraria reddit xii denarios per annum*.

The prefatory matter, before the account of the *Terra Regis* begins, contains a reference to a due received by the king (64b): 'From Wiltshire, the king has £10 instead of a falcon (*pro accipitre*) and 20*s.* for a sumpter horse (*pro summario*) and 100*s.* and 5 *orae* for fodder (*pro feno*).'

An unusual feature is the statement of the rents (*redditus*) received by the sheriff which precedes the account of his fief (69):

Edward the sheriff has yearly from the profits (*de denariis*) which belong to the shrievalty, 130 porkers (*porci*) and 32 bacon-hogs (*bacons*); 2 pecks (*modios*) and 8 sesters of wheat; and as much barley; 5 pecks and 4 sesters of oats; 16 sesters of honey, or instead of honey 16*s.*; 480 hens; 1,600 eggs; 100 cheeses; 52 lambs; 240 fleeces (*vellera ovium*); 162 acres of unreaped corn (*annonae vel Bled*). He also has £80 in value between the reeveland and what he receives thence (*inter reveland et quod inde habet*). When the rent (*firma*) fails with the reeves (*quando prepositis firma deficit*), then it is incumbent on Edward to supply the deficiency himself (*necesse est Edwardo restaurare de suo*).

REGIONAL SUMMARY

Wiltshire is traditionally divided into two main areas—upland and vale. The recognition of this distinction between 'chalk' and 'cheese' is of long standing. It was mentioned by Camden and by Speed and also by John Aubrey, who, in his *Natural History of Wiltshire* (1685), described the contrast in some detail. For our purpose, two other regions must be recognised. The Cotswolds overlap the county boundary in the north-west, and along the south-eastern margin lay some of the royal forests that were such a feature of Norman England. Each of these four areas was far from uniform, but the division will serve to give a broad view of the county in the eleventh century (Fig. 17).

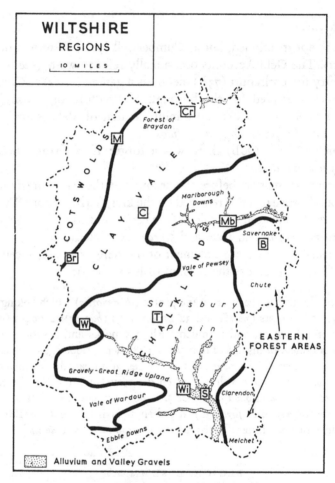

Fig. 17. Wiltshire: Regional subdivisions.

Domesday boroughs are indicated by initials: B, Bedwyn; Br, Bradford on Avon; C, Calne; Cr, Cricklade; M, Malmesbury; Mb, Marlborough; S, Salisbury; T, Tilshead; W, Warminster; Wi, Wilton.

(1) *The Cotswolds*

The dip-slope of the Cotswolds in north-west Wiltshire is an upland region which consists of Great Oolite and Cornbrash outcrops, and which rises in the south to over 400 ft above sea-level. It is crossed by various tributaries that flow to join the Bristol Avon, the largest being By Brook, which is deeply incised some 200 ft below the general surface. Villages were fairly numerous but they were mostly small, and the densities of plough-teams (under 2) and population (about 5) were low. Substantial amounts of wood were entered for some villages and small amounts of meadow for almost all villages. Pasture, on the other hand, seems to have been rare. Mills were common.

(2) The Clay Vale

The clayland consists very largely of stiff heavy clays—Oxford, Kimmeridge and Gault—lying mostly under 300 ft above sea-level. It is well watered by a close network of streams that flow either into the Bristol Avon or into the upper Thames. The densities of plough-teams (under 2) and population (under 7) were low in the northern part of the Vale, which included an area devoid of villages known later as the royal forest of Braydon or Braden. Southwards to Warminster, the Vale was well occupied, and here were some of the highest densities in the county: over 3 for plough-teams and over 8 for population. That these densities should be so high is surprising in view of the large amounts of wood entered for the manors of Chippenham and Melksham, each with woodland 4 leagues by 4 leagues, and each giving its name to a royal forest of which we hear in post-Domesday times. Further south, towards what was later known as the Forest of Selwood, where the land rises to above 400 ft, the densities decreased to below 1 for plough-teams and below 4 for population, although the vales of Warminster and Wardour seem to have been well occupied and tilled. Meadow was widely distributed throughout all the Clay Vale, and there were substantial quantities along the Bristol Avon and the Thames. There were also fair quantities of pasture, and the majority of vills, especially those along the Bristol Avon, had a mill or mills apiece.

(3) The Chalklands

The Chalklands lie for the most part over 600 ft above sea-level and in places they reach to over 900 ft. They are bounded towards the north and west by an escarpment that overlooks the Clay Vale below. From this outer limit the chalk beds dip gently to the south-east. Their sloping surface has been deeply incised by the Kennet, and more especially by the Salisbury Avon and its major tributaries, in such a way that the area as a whole is divided into a number of distinct upland blocks. The broad Vale of Pewsey, drained by the head-streams of the Salisbury Avon, separates the Marlborough Downs from the more southerly chalk masses which in turn are separated by the Avon itself and by the Wylye, the Nadder, the Ebble and the Bourne.

Strung along the valleys were the settlements of Domesday times, converging in a striking fashion upon Salisbury itself. Between these lines of villages were the relatively empty chalk uplands about which the Domesday text tells us nothing, that is, unless the entries for pasture refer to these areas; there is certainly a great deal of pasture recorded for southern Wiltshire. In spite of the large stretches of downland, the densities in the area bordering the Salisbury Avon, from the Vale of Pewsey southwards, were among the highest in the county—about 2·5 for plough-teams and over 10 for population. They were also high in the Kennet area east of Marlborough (3·5 and nearly 10 respectively) but it is possible that these high figures are due to the fact that here were the large manorial centres of Aldbourne and Ramsbury, with 'appendages' that may have been elsewhere. Over the remainder of the Chalklands the densities were more usually of the order of 2 or less for plough-teams and 7 or less for population. The figures were especially low for the bare area in the north-west of Salisbury Plain (1·6 and 4) and for the Grovely–Great Ridge area (1·6 and 4·8), in which was Grovely Forest. Almost every village in the river valleys had meadow and a mill or mills entered for it. It is difficult to be clear about the amount of wood over the Chalklands. The entries for many of the valley villages record some wood; that in the large manor of Amesbury was as much as 6 leagues by 4 leagues; but Amesbury included 'appendages', and we cannot say whether some of this wood was situated in the forest areas to the east.

(4) The Eastern Forest Areas

Along the south-eastern border of the county, the Chalk outcrop is covered with patches of Eocene sands and clays, Clay-with-flints and Plateau Gravels—all very unfavourable for agriculture. The densities of plough-teams (under 1·5) and population (under 4) were the lowest in the county. Various entries mention or imply the presence of forest in 1086, and in later times here were the royal forests of Savernake and Chute in the north and Clarendon and Melchet in the south. Amounts of meadow and pasture were small and there were relatively few mills. The quantities of wood entered under the names of the vills were also small, but there must have been a great deal of unrecorded wood in the forests and outside the manorial economy of the region.

BIBLIOGRAPHICAL NOTE

(1) An early edition of the Wiltshire section of the Domesday Book in H. P. Wyndham's *Wiltshire, extended from Domesday Book: to which is added a translation of the original Latin into English* (Salisbury, 1788). This is now of interest only as an early Domesday translation.

(2) There were two early studies by Henry Moody: (*i*) 'Notices on the Domesday Book for Wiltshire', *Proc. Archaeol. Inst. Salisbury, 1849*, v (London, 1851), pp. 177–81; (*ii*) 'The Domesday Book: Hampshire and Wiltshire', being the second essay (pp. 12–24) of *Notes and Essays, archaeological, historical and topographical, relating to the counties of Hants and Wilts*. (Winchester, London and Salisbury, 1851).

(3) In the mid-nineteenth century came W. H. Jones, *Domesday for Wiltshire* (Bath and London, 1865). Considering that it was written before the work of J. H. Round and F. W. Maitland was available, this is an outstanding study. As well as an extended text, translation and notes, it includes a long introduction.

It has, however, been superseded by the work of R. R. Darlington in *V.C.H. Wilts.* II (London, 1955). The translation of the Domesday text (pp. 113–68) is preceded by a long and valuable introduction (pp. 42–112) and is followed by a section dealing with the Wiltshire Geld Accounts—introduction, text and translation (pp. 169–217). An earlier section on 'Anglo-Saxon Wiltshire' (pp. 1–34) also contains much that is relevant to a study of the Domesday county.

(4) The following papers in the *Wiltshire Archaeological and Natural History Magazine* (Devizes) deal with various aspects:

W. H. JONES, 'Gleanings from the Wiltshire Domesday: I. Evidence as to the boundaries of the county being the same now as in the time of Domesday', no. XXIX (1867), vol. X, pp. 165–73.

W. H. JONES, 'Gleanings from the Wiltshire Domesday: II. On the names of owners or occupiers still preserved in those of persons or places in Wiltshire', no. XXXVIII (1872), vol. XII, pp. 42–58.

JOHN BEDDOE, 'A contribution to the anthropology of Wiltshire', no. CIII (1906), vol. XXXIV, pp. 15–41 (includes Domesday evidence).

F. W. MORGAN, 'Woodland in Wiltshire at the time of the Domesday Book', no. CLXII (1935), vol. XLVII, pp. 25–33.

F. W. MORGAN, 'The Domesday Geography of Wiltshire', no. CLXVII (1936), vol. XLVIII, pp. 68–81.

G. B. GRUNDY, 'The ancient woodland of Wiltshire', no. CLXXI (1939), vol. XLVIII, pp. 530–98.

H. C. BRENTNALL, 'The hundreds of Wiltshire', no. CLXXIX (1943), vol. L, pp. 219–29.

R. WELLDON FINN, 'The assessment of Wiltshire in 1083 and 1086', no. CLXXX (1946), vol. L, pp. 382–401.

R. WELLDON FINN, 'The making of the Wiltshire Domesday', no. CLXXXIX (1948), vol. LII, pp. 318–27.

M. W. HUGHES, 'The Domesday boroughs of Wiltshire', no. CXCI (1952), vol. LIV, pp. 257–78.

(5) Other works of interest in the Domesday study of the county are:

MARY BATESON, 'The burgesses of Domesday and the Malmesbury wall', *Eng. Hist. Rev.* XXI (London, 1906), pp. 709–22.

F. W. MORGAN, 'Domesday woodland in south-west England', *Antiquity*, X (Gloucester, 1936), pp. 306–24. This contains a map of the Domesday woodland of Wiltshire.

H. DE S. SHORTT, 'The mints of Wiltshire', *Numismatic Chronicle*, 6th ser., VIII (London, 1948), pp. 169–87.

H. DE S. SHORTT, 'The mints of Wiltshire from Eadgar to Henry III', *Archaeol. Jour.* CIV (London, 1948), pp. 112–28.

(6) A valuable aid to the Domesday study of the county is *The Place-Names of Wiltshire*, edited by J. E. B. Gover, Allen Mawer and F. M. Stenton (Cambridge, 1939).

CHAPTER II

DORSET

BY H. C. DARBY, LITT. D., F. B. A.

R. W. Eyton's study of the Dorset folios appeared in 1878, and was the first of his three Domesday studies on a county basis.[1] Although we cannot now accept many of his identifications or his ideas of Domesday mensuration, his partial 'analysis and digest' will always be of interest as a pioneer attempt at Domesday tabulation. The Domesday Book gives no hundredal rubrication for Dorset, and Eyton therefore used the Geld Accounts to reconstruct the hundreds of the county as they were at or near the time of the Inquest.[2] His interpretation of the evidence is sometimes open to criticism, but his work has proved of enormous value in making the material more manageable.

Dorset is one of the counties for which there is another text—although an incomplete one—of the Domesday survey. This text, in the *Liber Exoniensis*, covers four boroughs and a dozen fiefs out of 58; it deals wholly or in part with only 140 widely scattered places out of the 319 named places in the county (Fig. 31). Where it survives, the Exeter text has been followed in the present analysis. The Exeter entries are more detailed than those of the Exchequer Domesday Book itself. They distinguish the hidage of the demesne land from that of the villagers, which the Exchequer version does not always do. What is more, they list the demesne livestock. They also provide much information not included in the Exchequer version, e.g. they record 8 cottars at Frome St Quintin (*29*, 75 b), wood on a holding at Chelborough (*49*, 81 b), and values for a number of holdings. Furthermore, the two versions sometimes disagree in detail, e.g. the Exchequer version (78) attributes 5 teams to the men of West Milton, whereas the Exeter figure (*38*) is only 3 teams. Or, again, the account of Affpuddle at the foot of fo. 77 b seems to end in the middle

---

[1] R. W. Eyton, *A key to Domesday ... an analysis and digest of the Dorset survey* (London and Dorchester, 1878). The subsequent studies were those for Somerset (1880) and Staffordshire (1881).

[2] The Exchequer text makes reference to only two hundreds, *Bochelande Hund'* (75 b) and *Hundret Porbiche* (82). The former appears in the Geld Accounts as *Bochene' hundret'* (*23 b*); the latter is unmentioned.

of the statement about wood and does not give a value, information which the Exeter version supplies (*36b*). These are typical of the discrepancies that exist between the texts, and they are set out in Appendix II. To what extent such differences are due to errors in one or the other text, we cannot say.

We can also perceive what seem to be errors in some Exchequer entries for which there are no Exeter versions. Thus an entry for Frome on fo. 79 jumps from demesne teams to meadow, so omitting any reference either to the peasantry or to their teams. Carelessness may, perhaps, likewise explain the curious entry for Farnham (83): *geldabat pro ii hidis quae ibi sunt*. On the analogy of many other entries, this surely should read: *geldabat pro ii hidis. Terra est n carucis quae ibi sunt*. Another example comes from the concluding entries on the Dorset folios. The last three entries on fo. 85 were added after they had been omitted from their correct place higher on the folio, but the final entry was not completed and even ends in the middle of a sentence. Some of these omissions and apparent omissions are noted in the appropriate places below.

The *Liber Exoniensis* includes a Summary of the Dorset possessions of Glastonbury Abbey, and gives details for the demesne, for the land of the knights, and for a 5-hide manor held by two thegns. It is difficult to make a comparison between this information and that of the corresponding Domesday entries. The Domesday Book does not appear to mention the 5-hide manor, and it omits any reference to the teams of Sturminster Newton. The cottars of the Domesday entries appear as bordars in the Summary. The details are set out in the table on p. 71.

The *Liber Exoniensis* also includes a copy of an abstract of the record of the payment of a geld levied at a time near that of the Inquest. These accounts give information about the fiefs within each hundred, but they rarely mention place-names. In spite of the latter fact, however, the information is of great service in allocating the place-names of the Domesday Book to their correct settlements, all the more so because the Domesday text does not contain hundredal rubrics.[1] The differences between the details of the Geld Accounts and those of the Domesday folios are very many, and may testify to the imperfection of the Domesday record as well as to our lack of knowledge.

It is possible that some Domesday entries duplicate parts of others.

---

[1] Conversely, the Domesday text (82) mentions a hundred of *Porbiche* (? Purbeck) which is not mentioned in the Geld Accounts.

The two entries for Blackmanston on fo. 84b almost certainly repeat one another, and we have made our count on this assumption.

But there are other and greater difficulties. One is presented by those large composite manors which cover a number of unnamed places. Thus

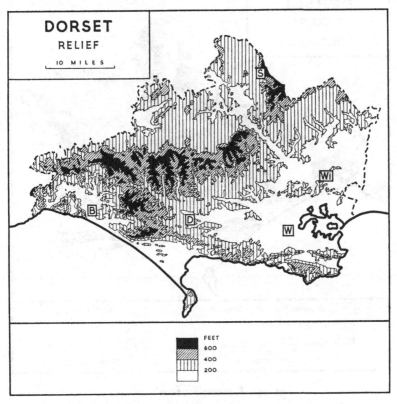

Fig. 18. Dorset: Relief.

Domesday boroughs are indicated by initials: B, Bridport; D, Dorchester; S, Shaftesbury; W, Wareham; Wi, Wimborne Minster.

the great manor of Sherborne (77), with anonymous sub-tenancies, and with a recorded population of 239, included a number of villages whose names do not appear in the Domesday Book. In the same way there must have been several villages or hamlets in the Isle of Portland with its 26 teams and its recorded population of 96; but we hear merely of

insula quae vocatur Porlanda and of its *pertinentes* (*26*, 75).[1] Puddletown
(*25*, 75) was valued *cum omnibus appendiciis* and Wareham (*28b*, 78b)
cum appendiciis suis, but again we are not told where the appendages

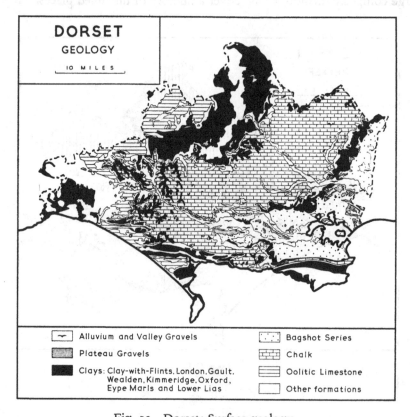

Fig. 19. Dorset: Surface geology.

Based on Geological Survey One-Inch Sheets (New Series)
312–14, 326–9, 341–3, and Quarter-Inch Sheet 19.

were.[2] The entries for some large manors do not mention sub-tenancies,
but their existence may be suspected from the large figures involved,
e.g. Canford (80b) with 84 recorded people and Sturminster Marshall

[1] Portland rendered £65 of white money *cum sibi pertinentibus.*

[2] The phrase *cum appendiciis suis* also occurs in connection with each of the five
groups of manors of the Ancient Demesne of the Crown (*27–8*, 75) but the main
manors in each group seem to be named.

(80) with 98. It is highly unlikely that such entries refer to a single village. Another major difficulty is the fact that there are no individual entries for those manors which were part of the Ancient Demesne of the Crown (*27-8*, 75). Although widely separated, they are grouped for the purpose of description in five entries, each with a single set of statistics; the constituents of each group are set out on p. 97. The problem raised by these and other composite entries are discussed as they arise below.

The present-day county of Dorset, in terms of which this study is written, is not exactly the same as the Domesday county. Thorncombe (*313*, 108) formed part of Devonshire in 1086, and was not transferred to Dorset until 1844. Seaborough (*154 bis*, *513*; 87b), nearby, was in Somerset until 1896. So were a group of other places to the north: Adber (*279*, 93; *355b*, 95b; *493b*,99) Goathill (*278b*, 92b), Poyntington (*279*, 93), Sandford Orcas (*466b*, *521b*; 99), Trent (*279*,93) and Weathergrove (*152*, *522b*; 89). Not far away, the parish of Holwell (not named in the Domesday Book) was until 1844 a detached portion of the Somerset hundred of Milborne. In the east of the county, Gussage St Michael (69) and one holding in Gillingham (73b) were mistakenly described in the

The Glastonbury Fief

Comparison of the Exchequer entries (77b) with those of the
Exeter Summary (*527b*)

	Demesne		Totals	
	D.B.	Summary	D.B.	Summary
Hides	$15\frac{5}{8}$	$22\frac{7}{8}$	58	$59\frac{1}{4}$
Non-gelding carucates	22	22	22	22
Demesne teams	4*	9	17	25
Villagers' teams	8*	20	23	44
Villeins	43	43	83	92
Bordars, etc.	72	72	117	144
Coliberts	13	13	13	13
Serfs	19	19	42	42
Value	£45	£45	£85 10s.	£78 10s.

The Summary adds, 'This land is sufficient for 105 teams';
the Domesday plough-lands total 100.

* D.B. omits the teams at Sturminster Newton, which the Summary implies were
5 demesne teams and 12 villagers' teams.

Wiltshire folios.[1] There have been losses as well as gains. Stockland (*44b*, 78) was transferred to Devonshire in 1844; Chardstock (77) and Hawkchurch (not described in the Domesday Book) went to the same county in 1869. Finally, Kinson (80b) was transferred to Hampshire in 1930.

SETTLEMENTS AND THEIR DISTRIBUTION

The total number of separate places mentioned in the Domesday Book for the area now forming Dorset seems to be at least 319, including the five boroughs of Bridport, Dorchester, Shaftesbury, Wareham and Wimborne Minster. This figure, for a variety of reasons, cannot accurately reflect the total number of settlements in 1086. In the first place, there are eleven entries in which no place-names appear, and it is possible that some of these refer to holdings at places not named elsewhere in the text.[2] One entry, for example, describes a 5-hide holding as being in three places (*in tribus locis*, 83) but there is no clue to their names. In the second place, as we have seen, the entries for some large manors cover not only the caput itself but also several unnamed dependencies, and again we have no means of telling whether these are named elsewhere in the text.[3]

In the third place, when two or more adjoining villages bear the same basic name today, it is not always clear whether more than one existed in the eleventh century. There is no indication in the Domesday text that, say, the East Lulworth and West Lulworth of today existed as separate villages; the Domesday information about them is entered under only one name (*Loloworde*, *28*; 75, 79b *bis*, 83), though they may well have been separate in the eleventh century as they certainly were by the end of the thirteenth.[4] When the holdings under such names were in different hundreds (as deduced from the Geld Accounts), we have counted them as

---

[1] R. R. Darlington in *V.C.H. Wilts.* II (London, 1955), pp. 135, 159, 175, 217. For a possible reason for the mistake, see R. Welldon Finn, 'The immediate sources of the Exchequer Domesday', *Bull. John Rylands Lib.* XL (Manchester, 1957), p. 62.

[2] The relevant entries are: William de Braiose *in Hundret Porbiche*, 7 hides less ½ virgate (82); Hugh de Boscherbert, *unum manerium* of 10 hides (83); Hugh d'Ivry, *terra in tribus locis* of 5 hides (83); *uxor Hugonis*, 1½ virgates (*56b*, 83b); Dodo, ½ hide (84); Alward, ¼ virgate (84); Godwin *venator*, 1 virgate and 4 acres (84); Ulvric *venator*, 1 hide (84); two bordars, ¼ virgate (84b); William d'Aumery, 3 hides and 2½ virgates (84b); Hugh Gosbert, 1 virgate (84b).

[3] See p. 69 above.

[4] For the history of these, and of all other names mentioned in this chapter, see A. Fägersten, *The Place-names of Dorset* (Uppsala, 1933).

separate settlements. Thus there are two entries for *Poleham*, and we have counted East Pulham in Hundredbarrow hundred (*48*, 81b) and West Pulham in Buckland hundred (79). This is not satisfactory, because if two or more such places lay in the same hundred, they have been counted as one in the present analysis. Moreover, the criterion of separate hundreds may not be infallible.

The problem is acute in Dorset because here so many groups of places took their respective Domesday names from the streams along which they were aligned. Such groups are associated with the Caundle, the Cerne, the Frome, the Iwerne, the Piddle, the Stour, the Tarrant, the eastern Winterborne, the western Winterborne, and the Wimborne (now called the Allen). Thus there are 15 entries for Tarrant, and today there are eight adjoining settlements with the same basic name. The entries apparently belong to only two hundreds, Badbury and Longbarrow, and the Tarrant holdings are accordingly represented on Fig. 20 by only two symbols. The table on pp. 74–6 shows how such groups of entries have been allocated on the basis of their respective hundreds as indicated by the Geld Accounts. Conjecture must enter into such allocations, but this hardly invalidates the general pattern of the distributions.

Very occasionally the text does distinguish between the units of a group with the same basic name. Thus we near of *Litelfrome* (*29*, 75b) as well as of the numerous Frome holdings; we also hear of Little Puddle (*Litelpidele, Litelpidre, 25b, 36*; 75b, 77b) and of Affpuddle (*Affapidele, 36b*, 77b) as well as of the Puddle holdings; and of Up Wimborne or *Opewinburne* (*27*, 75) as well as of the Wimborne holdings.[1] Then again a distinction is drawn between Littlebredy (*Litelbride, 37*, 78) and Long Bredy (*Langebride, 37b*, 78).[2] On the other hand, Seaborough, now in Dorset, appears in the Somerset folios as *Seveberge* (87b) and *alia Seveberge* (87b), but the Exeter text speaks only of one Seaborough (*154 bis, 513*), and there is no later evidence for two villages here; we must conclude that the reference is to two holdings at one place.[3]

Quite different are those groups of places bearing the same Domesday name but not lying adjacent. Thus, of two entries bearing the name *Bradeford*, one refers to Bradford Abbas (77) and the other to Bradford

---

[1] For the identification of *Opewinburne* with All Hallows Farm, see p. 76 below.

[2] There is a third Bredy (*Bridie*, 82b), not adjoining these, and lying 4 miles or so to the west in Burton Bradstock.

[3] For the use of *alia* in Essex, see H. C. Darby, *The Domesday Geography of Eastern England* (Cambridge, 2nd ed. 1957), pp. 212–13.

Ten Groups of Dorset Settlements

Caundle Brook

Bishop's Caundle, Purse Caundle, Stourton Caundle and Caundle Wake	*Candel, Candel(l)e, 41, 280b; 78b, 80 bis, 82 ter, 84b, 93*	Brownshall hundred
Caundle Marsh	*Candele, 84b*	Sherborne hundred

Cerne

Forston, Herrison and Pulston	*Cerne(l), 77, 79 quin., 83*	Dorchester hundred
Charminster	*Cerminstre, 75b*	Dorchester hundred
Godmanstone	*Cerna, Cerne(l), 45; 78, 79*	Modbury hundred
Up Cerne	*Obcerne, 75b*	Sherbourne hundred
Cerne Abbas	*Cernelium, Cerneli, 36, 77b*	Stone hundred

Frome

Frome Billet and Frome Whitfield	*Frome, Frome, 54; 79 bis, 83b, 84b*	Dorchester hundred
Frampton	*Frantone, 78b*	Frampton hundred
Frome St Quintin	*Litelfroma,Litelfrome, 29,75b*	Pimperne hundred
Chilfrome, Frome Vauchurch and Cruxton	*From(m)a, Frome, 27b, 48b; 75, 80b, 81b bis*	Tollerford hundred

Iwerne

Iwerne Courtney	*Werne, 81*	Farringdon hundred
Stepleton Iwerne	*Werne, 49, 81b*	Longbarrow hundred
Lazerton (in Stourpaine)	*Werne, 84*	Pimperne hundred
Ranston (in Iwerne Courtney)	*Iwerne, 80b*	Pimperne hundred
Iwerne Minster	*Euneminstre, 78b*	Sixpenny Handley hundred

Piddle

Affpuddle	*Affapidela, Affapidele, 36b, 77b*	Bere hundred
Briants Puddle and Turners Puddle	*Pidela, Pidele, 56b; 83b, 84b*	Bere hundred
Piddlehinton	*Pidele, 79*	Puddletown hundred
Tolpuddle, Athelhampton and Burleston	*Pidela, Pidele, 39, 43b; 77, 78, 78b*	Puddletown hundred

Piddle (*cont.*)

Little Puddle	*Litelpidel, Litelpidra,*	Puddletown hundred
	Litelpidele, Litelpidre,	
	Pidra, Pidre, 25 b, 36, 44 b;	
	75 b, 77 b, 78	
Puddletown,	*Pidele, Pidere, Pidra,*	Puddletown hundred
Bardolfeston	*Piretona, Pidredone,*	
	Piretone, Pitretone, 25,	
	28, 53; 75, 77, 79, 79 b, 82 b	
Piddletrenthide	*Pidrie,* 77 b	Stone hundred

Tarrant

Tarrant Crawford	*Tarente,* 77 b	Badbury hundred
Tarrant Gunville,	*Tarenta, Taerenta, Tarente,*	Longbarrow
Tarrant Hinton,	*Terente, 31, 31 b, 32, 58,*	hundred
Tarrant Keynston,	*59 bis;* 75 b *ter,* 76, 77 b *bis,*	
Tarrant Launceston,	78 b, 79, 80 b, 82 b, 83,	
Tarrant Monkton,	83 b *ter*	
Tarrant Rawston and		
Tarrant Rushton		

Wey

Broadwey, Upwey and	*Wai(a), 31, 54 b, 55;* 75 b,	Cullifordtree
Weymouth	79 *ter,* 83, 83 b *bis,* 84	hundred

Winterborne (*Eastern*)

Winterborne Kingston	*Wintreburne,* 79 b, 84, 85	Bere hundred
Winterborne Clenston,	*Wintreborna, Wintreburne,*	Combsditch
Winterborne Houghton	*28 bis, 44, 47 b, 56 bis, 57,*	hundred
Winterborne Muston,	*58 b bis;* 75, 77 *bis,* 78, 79 b	
Winterborne Tomson,	*quin.,* 82 *ter,* 83 b *sexiens,*	
Winterborne Whit-	84, 84 b, 85	
church, Winterborne		
Zelston, and Anderson		
Winterborne Stickland	*Wintreburne,* 79	Hundredbarrow
		hundred

Winterborne (*Western*)

Winterborne Came,	*Wintreborna, Wintreburne,*	Cullifordtree
Winterborne Herring-	*33, 55, 55 b;* 79 *bis,*	hundred
ston, Winterborne	83 b *bis,* 84 *bis,* 84 b,	
Monkton and Winter-	85	
borne Steepleton		

Winterborne (Western) (cont.)

Winterborne St Martin	*Wintreborna, Wintreburne,* 54; 79, 83b	Dorchester hundred
Winterborne Abbas	*Wintreborna, Wintreburne,* 37b, 78	Eggardon hundred
Wimborne (now Allen)		
Wimborne St Giles	*Winburna, Winburne, 56;* 77b, 83b, 84, 85	*Albretesberga* hundred
Wimborne Minster	*Winborna, Winborne, Winburne, 27, 30b; 75,* 75b, 76 *bis,* 78b, 79b, 80b	Badbury hundred
Monkton Up Wimborne	*Winburne,* 79	Knowlton hundred
All Hallows Farm	*Obpewinborna, Opewinburne, 27,* 75	Extra-hundredal

Note: Monkton Up Wimborne has been equated with *Winburne* and not with *Obpewinborna/Opewinburne.*

Peverell (80b); of three bearing the name *Come,* one refers to Coombe in Langton Matravers (84b) and two to Coombe Keynes (*62b;* 77b, 82b); and of two bearing the name *Westone,* one refers to Buckhorn Weston (79) and the other to Stalbridge Weston (77). The absence of hundredal rubrication increases the difficulty of identifying each holding, but the Geld Accounts and later manorial history help to show which holding belongs to which village.

The total of 319 includes a dozen or so places about which very little information is given. The only thing we are told about Whitchurch Canonicorum is that its church was held by the abbey of St Wandrille in Normandy (*28,* 78b). We hear nothing of Stoborough (79b) except that it was rated at ½ hide, and had a mill and three bordars and was worth 40s. What did the three bordars do? And what accounted for an annual value of as much as 40s? Then there are a number of places for which neither teams nor population are recorded, e.g. Blackmanston (84b *bis*), Chettle (83), Herston (*52;* 82b, 85), *Hiwes* (80b), Holton (82), Holwell in Radipole (79), Preston in Tarrant Rushton (77b), Rollington (*51b,* 82b), and Studley (84b); the entries for these places give details merely of assessments, plough-lands, values and, sometimes, meadow and pasture. Here might also be mentioned those other villages which have no population or no teams or neither entered for them.[1] Finally, the phrase *castellum Warham* (78b) refers to Corfe Castle; we hear of no settlement growing around, but it has been included in the total of 319.

About 120 of the total of 319 Domesday place-names are represented today not by parish names but merely by those of hamlets or of individual

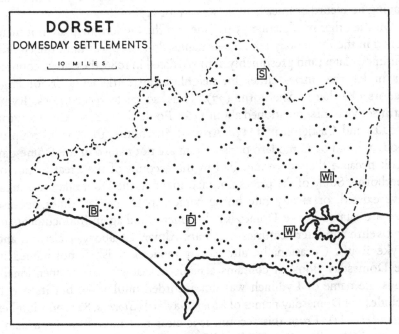

Fig. 20. Dorset: Domesday place-names.

Domesday boroughs are indicated by initials: B, Bridport; D, Dorchester; S, Shaftesbury; W, Wareham; Wi, Wimborne Minster.

farms and houses. Thus *Netelcome* (*38*, 78) is now the hamlet of Nettle-combe in Powerstock, and *Silfemetone* (*39b*; 78b, 80, 84b) is that of Shilvinghampton in Portisham. *Romescumbe* (*38b*, 78), for example, appears in the name of Renscombe Farm in Worth Matravers, and *Plumbere* (84) in that of Plumber Farm in Lydlinch. Some names have disappeared entirely from the map. *Stoches* (80b *bis*, 84b) and *Stodlege* (84b) became the lost places of Stoke Wallis and Studley in Whitchurch Canoni-

[1] See pp. 88 and 90 below.

corum; and *Levetesford* (77b) may have become the lost Leftisford in Cranborne. Other settlements appear to have left no trace of their names or sites, and the following names have remained unidentified in this analysis: *Hiwes* (80*b*), *Langeford* (77b), *Odeham* (77b), *Scetre* (*31 b*, 75 b), *Selavestune* (82b) and *Slitlege* (84b). Whether these will yet be located, or whether the places they represent have completely disappeared, leaving no record or trace behind, we cannot say.

On the other hand, some 55 parishes on the modern map are not mentioned in the Domesday text. Their names do not appear until the twelfth century or later; and presumably, if they existed in 1086, they are accounted for under the entries either for neighbouring settlements or for large manors like that of Sherborne (77). Thus, so far as record goes, Ryme Intrinseca was first mentioned about 1160, Poole about 1179, Cheddington in 1244 and Castleton in 1333.[1] Around Sherborne, there is a group of about a dozen or so modern parishes that are not named in the Domesday Book because they were included in the account of the great manor of Sherborne. Some of the places not mentioned in the Domesday text must have existed, or at any rate borne names, in Domesday times, because they are named in pre-Domesday documents and again in documents of the twelfth and later centuries. Such are Almer, Chalbury, Halstock, and Wyke Regis. Occasionally, although a modern parish is not named in the Domesday Book, it contains a name or names that are mentioned. Thus the name of Lydlinch was not recorded until 1166, but its parish includes the Domesday names of Stock Gaylard (*Stoches*, 82) and Plumber (*Plumbere*, 84). From this account it is clear that there have been many changes in the village geography of the county, and that the list of Domesday names differs considerably from that of present-day parishes.

The distribution of Domesday settlements over the county was fairly even, especially when the blank spaces that result from large composite manors are ignored, e.g. around Sherborne (Fig. 20). Three general features stood out. The first was the relative sparseness of villages in the sandy Tertiary Basin around Poole Harbour. The second was the alignment of villages along such rivers as the Frome, the Piddle and the Stour which flow across the chalkland and the Tertiary sands. Thirdly, there was the more irregular distribution of villages over the claylands of the north and west.

[1] The dates of the place-names in this section are from E. Ekwall, *The Concise Oxford Dictionary of English Place-Names* (Oxford, 4th ed. 1960).

THE DISTRIBUTION OF PROSPERITY AND POPULATION

Some idea of the nature of the information in the Domesday folios for Dorset, and of the form in which it is presented, may be obtained from the account of Powerstock. It is situated in the west of the county, 5 miles north-east of Bridport. The village was held entirely by Roger Arundel, and so it is described in a single entry. There is also a version in the description of Roger's fief in the *Liber Exoniensis*, and both entries are set out below.

Exeter Domesday (*5o b–5 1*)

Roger holds one manor which is called Powerstock [and] which Ailmar held on the day that King Edward was alive and dead [and] which paid geld for 6 hides. Now Hugh holds it of Roger. Six plough-teams can plough these [hides]. Thereof Hugh has in demesne 3 hides and 2½ plough-teams. There are 5 villeins and 9 bordars and 5 serfs, and 2 rounceys and 4 animals and 13 swine and 158 sheep and 16 goats, and 2 mills which render 3s., and 11 furlongs of wood long and 2½ furlongs in breadth, and 13 acres of meadow, and 15 furlongs of pasture long and 2 in breadth. [It is worth] £6, and when he received it, £4.

Exchequer Domesday (82b)

Hugh holds Powerstock of Roger. Ailmar held it in the time of King Edward, and it paid geld for 6 hides. There is land for 6 plough-teams. In demesne are 2½ plough-teams and 5 serfs, and 5 villeins and 9 bordars with 2½ plough-teams. There, 2 mills render 3s.; and 13 acres of meadow; and 15 furlongs of pasture in length and 2 in breadth. Wood 11 furlongs long and 2½ furlongs broad. It was worth £4; now £6.

This account does not include all the kinds of information that appear elsewhere in the folios for the county: there is no mention, for example, of the categories of population known as coliberts and coscets. But it is a fairly representative and straightforward entry, and it sets out the recurring standard items that are entered under most place-names. These are five in number: (1) assessment, (2) plough-lands, (3) plough-teams, (4) population, and (5) values. The bearing of these five items of information upon regional variations in the prosperity of the county must now be considered.

(1) *Assessment*

The Dorset assessment is stated in terms of hides, of virgates, and of what appear to be geld acres.[1] Fractions of both hides and virgates are common. The usual Exchequer formula is *geldabat pro n hidis*,[2] but there are frequent variations, as the following examples show:

Ringstead (84b): *Idem Brictuinus tenet i hidam in Ringestede. Ipse tenuit T.R.E.*

Woolcombe in Toller Porcorum (84b): *Idem Hugo tenet iii virgatas terrae in Wilecome; Dodo monachus tenuit T.R.E. et pro tanto geldabat.*

Afflington (82): *Idem Walterus tenet de Willelmi de Braiose iii virgatas terrae et dimidiam.*

Caundle (82): *Ibi sunt v hidae.*

The Exchequer version frequently states also the assessment of the demesne portion of an estate; the Exeter text not only almost invariably does this, but, in addition, it states explicitly the assessment of the portion in the occupation of the peasantry. The sum of the demesne and other hidage usually agrees with the total given for a holding as a whole, and there are only occasional discrepancies. Thus at Renscombe (*38b*) there were 2 h 3 v in demesne and 3 h 1 v with the peasantry, making a total of 6 hides as compared with a stated total of 5 h 1 v. There is an unusual phrase in the account of Horton (78b), where we read that the king held the two best hides (*duae meliores hidae*) of the manor in the forest of Wimborne. Occasional entries omit any reference to assessment, e.g. one for the manor of Portland (*26, 75*), another for Gillingham (77b), and another for Milborne Stileham (84b); the Geld Accounts indicate, however, that the missing figure for the latter must be 1½ hides.[3]

Eyton believed that the Dorset hide comprised 48 acres, but he gave no reason for this view.[4] Round, in criticising Eyton, declared himself strongly in favour of a 120-acre hide in Dorset as elsewhere.[5] So did Maitland, who discussed that Dorset entry in the Exeter text which seems to imply that a virgate comprised 10 acres: we are told that the 10-hide manor of East Pulham consisted of 4 h 1 v 6 ac in demesne and

---

[1] The Exeter folios mention the ferling in the account of Sandford Orcas (since transferred from Somerset to Dorset): *De his habet Hunfridus iii hidas et dimidiam i fertinus minus (466b)*.

[2] The corresponding Exeter formula runs: *reddit gildum pro n hidis*.

[3] R. W. Eyton, *op. cit.* p. 115 n. The 1½ hides have been included in the total on p. 84.   [4] *Ibid.* p. 14.

[5] J. H. Round in P. E. Dove (ed.), *Domesday Studies*, I (London, 1888), p. 213.

5½ h 4 ac with the villeins (*48*). Commenting on this, Maitland wrote: 'Now three or four such entries would certainly set the matter at rest; but a single entry can not.'[1] He pointed out that immediately following the account of East Pulham is that of the 5-hide manor of Hammoon (*48*) composed, so we are told, of 3 h 8 ac in demesne and 2 h less 12 ac with the villeins. Should we then assume, he asked, that 5 h = 5 h less 4 ac? Arithmetical error apart, Maitland could only suspect 'that here, as in Middlesex, geldable units and actual areal units have already begun to perplex each other'. Acres are very infrequently mentioned and, when they are, their numbers are small. Both these facts led Tait to favour not a 120-acre hide but 'a little hide' of 40 or 48 acres.[2] He pointed to a reckoning in the Wiltshire Geld Accounts that supported the East Pulham entry in indicating a 40-acre hide;[3] but, in general, he thought that the margin of error involved in any calculation was so great as to make it impossible to choose between a 40- and a 48-acre hide. We may, however, doubt whether 40 acres normally made up a hide, because we hear quite often of a third or two thirds of a virgate or of a hide.[4] One of the Tarrant entries even mentions 'a third part of half a hide' (*77* b). Less often we hear of quarter hides.[5]

The account of the Ancient Demesne of the Crown (*27–28, 75*) contains five composite entries, each of which includes the following statement or some variant of it: *Nescitur quot hidae sint ibi quia non geldabat T.R.E.*; each of the five groups of places was responsible for the ancient render either of one night's *firma* or of half a night's *firma*.[6] At four other places we hear also of hidated holdings that had never paid geld:

Swyre (80 b): *In ista villa tenet Willelmus quandam partem terrae quae nunquam geldavit T.R.E. sed erat in dominio et in firma regis.*

Warmwell (80): *Duo taini tenuerunt T.R.E. et geldabat pro ii hidis et una virgata terrae. Praeter has est ibi una virgata terrae quae nunquam geldavit.*

Wimborne Minster (*30 b, 75* b): *Ibi est dimidia hida et nunquam geldavit.*

Witchampton (79 b): *Ibi habet Hubertus unam virgatam terrae et terciam partem unius virgatae de qua nunquam dedit geldum.*

---

[1] F. W. Maitland, *Domesday Book and Beyond* (Cambridge, 1897), p. 485.

[2] J. Tait, 'Large hides and small hides', *Eng. Hist. Rev.* XVII (1902), pp. 280–2.

[3] See p. 14 above.

[4] E.g. on fo. 84 there are three examples—at Wilkswood (*60* b), at Hampreston, and on Alward's anonymous holding.

[5] E.g. at Rushton (84 b) and at Rollington (*51* b, 82 b).

[6] The composite entries are those associated with Burton Bradstock, Wimborne Minster, Dorchester, Pimperne and Winfrith Newburgh. See p. 97 below.

A marginal note against the account of Catsley on fo. 80 says: *Ibi est una virgata terrae de qua celatum est geldum T.R.W.*; and the Geld Accounts tell us that from this virgate *nunquam habuit rex geldum* (*19b*). The Exeter Domesday mentions half a hide or so at Puddle *quae nunquam gildavit sed celatum est* (*56b*) but there is no reference to this in the Exchequer version (83b). Furthermore, the Exeter text alone says that a virgate at Winterborne (*58b*. 83b) never paid geld to King William: *nunquam habuit W. rex geldum suum.*[1]

There are eleven other entries that refer to land which did not pay geld, and they are set out on p. 83. Five of these speak not of hides but of *carucatae terrae*; they all come on fo. 77 in the account of the lands of Sherborne Abbey.[2] In the other six entries, carucates are not mentioned by name but are implied. The total for all eleven entries amounts to 64½ carucates. Some of these are also mentioned in the Geld Accounts for the hundreds of Sherborne (*23b*) and Yetminster (*17*) and in the Summary of the lands of Glastonbury Abbey (*527b*). 'They are,' in the words of R. Welldon Finn, 'it seems, unassessed land, probably acquired since the shires were first or last assessed, intakes from the former waste, or attempts to reconcile actual possession with existent unsanctioned documentary evidence or rights of ownership.'[3]

A glance through the Dorset folios reveals many examples of the 5-hide unit. Thus Nettlecombe (*38*, 78) was assessed at 5 hides, Melbury Abbas at 10 (78b) and Yetminster at 15 (75b). Three Dorset manors are called *Fifehide*; each is assessed at 5 hides, and they are represented today by Fifehead Magdalen (80), Fifehead Neville (82) and Fifehead St Quintin (78b). When a village was divided amongst a number of owners, the same feature can sometimes be demonstrated. Thus Edmondsham (*29b*; 75b, 83 *bis*), with three holdings, was assessed at 5 hides (2 + 1½ + 1½); so was Creech (*60b*; 79b, 80, 82, 84) with four holdings (2 + 2 + ½ + ½). Six places in the modern county were assessed at 2½ hides each; 37 places

---

[1] The non-gelding lands at Puddle and Winterborne are also mentioned in the Geld Accounts (*20, 22*). See V. H. Galbraith, *The Making of Domesday Book* (Oxford, 1961), pp. 225–6.

[2] The word *carucatae* also occurs in the entry for Wootton (83), but this is obviously an error for *hidae: geldabat pro xii hidis. Terra est xvi carucis. De ea sunt in dominio iiii carucatae et ibi iii carucae.*

[3] R. Welldon Finn, *The Domesday Inquest* (London, 1961), p. 132. For earlier discussions of *carucatae terrae*, see (1) R. W. Eyton, *op. cit.* pp. 16–22; (2) J. H. Round, in P. E. Dove (ed.), *op. cit.* I, pp. 105–7.

in the modern county were assessed at 5 hides each, fifteen at 10 hides, and five at 15 hides. There were also a number of other places assessed at 20, 25 and even 30 hides. We cannot be far wrong in saying that the 5-hide principle is readily apparent in about one fifth of the Dorset villages. It is also possible that some villages were grouped in blocks for the purpose of assessment. Thus Littlebredy (*37*, 78) with 11 hides and Long Bredy (*37b*, 78) with 9 hides formed a 20-hide unit. In the absence of information about such possible groups, we cannot be certain about the full extent of the 5-hide unit in Dorset.

Dorset: *Carucatae terrae nunquam geldantes*

A. *Carucatae terrae* mentioned

Sherborne (77): *In hoc manerio Scireburne, praeter supradictam terram, habet episcopus in dominio xvi carucatas terrae. Haec terra nunquam per hidas divisa fuit neque geldavit...De hac quieta terra tenet Sinod de episcopo i carucatam terrae et Eduuardus aliam.*

Sherborne (77): *In hac eadem Scireburne tenent monachi ejusdem episcopi ix carucatas terrae et dimidiam quae nec per hidas divisae fuerunt nec unquam geldaverunt...De hac terra monachorum tenet Lanbertus de eis i carucatam terrae.*

Stoke Abbott (77): *Terra est vii carucis. Praeter hanc sunt ibi ii carucatae quae nunquam divisae sunt per hidas.*

Beaminster (77): *Terra est xx carucis. Praeter hanc terram habet in dominio ii carucatas terrae quae nunquam geldaverunt.*

Netherbury (77): *Terra est xx carucis. Praeter hanc habet in dominio ii carucatas terrae quae nunquam geldaverunt.*

B. *Carucatae terrae* implied

Charminster (75b): *In ipso manerio habet episcopus tantum terrae quantum possunt arare ii carucae. Haec nunquam geldavit.*

Alton Pancras (75b): *Terra est vi carucis. Praeter hanc habet terram ii carucis in dominio quae nunquam geldavit.*

Yetminster (75b): *Terra est xx carucis. Praeter hanc habet terram vi carucis quae nunquam geldavit.*

Lyme Regis (75b): *Terra est i caruca. Nunquam geldavit.*

Sturminster Newton (77b): *Terra est xxxv carucis. Praeter hanc est terra xiiii carucis in dominio ibi quae nunquam geldavit.*

Buckland Newton (77b): *Terra est xxiiii carucis. Praeter hanc est in dominio terra viii carucis quae nunquam geldavit.*

It is clear, however, that the assessment was largely artificial in character, and bore no constant relation to the agricultural resources of a vill. A striking example of low hidation is that of Puddletown (25, 75). It answered for only half a hide, yet there were 14 teams at work and there was a recorded population of 55. The variation among a representative selection of 5-hide vills speaks for itself:

	Plough-lands	Plough-teams	Popula-tion	*Valuit*	*Valet*
Boveridge (77 b)	7	5	24	100s.	100s.
Corton (80)	4	2	12	£9	£7
Ibberton (25 b, 75 b)	5	5	19	£10	£10
Littleton (79 b)	3	1	8	£4	40s.
Manston (82)	8	4	19	£6	100s.

The total assessment, including non-gelding hides, amounted to 2,349 hides, 3½ virgates and 65¼ acres.[1] This is for the area included within the modern county, and so is not strictly comparable with the estimates of Eyton[2] and Maitland[3] for the Domesday county. The nature of some of the entries, and the possibility of unrecognized duplicate entries, make exact calculation very difficult. All that any estimate can do is to indicate the order of magnitude involved. We must also remember the five groups of places of which it was said, *nescitur quot hidae sint ibi*, and also those 64¼ carucates of land that had never paid geld or never been assessed.

(2) *Plough-lands*

Plough-lands are systematically entered for the Dorset holdings, and the Exchequer formula normally runs: 'There is land for *n* plough-teams' (*Terra est n carucis*). The Exeter text usually says: *hanc terram possunt arare n carrucae.*[4] Among the variants is the form *potest arari cum n*

---

[1] In view of the uncertainty about the number of acres to the hide, the totals for hides and virgates, on the one hand, and for acres on the other, have been kept separate. Four acres have been taken away from the total of acres to allow for the Woolgarston entry (82), which speaks of 4 hides less 4 acres.

[2] R. W. Eyton, *op. cit.* p. 144. Eyton makes various adjustments to the figures, but it would seem that the total corresponding to the present count is 2,366 hides, 2 virgates, 11 acres.

[3] F. W. Maitland, *op. cit.* p. 400, reckoned 2,277 hides.

[4] For a similar phrase in the Exchequer entry for Charminster (75 b), see p. 83 above.

carrucis; and something similar to this is encountered once in the Exchequer folios—in the entry for Sturminster Newton (77b).[1] A reckoning in terms of half-teams is not unusual; but there are only very occasional references to oxen, such as that for Woodstreet (85): *Terra est vi bobus.* In two entries the phrase *Terra est* is followed by a blank space in which a figure was never inserted (Cheselbourne, 78b; and Nettlecombe, *38*, 78).[2] There are also nine entries which make no reference to plough-lands, although they state the number of teams at work, e.g. that for Wimborne St Giles (84); these eleven entries account for a total of 54 teams. There are, moreover, fifteen other entries which mention neither plough-lands nor teams; this is so, for example, in the first entry for Farnham on fo. 83, where something seems to have been omitted.

The relation between plough-lands and plough-teams varies a great deal in individual entries. They are equal in number in about 46 per cent of the entries which record both. Occasionally we are explicitly told that this is so, as in the entry for Stoke Wallis (80b): *Terra est i caruca quae ibi est.*[3] A deficiency of teams is also quite common, and is encountered in another 46 per cent or so of the entries which record both plough-lands and teams. The values of some deficient holdings had fallen in the years prior to 1086, but on others they had remained constant, and on yet others they had even increased. No general correlation between plough-team deficiency and decrease in value is therefore possible, as the following table shows:

		Plough-lands	Teams	*Valuit*	*Valet*
Decrease in value	⎧ Manston (82)	8	4	£6	£5
	⎩ Stour (80)	9	5	£9	£8
Same value	⎧ Chilfrome (*48b*, 81b)	6	4	£6	£6
	⎩ Yetminster (75b)	20	12	£22	£22
Increase in value	⎧ Buckhorn Weston (79)	6	3½	£4	£7
	⎩ Wootton[4] (83)	16	12	£10	£12

These figures are merely indications of unexplained changes on individual holdings. The entry for Swyre (80b) contains a most unusual phrase,

[1] The sentence runs: *Hae viii hidae possunt arari xi carucis.*
[2] An entry for Cheselbourne does not appear in the surviving Exeter text, and the Exeter entry for Nettlecombe (*38*) entirely omits the plough-land formula.
[3] The Exeter version of a similar entry for Bere Regis (*56b*, 83b) uses a different form of words: *ibi habet Willelmus dimidiam carrucam quia pro tanto arari potest.*
[4] Equated here with Abbots Wootton and Wootton Fitzpaine.

which implies an extension of the arable: *Prius erat pascualis, modo seminabilis.*[1]

An excess of teams is a good deal less frequently encountered, being

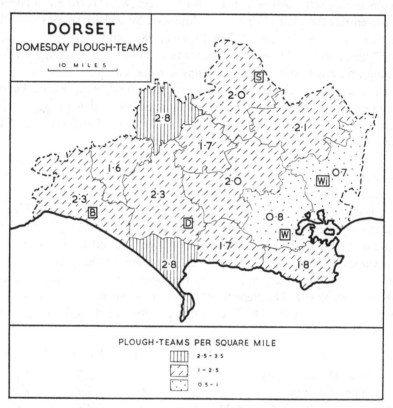

Fig. 21. Dorset: Domesday plough-teams in 1086 (by densities).
Domesday boroughs are indicated by initials: B, Bridport; D, Dorchester; S, Shaftesbury; W, Wareham; Wi, Wimborne Minster.

found on only 8 per cent of the entries which record both plough-lands and teams. The values on some of these holdings had risen, e.g. at Nutford (85) and Winterborne (84b), but very frequently this was not so. Other holdings with excess teams were valued the same in 1086 as at an earlier date, e.g. at Clifton Maybank (80) and Littlebredy (37, 78). The value of yet other holdings with excess teams had even dropped, e.g. at East

[1] See p. 89 below.

Pulham (*48*, 81b) and West Milton (*38*, 78). Our attention is sometimes specifically drawn to excess teams. An entry for Frome (84b), for example, reads: *Terra est ii carucis. Tamen sunt ibi iii carucae.* This phrase, or

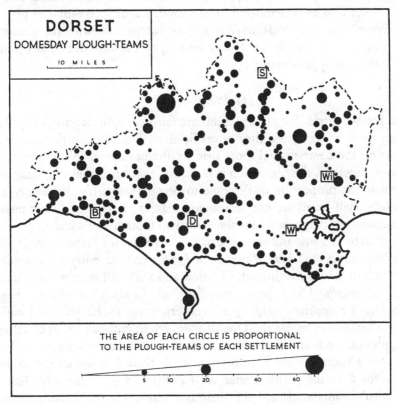

DORSET
DOMESDAY PLOUGH-TEAMS
10 MILES

THE AREA OF EACH CIRCLE IS PROPORTIONAL
TO THE PLOUGH-TEAMS OF EACH SETTLEMENT
5   10   20   40   60

Fig. 22. Dorset: Domesday plough-teams in 1086 (by settlements).

Domesday boroughs are indicated by initials: B, Bridport; D, Dorchester; S, Shaftesbury; W, Wareham; Wi, Wimborne Minster.

something like it, occurs altogether in eight entries, but no entry hints at a reason for the overstocking.[1]

[1] The eight entries are: Dodo's anonymous ½ hide (84), Crawford (84), Farnham (*57b*, 83b), Frome (84b), Nutford (85), *Odeham* (77b), Seaborough (*154*, 87b), Winterborne (84b). There is an Exeter version of the Farnham entry (*57b*) which records the presence of half a plough-land and one team without drawing attention to the excess.

The total number of plough-lands recorded for the area within the modern county is 2,220¾. If, however, we assume that the plough-lands equalled the teams on those holdings where a figure for the former is not given, the total becomes 2,274¾. The presence of 64½ non-gelding carucates must also be remembered.[1] For various reasons the estimates given by Eyton[2] and Maitland[3] are not strictly comparable with the present total. In view of the nature of some entries, no estimate can be anything other than an approximation.

(3) Plough-teams

The Dorset entries for plough-teams are fairly straightforward, and, like those for other counties, they normally draw a distinction between the teams on the demesne and those held by the peasantry. In a number of entries, however, the distinction is not clear, and this makes it difficult to calculate the proportion of demesne to peasant teams for the county as a whole. Half-teams are not uncommon. About 120 entries do not mention teams although they sometimes refer to plough-lands and to people as well as to meadow and wood. Such are the entries for Brenscombe (*61b*, 84), Smedmore (82), and Watercombe (*31b*, 75b). An entry for Caundle (83) tells of half a plough-land, 2 borders and 2 acres of meadow, and then adds *Nil amplius*.[4] The Exeter entry for Ower (*44b*), where the recorded population comprised only 13 saltworkers, says: *In his [iii hidis] nulla caruca est, nec arare potest*.[5] One of the Tarrant entries (*32*, 75b) refers to 3 plough-lands and to one team on the demesne, but leaves a gap where the men's teams should have been entered; there is a similar gap in the entry for Coombe (84b). Under West Pulham (79), on the other hand, the men's teams are stated and there is a gap where the demesne teams should have been entered. The omission of the full tale of teams for the

[1] See p. 82 above.
[2] R. W. Eyton, *op. cit.* p. 146. The total of 2,333 was for the Domesday county, and it involved assumptions about the *carucatae terrae*. Eyton did not specifically state the number of plough-lands, but their equivalent in acres at the rate of 120 acres per plough-land.
[3] F. W. Maitland, *op. cit.* pp. 401, 410. The total of 2,333 was for the Domesday county but, as in the present estimate, allowance was made for the missing plough-land figures.
[4] No people are recorded for the two holdings at Thornham (84 *bis*), and the Exeter entry says of one of the holdings: *Ibi habet R[obertus] i carucam et nichil amplius (61b)*.
[5] The Exchequer entry for Ower (78) reads: *Ibi nulla caruca habetur*.

large manor of Sturminster Newton (77b) is a striking one.[1] The unusual
entry that speaks of the conversion from pasture to arable at Swyre (80b)
does not say how many plough-teams were involved—*Prius erat pascualis, modo seminabilis.* All these omissions illustrate the uncertainty that
must always be remembered when dealing with Domesday statistics.
Occasional vills seem to have possessed a number of teams beyond the
capacity of their inhabitants. Thus one team would seem to have been
too much for the solitary inhabitant of Moulham (85) to manage without
help.

The total number of plough-teams amounted to 1,841¾, but this refers
to the area included in the modern county, and, in any case, a definitive
total is hardly possible.[2]

(4) *Population*

The bulk of the population was comprised in the five main categories of
bordars, villeins, serfs, cottars and coscets. In addition to these were the
burgesses and a miscellaneous group that included salt-workers, coliberts, rent-payers and others. The details of the groups are summarised
on p. 90. There are two other estimates, by Sir Henry Ellis[3] and by R. W.
Eyton[4] respectively, but the present estimate is not strictly comparable
with these because it is in terms of the modern county. In any case, an
estimate of Domesday population can rarely be definitive, and all that
can be claimed for the present figures is that they indicate the order of
magnitude involved. These figures are those of recorded population,
and must be multiplied by some factor, say 4 or 5, in order to obtain the
actual population; but this does not affect the relative density as between
one area and another.[5] That is all that a map such as Fig. 23 can roughly
indicate.

It is impossible to say how complete were these Domesday figures. We
should have expected to hear more, for example, of priests. Then there are

[1] But they are accounted for in the Summary for Glastonbury Abbey—see p. 71
above.
[2] F. W. Maitland's total for the Domesday county was 1,762 teams (*op. cit.* p. 401).
R. W. Eyton's analysis does not cover plough-teams.
[3] Sir Henry Ellis, *A General Introduction to Domesday Book*, II (London, 1833),
pp. 438–40. His total for the Domesday county (excluding tenants and sub-tenants)
came to 7,466.
[4] R. W. Eyton, *op. cit.* p. 151. His total for the county (excluding tenants and sub-
tenants), came to 7,275.
[5] But see p. 368 below for the complication of serfs.

Recorded Population of Dorset in 1086

A. *Rural Population*

Bordars	3,032
Villeins	2,569
Serfs	1,243
Cottars	196
Coscets	182
Miscellaneous	122
Total	7,344

There were also 3 bondwomen (*ancillae*).

Details of Miscellaneous Rural Population

Salt-workers . .	56	Priests . . .	4
Coliberts . . .	33	*Servientes francigenae* .	2
Rent-payers (*censores*)	11	Smith (*unus faber*) .	1
Fishermen . .	4	*Rusticus* . . .	1
Men (*homines*) . .	10	Total . . .	122

B. *Urban Population*

BRIDPORT
101 *domus*; 20 *domus destitutae* or *penitus destructae* (see p. 119 below).

DORCHESTER
89 *domus*; 1 burgess; 100 *domus penitus destructae* (see p. 118 below).

WAREHAM
142 *domus*; 4 burgesses; 1 bordar; 150 *domus destructae*, *penitus destructae* or *vastae* (see p. 119 below).

WIMBORNE MINSTER
14 *domus*; 8 burgesses; 3 bordars (see p. 122 below).

SHAFTESBURY
177 *domus*; 80 *domus destructae, omnino destructae* or *penitus destructae* (see p. 121 below).

Note; The bordars are also included in Table A.

some 65 or so entries, involving about 50 places, which do not mention population although they sometimes refer to plough-lands and even plough-teams and to other resources such as wood or meadow. Such is the Exchequer entry for Todber (82), which answered for two hides:

Terra est ii carucis quae ibi sunt in dominio, et molinum reddit x solidos, et x acrae prati. Silva dimidia leuua longa et una quarentena lata. Valuit iii libras. Modo iiii libras.

We hear from the Exeter version (47) that there were also stock there. One wonders who looked after the animals, and worked the teams and the mill. Could it have been unrecorded serfs? Or occupiers of neighbouring settlements? Or were the inhabitants of Todber accidentally omitted from the record? Other examples of entries with mention of teams but not of population are those for Brockington (79b), Ilsington (80), Thornham (61, 61b; 84 bis), and the unidentified Odeham (77b). Then there are those other place-names for which hardly any information is given.[1] We cannot be certain about the significance of these omissions, but they (or some of them) may imply unrecorded inhabitants.

Unnamed thegns are mentioned quite frequently as holders of estates in 1066, and they are usually entered in small groups, e.g. the 3 at Buckham (77), the 5 at Rampisham (77), and the 9 at Rollington (51b, 82b).[2] Altogether, some 340 unnamed thegns are enumerated for 1066. Unnamed thegns are not mentioned for 1086 with but four exceptions: ten thegns still held Lower Kingcombe (84b); six held part of Sherborne (77); two held part of Stoke Abbott (77); and one held part of Long Bredy (37b, 78). It is impossible to say what had become of the earlier groups. There had been 7 thegns at Poorton (80b) in 1066, and there were 7 villeins there in 1086. There had been 5 thegns at Higher Kingcombe (80b) in 1066, and there were 5 villeins there in 1086.[3] Can we suppose that at both places the villeins of 1086 were the thegns of 1066 or their descendants?[4] But usually the figures do not so conveniently lend themselves to such suggestions. Thus we have no means of knowing whether the 2 thegns at Leigh (80b) in 1066 were represented among the 3 villeins there in 1086. The thegns at Lower Kingcombe,[5] Sherborne,[6] Stoke Abbott,[7]

---

[1] See p. 76 above.

[2] For the groups of thegns in Bedfordshire and Buckinghamshire, see H. C. Darby and Eila M. J. Campbell, *The Domesday Geography of South-east England* (Cambridge, 1962), pp. 24, 157 and 595.

[3] In the same way Galton (85), held by 4 freemen (*liberi homines*) in 1066, was occupied in 1086 by 4 men (*homines*) rendering 12s. 4d. It also seems possible that the 2 bordars in possession of an anonymous holding had once been freemen: *Duo bordarii tenent quartam partem unius virgatae terrae. Ipsi libere tenuerunt T.R.E.* (84b).

[4] In the section dealing with *Terrae Tainorum Regis*, there are several entries which name the 1086 holder and then add: *Pater eius tenuit*.

[5] Lower Kingcombe (84b): *Decem taini tenent Chimedecome. Ipsi tenuerunt T.R.E. pro uno manerio.*

[6] Sherborne (77): *De eadem etiam terram tenent vi taini viii hidas.*

[7] Stoke Abbott (77): *et ii taini tenent ii hidas et dimidiam.*

and Long Bredy[1] in 1086 have been excluded from our totals because they were landholders; so likewise have the named thegns. So have the *Angli* at Handley,[2] the knight at Nettlecombe,[3] and the knight and widow at Piddletrenthide.[4] The decision whether to include or not is often difficult.[5]

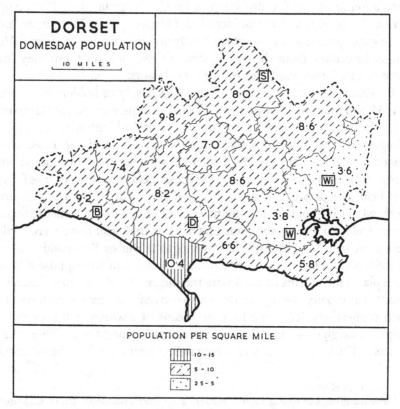

DORSET

DOMESDAY POPULATION

10 MILES

POPULATION PER SQUARE MILE

|||||| 10 – 15

///// 5 – 10

:::: 2·5 – 5

Fig. 23. Dorset: Domesday population in 1086 (by densities).

Domesday boroughs are indicated by initials: B, Bridport; D, Dorchester; S, Shaftesbury; W, Wareham; Wi, Wimborne Minster.

[1] Long Bredy (*37b*, 78): *et unus tainus habet i hidam.*
[2] Handley (78b): *De hac terra tenent ii Angli libere iiii hidas.*
[3] Nettlecombe (*38*, 78): *De eadem terra tenet unus miles ii hidas.*
[4] Piddletrenthide (77b): *De eadem terra tenent i miles et quaedam vidua iii hidas.*
[5] We have included the 6 men at Ringstead (84b) because (1) they were rent-payers although not called so; (2) there was no other population on the holding: *Sex homines tenent eam ad firmam.*

Bordars constituted the most important element numerically in the population, and amounted to over 40 per cent of the total. On one holding at Langton Herring (*31*, 75 b) one of the eight bordars rendered 30*d.* per annum. The Exeter version of an entry for a *Waia* manor tells us that

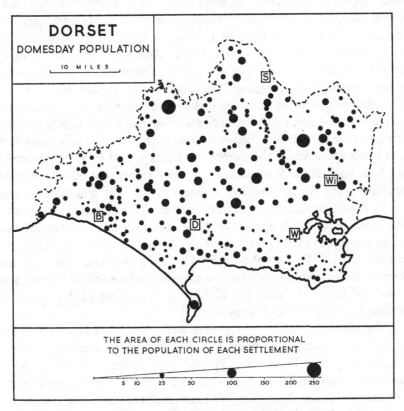

DORSET

DOMESDAY POPULATION

10 MILES

THE AREA OF EACH CIRCLE IS PROPORTIONAL
TO THE POPULATION OF EACH SETTLEMENT

5   10   25   50   100   150   200   250

Fig. 24. Dorset: Domesday population in 1086 by (settlements).

Domesday boroughs are indicated by initials: B, Bridport; D, Dorchester;
S, Shaftesbury; W, Wareham; Wi, Wimborne Minster.

6 bordars alone had half a hide (*54b*, 83 b); and an Exeter entry for a Cerne manor (*45*, 78) likewise says that 7 bordars had 1 virgate and 5 acres.[1] The reference to 2 bordars who held a ¼ virgate on an anonymous holding (84 b) is interesting. They may have been freemen reduced to bordars: *Ipsi libere tenuerunt T.R.E.* Next in numerical order came

[1] The Exchequer version mentions 5 bordars.

villeins, who accounted for nearly 35 per cent of the total. A common Exeter formula states the hidage and plough-teams held by the villeins before going on to enumerate the categories of people on a holding; and the word *villani* is sometimes used in this connection even when no villeins appear in the subsequent enumeration,[1] e.g. in the entry for Durweston (*58b*, 83 b).[2] The Exeter folios sometimes supply additional information. Thus we hear that the 4 villeins who paid £3 at Wraxall did so *de gablo* (*51*, 82 b); that one villein at Adber occupied one virgate (*355b*, 95 b); and that on one of the Tarrant holdings there was *i villanus qui manet ibi habet aliam virgatam et dimidiam carucam* (*58b*, 83 b). One of the Exeter entries for Morden (*56*) tells of villeins with 1½ virgates on a holding of 5 virgates, but we are not told how many they numbered; the corresponding entry in the Exchequer text (83 b) does not help because it has a blank space where the number of teams and of inhabitants should have been entered. Serfs amounted to nearly 17 per cent of the recorded population, but the distribution as between one fief and another was very variable. Cottars and coscets each accounted for 3 per cent of the total. Each group occasionally appears in entries alongside bordars, e.g. in those for Beaminster (77) and Frampton (78 b), but they never appear together in the same entry.[3] Watercombe (*31b*, 75 b) had a population of only one coscet on its hide and plough-land; it also had pasture and half a mill, and it rendered 15s. The Exeter text alone tells us that two of the coscets at Nutford (*31b*, 75 b) held 8 acres.

In the miscellaneous category, there were 56 salt-workers at 3 places along the coast, but salt-workers are not mentioned for 2 other places said to have salt-pans.[4] There were 33 coliberts—20 on some of the royal manors described on fo. 75, and another 13 at Sturminster Newton (77 b). Nine rent-payers at Allington (80b) rendered 11s., and 2 at Askerswell (78 b) rendered 15s.; the latter appear in the Exeter version as *gablatores* (*42*). There were also 6 men (*homines*) holding *ad firmam* at Ringstead (84 b), and 4 rendering 12s. 4d. at Galton (85). Four fishermen are recorded for Bridge (83, 84 b), but one of the entries for Lyme Regis (75 b) also speaks

[1] In the entry for Long Bredy (*37b*, 78), the Exeter version uses *rustici* instead of *villani*, but *rustici* are not mentioned in the detailed enumeration of people that follows.
[2] The total recorded population for Durweston amounts to 3 bordars.
[3] To this there is one exception in the Dorset folios—in the entry for Kinson (80 b), which was transferred to Hampshire in 1930. Here were 7 serfs, 18 villeins, 14 coscets and 4 cottars.
[4] See p. 112 below.

of an unspecified number of fishermen who rendered 15s. to the monks. There were 4 priests.[1] There were 2 *servientes francigenae* on a Cerne holding (79), and a smith at Melbury (79b). Finally, the villein on a holding at Uploders (83b) appears in the Exeter text as *rusticus* (58).[2] Not included in the count are the 3 bondwomen (*ancillae*) at Crichel (80b).

(5) *Values*

The value of an estate is generally given for two dates—for 1086 and for some earlier year. Some entries give a figure only for 1086, and a very small number give no value at all.[3] The following examples illustrate some of the variations to be found in the Exchequer version:

Nutford (75 b): *Valuit et valet xxv solidos.*
Fontmell Magna (78b): *Valuit x libras. Modo xv libras.*
Cranborne (75 b): *Valuit xxiiii libras. Modo reddit xxx libras.*
Frampton (78b): *Totum valuit et reddit xl libras.*
Osmington (78): *Valet viii libras.*
Hampreston (75 b): *Reddit l solidos.*
Affpuddle (77b): No value stated.

The general use of the word *valuit* in the Exchequer folios leaves the exact date of the earlier valuation in doubt, but two entries are more explicit and speak of the day on which the existing holders entered into possession:[4]

Sturminster Marshall (80): *Valebat lxvi libras quando recepit. Modo lv libras.*
Wraxall (82b): *Inter totum valet manerium ix libras. Quando recepit iiii libras.*

In contrast to the Domesday Book, the Exeter text almost invariably gives the earlier date as *quando recepit*. Here are the two versions of the valuation of Bexington:

E.D. (*50b*): *haec mansio valet vi libras, et quando recepit iiii libras.*
D.B. (82b): *Valuit iiii libras, modo vi libras.*

[1] See p. 116 below.
[2] *Rustici* is also used interchangeably with *villani* in the Exeter account of Long Bredy (*37b*, 78)—see p. 94 above.
[3] The entry for Adber (*493b*, 99), in the Somerset folios, follows a common Somerset form: *Olim et modo valet xx solidos.* The entry for Sandford Orcas, also in the Somerset folios (*466b*, 99), likewise uses the word *olim*. So does the entry for Thorncombe (*313*, 108) in the Devonshire folios.
[4] So does a third entry—for Kinson: *quando recepit valebat l libras. Modo lxx libras.* But Kinson was transferred to Hampshire in 1930.

Occasionally, when the Domesday Book states a value for only one date, the Exeter text shows that the same figure applied to both dates, e.g. in the entry for Bockhampton:

E.D. (*33*): *valet per annum iii libras, et quando comitissa [Boloniensis] recepit tantundem.*

D.B. (85): *Valet iii libras.*

A few Exeter entries place the earlier date at another time, and one entry for a holding at Chaldon Herring gives its value at three dates.

Ashmore

E.D. (*29*): *haec mansio reddidit xv libras vivente regina, et modo reddit tantundem.*

D.B. (75 b): *Valuit et valet xv libras.*

Cranborne

E.D. (*29*): *haec mansio reddidit xxiiii libras vivente regina, et modo reddit xxx libras.*

D.B. (75 b): *Valuit xxiiii libras. Modo reddit xxx libras.*

Littlebredy

E.D. (*37*): *haec mansio reddit xvi libras, et tempore E. abbatis valebat tantundem.*

D.B. (78): *Valuit et valet xvi libras.*

Chaldon Herring

E.D. (*59 b*): *valet per annum viii libras, et quando Hugo recepit eam valebat x libras, et in vita Hugonis reddidit xi libras.*

D.B. (83 b): *Valuit x libras. Modo viii libras.*

In the Exeter text (e.g. on fo. *36 b*), the word *valet* has sometimes been struck out and *reddit* written above.

The amounts are usually stated in round numbers of pounds and shillings, e.g. £6 or 70s. Occasional precise figures, however, suggest either a sum of rents or careful appraisal, e.g. the 21s. 3d. at Moreton (84b), and the rise from 40s. to 65s. 8d. at Crichel (83). We hear of assayed money in the entry for the Isle of Portland:

E.D. (*26*): *Haec reddit lxv libras candidos* [sic] *per annum.*

D.B. (75): *Hoc manerium cum sibi pertinentibus reddit lxv libras albas.*

Honey formed part of the render at two places:

Holworth (*44 b*, 78): *Valet iii libras et sextarium mellis.*

Rushton (82): *Reddit xxx solidos et iiii sextaria mellis.*

A few valuations give us a glimpse of an archaic system that had almost disappeared. This was the arrangement of the manors of the Ancient Demesne of the Crown in groups from each of which the king received a fixed rent in kind—the *firma unius noctis*. The five groups (*27-8*, 75) each comprised widely separated places, and were as follows:

(1) Burton Bradstock, Bere Regis, Colber, Shipton Gorge, Bradpole, Chideock.

(2) Wimborne Minster, Shapwick, Crichel, All Hallows Farm (*Opewinburne*).

(3) Dorchester, Fordington, Sutton Poyntz, Gillingham, Frome.

(4) Pimperne, Charlton Marshall.

(5) Winfrith Newburgh, Lulworth, Winterborne, Knowlton.

The first three groups each rendered one night's *firma*, and the last two groups each rendered half a night's *firma*.[1] The statement for the first group is typical: *Hoc manerium cum suis appendiciis et consuetudinibus reddit firmam unius noctis*. The ancient renders had apparently not been commuted for money payments. None of the groups seems to have been assessed in hides, and we are explicitly told in each of the relevant entries that no geld had been paid in 1066.[2] The *firma noctis* is also mentioned in the accounts of the boroughs of Dorchester, Bridport and Wareham (*11b–12b*, 75).[3]

Values frequently remained the same; but they had sometimes risen, and, less often, they had fallen. Occasionally, the variations were considerable, e.g. Wynford Eagle (80b) had risen from £12 to £19, and Iwerne Courtney (81) had fallen from £15 to £10. The rise at Crawford (84) from 30*d.* to 40*s.* is striking; so is that at Moulham (83) from 5*s.* to 30*s.* Whether scribal errors were involved in such figures, we cannot say.[4] A holding of one plough-land at Nyland (79) was waste—*Vasta est*. Another holding of half a plough-land at Hurpston rendered 12*s.* 6*d.*;

---

[1] Similar arrangements were to be found occasionally in the adjacent counties of Hampshire, Somerset and Wiltshire. See p. 15 above and p. 150 below. See also J. H. Round, *Feudal England* (London, 1895), pp. 109–15. There was another royal holding at Wimborne Minster (in the second group above) which was valued in the usual way and which (we are specifically told) did not form part of the *noctis firma* of Wimborne (*30b*, 75b).

[2] See p. 81 above.

[3] Bridport and Wareham, however, unlike Dorchester, are not mentioned in any of the five groups of royal estates rendering a *firma unius noctis*.

[4] Could the figures be errors for 30*d.* and 40*d.* and for 5*s.* and 30*d.*?

the Domesday Book merely records the fact, but the Exeter Book explains that the holding rendered this amount in spite of the fact that it had been entirely devastated:

> E.D. (60): *haec terra omnino devastata est et tamen valet xii solidos et vi denarios.*
> D.B. (84): *Valet xii solidos et vi denarios.*

The contrast between these two entries illustrates what may so often lie behind the terse Domesday summaries.

Generally speaking, the greater the number of teams and men on a holding, the higher its value; but there is much variation, and it is impossible to discern any consistent relationship, as the following figures for five holdings, each rendering £3 in 1086, show:

	Teams	Population	Other resources
Bockhampton (33, 85)	3	6	Mill, meadow, wood
Bradle (83)	1½	5	Mill, pasture, wood
Elworth (80b)	2	8	Meadow, pasture
Hemsworth (83)	1	4	Meadow, pasture
Plumber (84)	2	13	Meadow, wood

There are, moreover, many perplexities. Why, for example, should a holding at Blandford St Mary (79b), with one plough-land but no team or anything else, render 12s.? Apart from 12 acres of meadow, Chettle (83) was in a similar position, yet it rendered 20s.

Conclusion

For the purpose of calculating densities, Dorset has been divided into thirteen units, a division based upon the following considerations: the main physical and geological features; the hundreds as deduced from the Geld Accounts; and the desirability of keeping within one unit the land that was probably included in one of the great manors, e.g. that of Sherborne (77). Furthermore, the places mentioned in each of the five composite entries that describe the Ancient Demesne of the Crown (27-8, 75) are so widely scattered that it is impossible to include any group within one unit area, and the totals within each of these entries have therefore been divided equally among its constituent holdings.[1] The thirteen units

[1] For the constituent holdings of each, see p. 97 above.

do not constitute an ideal division, and the fact that their boundaries have been drawn to coincide with those of civil parishes increases their artificial character; many parishes (perhaps most) extend across more than one kind of soil and country. Even so, the division does provide a basis for distinguishing broad variations over the face of the countryside.

Of the five standard formulae, those relating to plough-teams and population are most likely to reflect something of the distribution of wealth and prosperity throughout the county. Taken together, certain common features stand out on Figs. 21 and 23. The main contrast on both maps is that between the Tertiary Basin with its light sandy soils and the rest of the county. The density of teams in the Tertiary area was under one as compared with between 1·5 and nearly 3 elsewhere; the density of population was under 4 as compared with between 5 and about 10 elsewhere. Taking a general view, there are no great differences between the densities of chalkland and clayland.

Figs. 22 and 24 are supplementary to the density maps, but it is necessary to make reservations about them. As we have seen on p. 72, some Domesday names covered two or more settlements, and several of the symbols should thus appear as two or more smaller symbols. This limitation, although serious locally, does not affect the impression conveyed by the maps. Generally speaking, they confirm and amplify the information on the density maps.

Types of entries

WOODLAND

The Dorset folios give information both for wood itself and for underwood. The amount of each is normally expressed in one of two ways—in terms of linear dimensions or in terms of acres. There are also a number of variant entries.

Woodland is called *silva* in the Exchequer folios and *nemus* in the Exeter folios. The linear dimensions are stated in terms of leagues and furlongs, and, very rarely, of perches. The following examples from the Exchequer text illustrate the kind of entry that is encountered:

Stalbridge (77): *Silva i leuua longa et iii quarentenis lata.*
Stock Gaylard (82): *Silva x quarentenis longa et iiii quarentenis lata.*
Coombe Keynes (77b): *Silva vi quarentenis longa et tantundem lata.*
Tarrant (77b): *Silva viii quarentenis longa et totidem lata.*
Tarrant (78b): *Silva l perticis longa et xl lata.*
Leigh in Colehill (80b): *Silva i quarentena longa et v virgis lata.*

The Exeter version occasionally uses *silva* instead of *nemus*, e.g. in the entry for Poorton (*51*, 82b). The Exeter entry for Puddletown (*25*) is unusual in that it describes two stretches of wood: one not mentioned in the corresponding Exchequer entry, and another (*in alio loco*) which is (*75*). When the quantities are in furlongs, the number is usually less than the twelve commonly supposed to make a Domesday league, but Broad-windsor (*85*), for example, had wood 30 furlongs by 8, and Okeford Fitzpaine (*83*) had wood 23 furlongs by 9. The exact significance of these linear dimensions is far from clear, and we cannot hope to convert them into modern acreages.[1] All we can do is to plot them diagrammatically as on Fig. 25, and we have assumed that a league comprised twelve furlongs. Occasionally, one set of measurements is found in a linked entry covering a number of widely separate places, e.g. the entries for the five groups of manors of the Ancient Demesne.[2] This seems to imply, although not necessarily, some process of addition whereby the dimensions of separate tracts of wood have been consolidated into one sum. On Fig. 25, such measurements have been plotted for the first-named place only, and the other places in the entry have each been indicated merely by a conventional symbol to show the possible places where wood may have been.[3]

Twelve entries state the amount of woodland to be 'in length and breadth', or, more rarely, as 'between length and breadth'; R. W. Eyton thought that these variant formulae indicated 'areal leagues' or 'areal furlongs'.[4] On the other hand, when the Domesday Book gives one form, the Exeter version occasionally gives another:

Adber

> E.D. (*493b*): *i quadragenaria nemoris in longitudine et in latitudine.*
> D.B. (99): *una quarentena silvae in longitudine et latitudine.*

Worth Matravers

> E.D. (*51b*): *vii quadragenariae nemoris in longitudine et totidem in latitudine.*
> D.B. (82b): *vii quarentenae silvae inter longitudinem et latitudinem.*

[1] See p. 372 below.
[2] See p. 97 above.
[3] The wood in a linked entry has always been plotted for the first-named place, in addition to any other wood that is mentioned in a separate entry for that place. If any place (other than the first-named) in a linked entry also has wood entered for it in a separate entry, that wood has been plotted instead of the conventional symbol for a possible place with wood.
[4] R. W. Eyton, *op. cit.* pp. 31–5.

One is tempted to assume that the variant formulae imply the same facts
as the more usual formula, and that they occur occasionally when the
length and breadth of woodland are the same. We have followed this
assumption in drawing Fig. 25.

Some twenty or so Exchequer entries give only one dimension, and
again Eyton thought that these indicated areal quantities. The Exeter
versions of such entries, however, occasionally follow the normal formula:

Hurpston
  E.D. (*60 b*): *i quadragenaria nemoris in longitudine et tantundem in latitudine.*
  D.B. (84): *i quarentena silvae.*

Wilkswood
  E.D. (*60 b*): *iiii quadragenariae nemoris in longitudine et latitudine.*
  D.B. (84): *iiii quarentenae silvae.*

It may be that, as with the other variants discussed above, all these
different expressions are merely alternative ways of saying the same thing
when the length and breadth of wood are the same. On Fig. 25, however,
such entries have been plotted as single lines—that is, in the absence of an
alternative expression in the Exeter folios.

Quite frequently, the amount of wood is stated in acres. The figures
range from half an acre on Dodo's anonymous holding (84) to 40 acres at
Sutton Waldron (82), but the great majority do not rise above 15 acres.
The figures are usually in round numbers that suggest estimates, e.g. 10 or
20 acres; but occasional entries seem to be precise, e.g. Steeple (80) had
3 acres of wood. No attempt has been made to convert these figures into
modern acres, and they have merely been plotted as conventional units on
Fig. 25. The wood on one holding in a manor might be measured in acres
and that on another holding in terms of linear dimensions. At Mappowder,
for example, there were 5 acres on one holding (80b), wood 2 leagues
by 1 on another (79b), and wood 4 by 3 furlongs on a third holding (84).
There are a few entries in which the acre seems to appear as a linear
measure:

Tarrant (83): *Silva i quarentena longa et iiii acris lata.*
Winterborne (77):[1] *Silva iii quarentenis et dimidia longa et iiii acris et ii lata*
[sic].
Witchampton (79b): *Silva una quarentena longa et viii acris lata.*

  [1] This is one of the eastern group of Winterbornes—see p. 75.

R. W. Eyton explained these as references to the 'lineal acre', comprising 4 perches.[1] On Fig. 25, they have been indicated under the category of 'other mention'.

Two other unusual entries must also be mentioned. The wood at Renscombe was said to be unproductive, and, in the Exeter text alone, so was that at Nettlecombe:

Renscombe (*38b*, 78): *Silva infructuosa v quarentenis longa et una quarentena lata.*

Nettlecombe (*38*, 78): *i leuga nemoris longitudine et viii quadragenae latitudine, quod nemus nullum fructum fert.*

We also hear of *Boscus de Havocumba* (*27*, 75): two thirds contributed to the royal *firma*, but 'the third oak' went to Earl Godwine, wrongly called Earl Edwin in the Domesday Book.[2] Another curiosity occurs in the account of Mapperton in Almer (78b), where there were *inter pasturam et silvam* eleven by eleven furlongs, plotted on Fig. 25 as 'other mention'.

A number of entries mention underwood (*silva minuta*), always alone and never with wood, although some of the places have other holdings with wood. The phrase *silva modica* appears in the entries for Stalbridge Weston (77) and Stoke Abbot (77). Neither of these entries appears in the Exeter text, in which *silva minuta* is always rendered as *nemusculus*. It is difficult to know whether the terms *silva minuta* and *silva modica* indicated the same thing, but they are shown by the same symbol on Fig. 25.[3] The amounts are usually in acres, and are all under 20 acres. The underwood on four holdings was measured in linear terms. One of these gives two dimensions, as normally; one gives dimensions 'in length and breadth'; and two give only one dimension—2 furlongs at Caundle (84b) and 6 at Milborne St Andrew (82b). There were also two references to *broca* or brushwood:

Wimborne Minster (80b): *ibi una leuua brocae.*[4]

Lytchett Matravers (80b): *Brocae i leuua in longitudine et latitudine.*

They appear on Fig. 25 under the category of 'other mention'; both holdings also had *silva* entered for them.

[1] R. W. Eyton, *op. cit.* pp. 25–7.
[2] *Boscus de Havocumbe* has not been plotted on Fig. 25 because its location is uncertain.
[3] But see p. 177 below for the Somerset evidence which equates the two phrases.
[4] This comes at the end of the account of Canford, and may possibly have been there.

The only reference to forest occurs in the account of Horton (78b), where we are told that the king held the two best of the seven hides of the Abbot of Horton's manor in the royal forest of Wimborne—*Duas meli-*

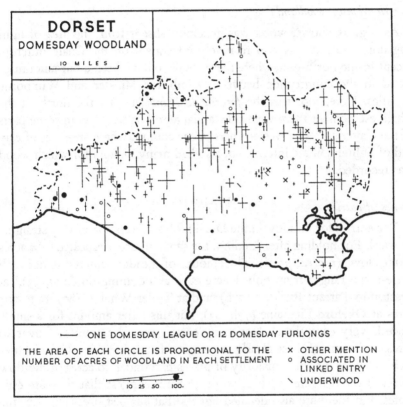

Fig. 25. Dorset: Domesday woodland in 1086.

When the wood of a settlement is entered partly in linear dimensions and partly in some other way, both are shown. When wood and underwood are entered for the same settlement, both are shown.

ores hidas de his vii tenet rex in foresta de Winburne. Could this have been part of Holt Forest of which we hear in later records? The earliest reference to Holt comes from the twelfth century in the form of *Win-burneholt.*[1] No reference is made to the forests of Blackmoor, Gillingham

---

[1] A. Fägersten, *op. cit.* p. 82.

and Powerstock or to Cranborne Chase. Some landholders are described as huntsmen in the Domesday entries, and still others in the Geld Accounts (Fig. 90).

Distribution of woodland

As Fig. 25 shows, wood was generally absent from the area of light Bagshot sands. It was also relatively infrequent on the Chalk outcrop, except in the north-east, where there seems to have been a fair amount of wood in the countryside between Wimborne Minster and Wimborne St Giles. Here, too, was the *foresta de Winburne*. To the north of the Chalk escarpment there were substantial amounts of wood in some parts of the clay vales. The western part of the county, where stretches of clay and of lighter soil are intermixed, also had many manors for which wood was recorded.

Types of entries   MEADOW

The entries for meadow in the Dorset folios are comparatively straight-forward. For holding after holding, the same phrase is repeated—'*n* acres of meadow' (*n acrae prati*). The amount of meadow entered under each place-name ranges from only ½ acre each at Poyntington (*279*, 93),[1] at Preston in Tarrant Rushton (77b), and at Toller Whelme (80) up to 204 acres at Okeford Fitzpaine (77b, 83), but this latter amount for a single place is very exceptional. Not many entries record amounts of over 50 acres, and some of these clearly are entries that cover a number of settlements. By far the great majority of places had under 30 acres of meadow apiece. Round figures such as 30 or 40 acres suggest that they are estimates, but there are also detailed figures that suggest precision, e.g. the 4½ acres at Bowood (77), the 27 acres at Sturthill (*57b*, 83b) and the 102 acres at Corfe Mullen (80b). The entry for Sherborne (77) records 130 acres 'of which', we are told, '3 acres are in Somerset near Milborne Port', that is, just across the county boundary. As for other counties, no attempt has been made to translate these figures into modern acreages, and they have been plotted as conventional units on Fig. 26. The meadow in a composite entry has been divided equally among the places involved.

There are very few unusual entries. Two measure meadow in terms of length and breadth, and one entry gives only a single dimension:

[1] Poyntington is described in the Somerset folios.

West Pulham (79): *Ibi viii quarentenae prati inter longitudinem et latitudinem.*
Waia (54b–55, 83b): pratum ix quarentenis longum et una quarentena latum.[1]
Puddle (*39, 78b*): *vi quarentenae prati.*

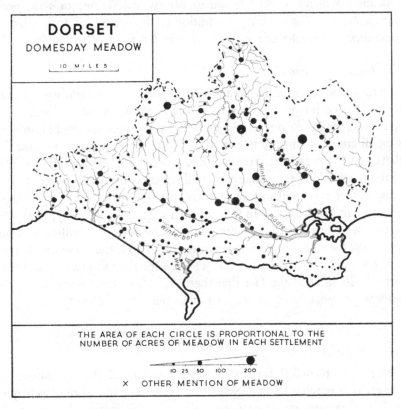

Fig. 26. Dorset: Domesday meadow in 1086.

When the meadow of a settlement is entered partly in acres and partly in some other
way, both are shown. Areas of alluvium are shown in grey; rivers passing through
these areas are not marked.

One entry for Mappowder (84) says, *Ibi aliquantum prati,* but *xvi* is
interlined above; maybe 16 acres are implied, but this has not been
assumed in drawing Fig. 26, and it has been plotted under the category of
'other mention'. An unusual reference to meadow in demesne occurs in
an Exeter entry for Rushton (*59b*) where we read that two knights held

[1] The Exeter version reads (*54b–55*): *ix quadragenariae prati et i in latitudinem.*

half a hide from the wife of Hugh fitzGrip, *exceptis xvi agris prati quos ipsa tenet in dominio*; the knights held 2 acres apiece, and the Exchequer version merely mentions a total of 20 acres. An entry for Melcombe (75 b) refers to meadow which had been leased to Wulfweard White: *xii acrae prati prestitae fuerunt Wlgaro Wit.* In addition to the entries for 1086, there is a reference to a render of hay in 1066 from 6 acres at Waddon (79).[1]

Distribution of meadowland

As might be expected, fair quantities of meadow—frequently more than 20 acres for each manor—were to be found along the alluvial valleys of the Frome, the Piddle and the Stour. Their tributaries, too, were bordered by small amounts (mostly under 20 acres), and this fact accounts for the relatively widespread distribution of meadow in the Chalk upland (Fig. 26). To the north, along the numerous small streams of the well-watered clay-lands, almost every village had meadow entered for it, and many villages had over 20 acres apiece. In the western part of the county, also well watered with small streams, there was likewise scarcely a village without some meadow. One interesting feature of Fig. 26 is the presence of small amounts—usually under 10 acres—entered for the villages of the Isle of Purbeck, in spite of the fact that there were but few streams here; the meadow presumably lay in the east–west belt of Wealden Clay.

PASTURE

Types of entries

Pasture is recorded for about three quarters of the settlements of Dorset. It is entered in the Exeter text as *pascua*, and in the Exchequer version as *pastura*. Its amount is usually recorded in one of two ways—either in linear dimensions or in acres. The linear dimensions are given in terms of leagues and furlongs and, very rarely, of perches. Some of the figures for furlongs are large, e.g. the 42 by 8 furlongs at Okeford Fitz-paine (83), and the 27 furlongs by 1 league 3 furlongs at Abbotsbury (*39b*, 78). The following examples from the Exchequer text indicate the kind of entry that is encountered:

Milton Abbas (78): *Pastura iii leuuis longa et una leuua lata.*
Ibberton (75 b): *Pastura vii quarentenis longa et iii quarentenis lata.*
Lulworth (83): *vi quarentenae pasturae in longitudine et tantundem in latitudine.*

[1] See p. 126 below.

Coombe Keynes (77b): *viii quarentenae pasturae in longitudine et totidem in latitudine.*

Poxwell (78): *Pasturae viii quarentenis et xxxvi virgis longa et iii quarentenis et xiiii perticis lata.*

Symondsbury (78): *Pastura v quarentenis longa et una quarentena lata x virgis minus.*

The exact significance of these linear dimensions is far from clear, and we cannot hope to convert them into modern acreages.[1] All we can do is to plot them diagrammatically as on Fig. 27, and we have assumed that a league composed 12 furlongs. Occasionally, one set of measurements is found in a linked entry covering a number of widely separated places, e.g. the entries for the five groups of manors of the Ancient Demesne.[2] This seems to imply, although not necessarily, some process of addition whereby the dimensions of separate tracts of pasture were consolidated into one sum. On Fig. 27 such measurements have been plotted for the first-named place only, and the other places in the entry have each been indicated merely by a conventional symbol to show the possible place where pasture may have been.[3]

Twenty-seven entries state the amount of pasture to be 'in length and breadth' or 'between length and breadth'. The formulae appear to be interchangeable; the Exeter entry for Afflington gives one, and the Exchequer entry gives the other:

E.D. (*62*): *iiii quadragenariae pascuae inter longitudinem et latitudinem.*

D.B. (82b): *Pasturae iiii quarentenae in longitudine et latitudine.*

As for wood, we need not invoke the 'areal leagues and furlongs' of Eyton.[4] All these entries have been plotted in the normal manner as intersecting lines. Some forty or so entries give only one dimension, but the Exeter versions of two of these give one of the other variants just discussed:

Ringstead

E.D. (*60*): *iiii quadragenariae pascuae inter longitudinem et latitudinem.*

D.B. (83b): *iiii quarentenae pasturae.*

---

[1] See p. 372 below.   [2] See p. 97 above.

[3] The pasture in a linked entry has always been plotted for the first-named place, in addition to any other pasture that is mentioned in a separate entry for that place. If any place (other than the first-named) in a linked entry also has pasture entered for it in a separate entry, that pasture has been plotted instead of the conventional symbol for a possible place with pasture.

[4] R. W. Eyton, *op. cit.* pp. 31–5. See p. 100 above.

108 DORSET

Tarrant

E.D. (*59*): *et in alia* [sic] *loco viii quadragenariae pascuae inter longitudinem et latitudinem.*

D.B. (83 b): *viii quarentenae pasturae.*

Eyton thought that these one-dimension entries also indicated areal quantities. But, especially in view of the evidence of the woodland measurements, one is tempted to assume that they imply the same facts as the more usual formula, and that they occur occasionally when the length and breadth of pasture are the same.[1] On Fig. 27, however, they have been plotted as single lines—that is, in absence of an alternative expression in the Exeter folios.

Quite frequently the amount of pasture is stated in terms of acres. Amounts are generally below 30 acres, but they range from one acre at Caundle Marsh (84 b) to 250 acres at West Knighton (82). The figures are often in round numbers that suggest estimates, but some seem to be precise, e.g. 19 acres in one of the Cerne entries (*45*, 78) and 25 acres in another for Uploders (79 b). Like the acreages for meadow, they have all been treated as conventional units. One of the entries for Farnham (83) speaks of 10 acres *inter longitudinem et latitudinem*, and the amount may be an error for 10 furlongs, but it is shown on Fig. 27 as 'other mention'. Occasionally, the acre seems to be used as a linear measure, as in an entry for Wootton (80): *vii quarentenae et iiii acras* [sic] *pasturae.*[2] Pasture for one holding in a settlement might be measured in acres and that on another holding in terms of linear dimensions. At Mapperton near Beaminster, for example, there were 12 acres on one holding (*49 b*, 80 b), and pasture 1 league by 4 furlongs on another (*25*, 75).

The Exchequer account of West Stafford (83 b) says that it was held by two men, and speaks, amongst other things, of 24 acres of meadow and of '16 furlongs of pasture and 8 acres'.[3] The Exeter version (*55 b*) separately surveys the portion held by each man, and so speaks twice of 12 acres of meadow and of '8 furlongs of pasture and 4 acres'. There are also parallel descriptions of part of Frome. The Exchequer account (81 b) says that it was held by two men, and speaks, amongst other things,

[1] See p. 101 above.
[2] Indicated under the category of 'other mention' on Fig. 27. For a similar use of the acre in measuring wood, see p. 101 above. The Wootton here is both Abbots Wootton and Wootton Fitzpaine.
[3] Plotted on Fig. 27 as 'other mention'.

of 20 acres of meadow, 9 acres of wood, and of pasture 17 furlongs by 17 furlongs (*pastura xvii quarentenis longa et tantundem lata*). The Exeter version (*48b*) again separately surveys the portion held by each man, and

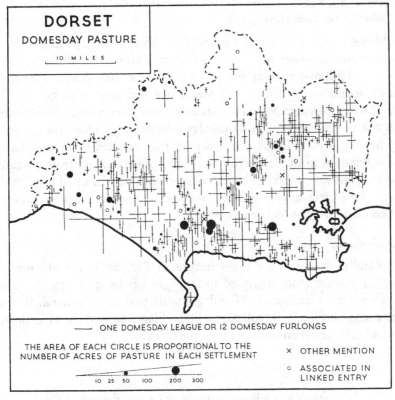

Fig. 27. Dorset: Domesday pasture in 1086.

When the pasture of a settlement is entered partly in linear dimensions and partly in some other way, both are shown.

speaks twice of 10 acres of meadow, 4½ acres of wood, and pasture 8½ furlongs by 8½ furlongs. But, in each case, pasture of the Exeter dimensions amounts only to one quarter of that entered in the Exchequer version. It is curious, but it is difficult to see whether it throws any light upon the obscure question of what exactly is implied by the linear dimensions of the Domesday Book.

Two holdings each comprised two separate quantities of pasture:

Spettisbury (*47b*, 82): *pastura v quarentenis et dimidia longa et ii quarentenis lata et alio loco super aquam pastura ii quarentenis longa et dimidia* [sic] *et i quarentena et dimidia lata.*

Tarrant (*59*, 83b): *Pastura iii quarentenis longa et ii quarentenis lata et in alio loco viii quarentenae pasturae.*

Could one of the Spettisbury pastures have been in the alluvial valley of the Stour and the other on the downs that stand above the village? A most unusual addition occurs at the end of the entry for Swyre (80b). It follows an account of the part of Swyre that had never paid geld in the time of King Edward: *Prius erat pascualis, modo seminabilis.* This constitutes one of the few explicit Domesday references to colonisation, and it may be compared with the description of the unnamed holding in Easewrithe hundred in Sussex (29) which had been reclaimed from pasture. Another curiosity occurs in the account of Mapperton in Almer (78b), where there were *inter pasturam et silvam* eleven furlongs by eleven, plotted on Fig. 27 as 'other mention'.

Distribution of pasture

Generally speaking, the main feature of Fig. 27 is the absence of recorded pasture from many of the villages of the northern clayland. Elsewhere, the distribution of villages with pasture is general. It was widespread on the Chalk upland, and the villages around the Heathlands also had substantial amounts entered for them.

FISHERIES

Fisheries (*piscariae*) as such are not mentioned in the folios for Dorset, and eel renders from mills appear only once—on a Tarrant manor where two mills rendered 30s. and 1,000 eels (77b); the small River Tarrant itself is a tributary of the Stour. The presence of fisheries was, however, implied at two other places (Fig. 28). The bishop of Salisbury's holding at Lyme Regis (75b) was occupied by an unspecified number of fishermen who paid a due in lieu of fish for the monks: *Piscatores tenent et reddunt xv solidos monachis ad pisces.* The other place was at Bridge (*57*; 83, 83b, 84b) where there was another teamless community of 1 villein and 4 fishermen.[1] The locality may have been situated near the isthmus that

[1] The spellings of this place-name are *Brige* (83, 84b) and *Briga* (*57*, 83b).

connects the Isle of Portland with the mainland, but, at any rate, some-where near Weymouth. These coastal groups of fishermen may have been engaged in sea-fishing. Clearly, neither this nor the catching of eels along the Tarrant can have comprised all the fishing activity of Dorset in the eleventh century.

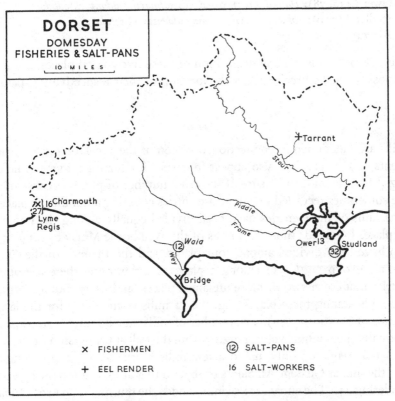

Fig. 28. Dorset: Domesday fisheries and salt-pans in 1086.

Waia is today represented by Broadwey, Upwey and Weymouth—see p. 75. The figures indicate the number of salt-pans or of salt-workers respectively. Lyme Regis had both fishermen and salt-workers.

SALT-PANS

The making of salt is implied in connection with five places. We hear of salt-pans themselves at two places: of 32 for Studland and of 12 for a *Waia* manor. For the other three places, salt-workers are entered, 27 for

Lyme Regis, 16 for Charmouth, and 13 for Ower. For two of the five places we hear also of money renders. The entries are as follows:

Charmouth (80): *Ibi xvi salinarii.*
Lyme Regis (77b): *Ibi xiii salinarii reddunt xiii solidos.*
Lyme Regis (85): *xiiii salinarii.*
Ower (*44b*, 78): *Ibi nulla caruca sed xiii salinarii reddunt xx solidos.*
Studland (79b): *Ibi sunt xxxii salinae reddentes xl solidos.*
Waia (79): *xii salinae.*

Fig. 28 shows the general disposition of these five coastal places, but it is impossible to be precise about the location of their respective salt-pans.

MILLS

Mills are mentioned in connection with 166 of the 319 Domesday settlements in Dorset; they also appear in entries for three anonymous holdings.[1] It is difficult to be sure of the exact number of places because mills are sometimes recorded in composite entries covering a number of places. The mills of these entries have been divided equally among the number of places involved; thus the 8 mills of the Wimborne Minster group (*27*, 75) have been divided among four places, and the 12 mills of the Dorchester group (*27b*, 75) among 5 places. Furthermore, there are other entries that cover two or more unnamed places, and we cannot apportion the mills among these places, e.g. the 12 mills in the entry for the large manor of Sherborne (77). The number of mills on a holding is stated, and normally their annual value, ranging from the mill at Catherston Leweston (80) that rendered only 3*d.* to a few mills worth over £1 per annum, e.g. the mill at Gussage All Saints (79b) and that at Povington (80b), each rendering 25*s.* The entry for Povington adds the unusual statement: *Hujus manerii molinus calumniatus est ad opus regis.* Occasionally, no render was entered, e.g. for mills at Ilsington (80), Poorton (83) and Rushton (82). On the other hand, 2 mills on one of the Tarrant manors (77b) rendered 1,000 eels as well as 30*s.*—the only eel render recorded for Dorset. Such sums as 5*s.*, 7*s. 6d.*, 10*s.*, 12*s. 6d.*, and 15*s.* were very frequent. Other common amounts were 20*d.* and 40*d.* and also 32*d.* Renders of uneven numbers of pence were rare, e.g. the 3 mills at Fontmell Magna (78b)

[1] One mill each on the holdings of Hugh de Boscherbert (83) and Dodo (84), and three parts of a mill on that of William d'Aumery (84b).

returned 11s. 7d. The account of Stoborough (79b) contains an unusual reference to a mill; the complete entry, and there is only one, merely says: *Ipse comes [Moritoniensis] habet in Stanberge i molinum cum dimidia hida et iii bordarios. Totum valet xl solidos.*

Fractions of mills are occasionally encountered. At Seaborough, described in the Somerset folios (*154 bis*, 87b *bis*), there were two holdings, each with half a mill rendering 10d.; presumably here was 1 mill. There were likewise two half-mills at Worgret; the value of one half was 10s. (82), but that of the other half was not stated (*37*, 78). Ringstead (83) had a half-mill rendering 4s., and at Watercombe (*31b*, 75b) nearby there was another half also rendering 4s., these also may possibly have been halves of the same mill. One royal holding at Okeford Inferior and Child Okeford (75) had 2 mills rendering 20s.; the Exeter version (25) adds 'of which the king has a half part'. The Domesday account (79) of the other holding (in the hands of the count of Mortain) tells us: *Ibi medietas ii molinorum reddit x solidos.* Could it be that these are the same 2 mills and that the account in the former entry is misleading? At Crawford (84) a quarter of a mill rendered 30d., but there is no mention of the remainder unless it was the 'three parts' of a mill rendering 9s. on the anonymous holding belonging to William d'Aumery (84b).[1] At Morden (84) we hear of 11d. *de parte molini*; there were also two other references to mills at Morden, one of 6s. 3d. (79b) and another of 45d. (*62b*, 82b). Does the reference to part of a mill imply a third mill? Or, after all, was there only one mill at Morden rendering 10s., a very usual amount? Similar doubts are raised by the entries for Wimborne St Giles. One speaks of 'a third part of a mill' rendering 15d. (*56*, 83b); another (85) says: *In molino villae xxii et dim'*, followed by a blank space. What does the latter entry mean? Can we put the two entries together by assuming five shares of 7½d. each and assigning two to the first entry and three to the second? The other entries for Wimborne St Giles do not mention mills (77b, 84).

Most Dorset villages with mills had one or two apiece. But the table on p. 114 cannot be very accurate because a group of mills in a large manor may have been distributed among its members in a way unknown to us. The group of 12 mills was entered for the large manor of Sherborne (77), which included a number of villages not named in the Domesday Book. The group of 11 mills belonged to the *Waia* manors along the River Wey,

---

[1] R. W. Eyton, *op. cit.* pp. 117–18, places the anonymous holding in the same parish as Crawford.

Domesday Mills in Dorset in 1086

Under 1 mill	3 settlements	5 mills	1 settlement
1 mill	101 settlements	6 mills	1 settlement
2 mills	37 settlements	7 mills	1 settlement
3 mills	16 settlements	11 mills	1 settlement
4 mills	4 settlements	12 mills	1 settlement

The table excludes the mills on 3 anonymous holdings.

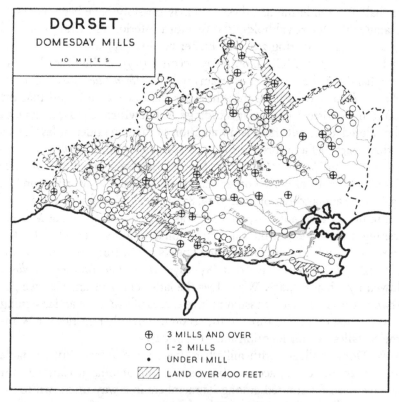

Fig. 29. Dorset: Domesday mills in 1086.

There were also mills on three anonymous holdings—see p. 112. Areas of alluvium are shown in grey; rivers passing through these areas are not marked.

all counted as one in this analysis.[1] The group of 7 mills belonged to those
Puddle manors probably in Puddletown hundred, and it must have covered
a number of settlements along the River Piddle.[2] The group of 6 belonged
to those Frome manors probably in Tollerford hundred, and is doubtful,
partly because it covered more than one settlement and partly because it
involves a linked entry.[3] The group of 5 belonged to those Tarrant manors
probably in Longbarrow hundred, and it covered a number of settlements
along the river Tarrant.[4] These examples illustrate the unsatisfactory
nature of the table above. We might go so far as to say that a large number
of mills entered under one name creates a suspicion that more than one
settlement is involved. Thus some of the groups of 4 mills, e.g. for Stour,
may well have involved more than one place.[5] So may some of the groups
of 3, e.g. for Wootton.[6] This is not to say that the table is without value,
because, at any rate, it suggests that there were some villages with more
than 1 mill at work.

Fig. 29 shows how the mills were associated with the rivers, and especially with the Stour, the Piddle, the Frome and their tributaries. There
were also many villages in the west of the county, along such small
streams as the Bride, the Brit and the Char. It is interesting to note that
there were mills associated with both the eastern Winterborne and the
western Winterborne. There were very few mills in the Isle of Purbeck.
Villages without any recorded mills were widespread throughout the
county, and we may well wonder whether the Domesday Book gives us
a complete picture of the eleventh-century mills of Dorset.

[1] Plotted at Broadwey. The folios for the five holdings are *54b, 55*; 79 *bis*, 83b *bis*
and 84.
[2] Plotted at Tolpuddle. The folios for the six holdings are *39, 43b, 53*; 77, 78, 78b,
79, 79b, and 82b.
[3] Plotted at Chilfrome. The folios for the four holdings are *27b, 48b*; 75, 80b
and 81b *bis*.
[4] Plotted at Tarrant Monkton. The folios for the four holdings are *32, 59*; 75b,
77b *bis* and 83b.
[5] The modern parishes are East Stour, West Stour and Stour Provost. The folios
are 78b and 80, and the mills are plotted at East Stour; Stourpaine (83) is separate
and some distance to the south.
[6] Abbots Wootton (in Whitchurch Canonicorum) and Wootton Fitzpaine. The
folios are 80 and 83, and the mills are plotted at Wootton Fitzpaine.

CHURCHES

Churches are not regularly enumerated in the folios for Dorset, but they are mentioned incidentally in connection with only 12 out of the 319 places recorded for the county, usually in groups and on royal manors. For Wimborne Minster there is not only a reference to the endowment of the *ecclesia* itself but also mention of an *ecclesiola*. Two separate churches were also mentioned for Wareham. We hear nothing, on the other hand, of a church at the borough of Shaftesbury, nor of churches on large manors such as that of Sherborne. Tithes are mentioned for the churches at Dorchester and Bere Regis.[1] Hugh fitzGrip gave a hide at Orchard in Church Knowle (84) to the *ecclesia* of Cranborne, but the Exeter text (*61 b*) makes clear that this was the abbey itself, and so it has not been counted here. The relevant entries are set out on p. 117. No priests are entered for any of these places, but they appear in entries for four other places,[2] and in two of these entries they are enumerated amongst the villeins and bordars.

Hinton Martell (76): *Presbyter vero hujus manerii habet alteram hidam et dimidiam, et ibi habet ii carucas cum iiii villanis et ii bordariis.*
Lutton Gwyle (80): *Ibi est presbyter et unus villanus et i bordarius.*
Langton Long Blandford (84): *iii villani cum presbytero et vi bordariis.*
Tarrant (76): *De ipsa eadem* [in Hinton Martell] *tenet alius presbyter manens in Tarente unam hidam et terciam partem i hidae.*[3]

Clearly the record of churches for Dorset—explicit or implied—is very incomplete. A number of Domesday place-names suggest the existence of unrecorded churches, e.g. Beaminster (77), Charminster (75b), Iwerne Minster (78b), Sturminster Marshall (80) and Yetminster (75b).[4] The name of Sturminster Newton appears only as *Newentone* (77b), but the minster is mentioned in pre-Domesday as well as post-Domesday documents.[5] Then, too, there were some churches not specifically mentioned, but which were named as tenants-in-chief: Abbotsbury, Cerne, Cranborne, Horton, Milton, Shaftesbury and Sherborne.

[1] For churchscot in 1066, see p. 126 below.
[2] The decision whether to include or not is frequently a difficult one. The priests at Hinton Martell and Tarrant were sub-tenants, but they have been counted because it looks as if they were resident.
[3] The word *manens* is interlined above. Two priests are said to have held (*tenuit*) portions of the substantial manor of Hinton Martell at some date previous to 1086.
[4] For all these names, see A. Fägersten, *op. cit.* pp. 262, 181, 26, 116 and 227.
[5] *Ibid.* pp. 47.

Domesday Churches in Dorset in 1086

Wimborne Minster (76): *De eadem ipsa terra* [in Hinton Martell] *pertinet ad ecclesiam de Winburne i hida et dimidia et dimidia virgata terrae.*

Wimborne Minster (78b): *Ad hanc ecclesiam* [Horton Abbey] *pertinent ecclesiola una in Winburne et terra duabus domibus.*

Burton Bradstock, Bridport, Whitchurch Canonicorum (28, 78b): *Ecclesia S. Wandregisili tenet ecclesiam de Bridetone et de Brideport et de Witcerce. His pertinent iiii hidae.*

Wareham (78b): *in Warham una ecclesia* [belonging to Horton Abbey].

Wareham (78b): *Ipsa ecclesia* [of St Wandrille] *tenet unam ecclesiam de rege in Warham ad quam pertinet i hida.*

Gillingham (78b): *De manerio Chingestone habet rex i hidam in qua facit castellum Warham et pro ea dedit S. Mariae* [Shaftesbury Abbey] *ecclesiam de Gelingeham cum appendiciis suis quae valent xl solidos.*

Dorchester, Bere Regis (27b, 79): *Bristuard presbyter tenet ecclesias de Dorecestre et Bere et decimas. Ibi pertinent i hida et xx acrae terrae.*

Winfrith Newburgh (28, 79): *Bollo presbyter habet ecclesiam de Winfrode cum una virgata terrae.*

Puddletown, Chaldon Herring, Fleet (28, 79): *Bollo presbyter ecclesiam* [sic] *habet de Pitretone et de Calvedone et de Flote. His adjacet i hida et dimidia.*

URBAN LIFE

There seem to have been five boroughs in eleventh-century Dorset, and the Dorset section of the Domesday Book opens with an account of four of them—Dorchester, Bridport, Wareham and Shaftesbury. There had been much destruction, and all four places had fewer houses in 1086 than twenty years earlier. The presence of moneyers in each of the four places is entered for 1066, but we are not told whether their activities had survived up to 1086. Markets are not mentioned for either date. The fifth borough, that of Wimborne Minster, is described in a number of entries in the body of the text. It is never specifically called a borough but it is said to have burgesses. It seems to have been a substantial agricultural settlement with an urban element. The information about all five boroughs is very meagre, and we can only conjecture about their economic life.[1]

[1] In order to complete the picture, one other reference to urban life must be mentioned. Goathill, now in Dorset, is described in the Somerset folios (92b); the Exeter version mentions 2 *masurae* in the Somerset borough of Milborne Port (278b)—see p. 200 below.

Dorchester

The account of Dorchester forms the first entry in the Exchequer folios for Dorset (75). In 1066, there had been 172 houses (*domus*) in the borough, but in 1086 there were only 88 *stantes*, to use the phrase of the Exeter text (*11b*). As many as 100 had lain waste from the time of Hugh the sheriff: *Modo sunt ibi quater xx et viii domus et c penitus destructae a tempore Hugonis vicecomitis usque nunc* (75).[1] The borough was rated at 10 hides. There had been two moneyers (*monetarii*) in 1066, each of whom paid to the king one mark of silver and 20s. whenever the coinage was changed (*quando moneta vertebatur*); but we are not told whether there were still moneyers there in 1086. Elsewhere (*27b*, 79), we hear of the church of Dorchester and its tithes. On two other folios of the Exchequer text only, urban properties in Dorchester are entered as belonging to Charminster and to Horton:

> Charminster (75 b): *In Dorecestre unus burgensis cum x acris terrae pertinet huic manerio.*
> Horton (78 b): *in Dorecestre una domus.*

Quite separate from the account of the borough, there is a description of the manor of Dorchester (*27b*, 75) linked in a composite entry with four other places, all five forming part of the Ancient Demesne of the Crown. The group rendered a *firma unius noctis* which is also mentioned in the account of the borough.[2] We hear also (*11b*, 75) of meadow, pasture, wood and mills, and of 56 teams at work on the same number of plough-lands; we hear, too, of a total population of 235. How much of this belonged to the community of Dorchester itself, we cannot say. If we divide this number of teams and of people equally among the five places, we obtain 11 or so teams and 47 recorded people for each place.[3] Such arithmetic, if it were valid, would mean that Dorchester comprised a recorded population of 137. Presumably this implies a total population of nearly 700. But all this is conjecture, and we are without any clear picture of the eleventh-century borough and its economic life.

---

[1] The arithmetic is curious, because 172 minus 88 *stantes* leaves 84, and not 100.
[2] See p. 97 above.
[3] The figure of 47 is a fifth of 235 (114 villeins, 89 bordars, 12 coliberts, 20 serfs).

Bridport

The account of Bridport is brief (*12*, 75). In 1066 there had been 120 houses (*domus*) in the borough, but in 1086 there were only 100 *stantes*, to use the phrase of the Exeter text. Twenty had been damaged (*viginti sunt ita destitutae quod qui in eis manent geldum solvere non valent*).[1] The borough was rated at 5 hides. There had been a moneyer in 1066 who paid to the king one mark of silver and 20*s.* whenever the coinage was changed, but we are not told whether he was still there in 1086. Elsewhere, there is a reference to a church (*28*, 78b);[2] and, furthermore, two Exchequer entries seem to indicate urban property appurtenant to manors not far away. In the first place, the account of Lyme Regis (75b) says that the bishop of Salisbury had one house rendering 6*d.*; Bridport is not named but it may well be implied. In the second place, on fo. 77 following the account of Netherbury, we are told that the bishop of Salisbury had half an acre in Bridport which rendered 6*d.*

The account of Bridport, like that of Dorchester, mentions the customary dues that pertained to a *firma noctis*. Unlike Dorchester, however, Bridport is not mentioned in any of the five groups of estates rendering a *firma unius noctis*.[3] Finally, we are not told anything about the general resources of the borough in arable, meadow, fisheries and the like. All we can do is to envisage it as a settlement of about 500 or so people, and very possibly more.

Wareham

At Wareham (*12b*, 75) in 1066 there had been 285 houses (*domus*) in the borough but in 1086 there were only 135 *stantes*, to use the phrase of the Exeter text. As many as 150 had been destroyed; some of these are described as *penitus destructae* from the time of Hugh the sheriff, some as *vastae*, and some as *destructae*. The borough was rated at 10 hides. There had been two moneyers (*monetarii*) in 1066, each of whom paid to the king one mark of silver and 20*s.* whenever the coinage was changed; but we are not told whether there were still moneyers there in 1086. The entry, like that for Dorchester, mentions the customary dues that

[1] The Exeter version reads: *Et xx domus sunt penitus destructae a tempore Hugonis vicecomitis usque nunc.*

[2] In the entry for Burton Bradstock, Bridport and Whitchurch Canonicorum—see p. 117.   [3] See p. 97 above.

pertained to a *firma noctis*. Unlike Dorchester, however, Wareham is not mentioned in any of the five groups of estates rendering a *firma unius noctis*.[1]

We also hear of a church in Wareham held by the abbey of St Wandrille in Normandy (*28 b*, 78 b). To it pertained one hide and one team with two

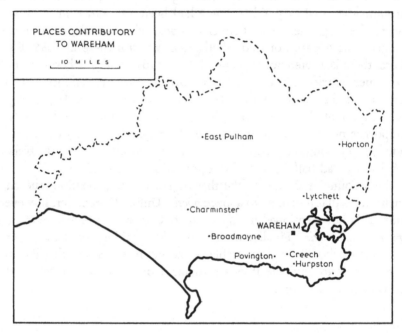

Fig. 30. Places contributory to Wareham.

bordars, and it was worth 70*s*. with its appendages (*cum appendiciis suis*), but we cannot say whether the team and the bordars were at Wareham itself. On other folios we read of burgesses and properties in Wareham but belonging to rural manors elsewhere, and these are set out below (Fig. 30).[2] The entry for Hurpston does not mention Wareham by name but it is probably implied; that for East Pulham appears only in the Exeter text. To what extent these holdings are included within the totals of the main

[1] See p. 97 above.
[2] The implications of this connection lie outside the scope of this chapter. For a discussion of the general problem see Carl Stephenson, *Borough and Town* (Cambridge, Mass., 1933), pp. 81 ff.

entry (*12b*, 75), we cannot say. Whether they are or not makes little difference to an estimate of the population of Wareham. A total of 147 recorded houses and people may imply a population of at least 700. We are given no hint of the economic activity that sustained this community.

Places contributory to Wareham

Broadmayne (80): *In Warham unus domus reddens v denarios.*
Charminster (75 b): *In Warham ii burgenses cum xii acris terrae.*
Creech (79 b): *una domus in Warham.*
East Pulham (*48*): *i ortus in Warham.*
Horton (78 b): *in Warham una ecclesia et v domus reddentes lxv denarios.*
Hurpston (84): *unus burgensis reddens viii denarios.*
Lytchett Matravers (80 b): *In Warham ii ortos et i bordarium.*
Povington (80 b): *In Warham unus burgensis reddens ii solidos.*

Shaftesbury

At Shaftesbury (*11*, 75) in 1066 there had been 257 houses (*domus*), but in 1086 there were only 177 *stantes*, to use the phrase of the Exeter text. As many as 80 had been destroyed; some of these are described as *destructae* from the time of Hugh the sheriff, some as *omnino destructae* or (in the Exeter text) as *penitus destructae*. The borough was rated at 20 hides. There had been three moneyers (*monetarii*) each of whom paid one mark of silver and 20*s.* whenever the coinage was changed. No reference is made to customary dues pertaining to a *firma noctis* as in the accounts of Dorchester, Bridport and Wareham (*11b–12b*; 75).[1]

Some houses were in the hands of the abbess of Shaftesbury, and one part of the account implies that two or more burgesses sometimes occupied the same house:

In parte abbatissae erant T.R.E. domus cliii. Modo sunt ibi cxi domus et xlii sunt omnino destructae.
Ibi habet abbatissa cl et unum burgenses et xx mansuras vacuas et i hortum.

It would thus seem as if the abbess had 151 burgesses in 111 houses.[2] On this analogy, a total of 177 houses would imply 240 people, or a minimum population of at least 1,000. There is no hint of the economic life of the town or of the activities of its people.

[1] See p. 97 above.
[2] For the relation of houses to burgesses, see A. Ballard, *The Domesday Boroughs* (Oxford, 1904), pp. 13, 56.

Wimborne Minster

The settlement of Wimborne Minster is described in three entries. One is a composite entry dealing with Wimborne Minster, Shapwick, Crichel and All Hallows Farm (*27, 75*).[1] The whole formed part of the Ancient Demesne of the Crown, and, with its appendages, it rendered a *firma unius noctis*.[2] The other two entries tell of another holding belonging to the king (*30b, 75b*) and of one belonging to the count of Mortain (*79b*); they are straightforward entries of the kind usually found in the Dorset folios, but the first tells us that the holding did not contribute to the *firma* of Wimborne. Taken together, the three entries present a picture of a large rural community with villeins, bordars, cottars and serfs, and with meadow, pasture, wood and mills. If we divide the teams and the people of the composite entry equally among the four places named, and add the results to the corresponding figures in the other two entries, we obtain totals of 10¾ teams and some 57 people for Wimborne as a whole.[3] None of the three entries calls Wimborne a borough or mentions burgesses.

The indication of an urban element at Wimborne Minster comes from entries relating to three places nearby, which also mention a church and a chapel:

Canford (80b): *Ad Winburne iii bordarii et una domus pertinent huic manerio.*
Hinton Martell (76): *in Winburne xi domos.*
Hinton Martell (76): *De eadem ipsa terra pertinet ad ecclesiam de Winburne i hida et dimidia virgata terrae. Mauricius episcopus [Lundoniensis] tenet et ibi habet vi bordarios et viii burgenses . . .*
Horton (78b): *Ad hanc ecclesiam pertinent ecclesiola una in Winburne et terra duabus domibus.*

The 6 bordars entered under Hinton Martell may have been there, and not at Wimborne Minster. The 22 burgesses and houses and the 3 bordars bring the recorded population of Wimborne Minster up to about 82. Presumably this implies a total population of over 400. We can only conjecture whether the urban element in this population was associated in any way with its minster which is mentioned in pre-Domesday documents.[4]

---

[1] All Hallows Farm is the modern equivalent of the Domesday *Opewinburne*.
[2] See p. 97 above.
[3] The figure of 57 is obtained by adding 4 villeins, 12 bordars and 3 serfs to a quarter of 153 (63 villeins, 68 bordars, 7 cottars and 15 serfs).
[4] A. Fägersten, *op. cit.* p. 87.

LIVESTOCK

Livestock are recorded for the demesne of those Dorset holdings surveyed in the Exeter text, i.e. for 140 out of the total of 319 Domesday settlements in the county. We cannot say how complete even these restricted figures were, and it may well be that some demesne animals were omitted. A number of entries for holdings with demesne land make no reference to livestock, e.g. those for Bockhampton (*33*) and Caundle (*41*). Conversely, stock are occasionally entered for holdings without recorded demesne, e.g. for Bere (*56b*) and Swanage (*61*). The table below shows the variety of the livestock mentioned and also the number of animals in each category.[1]

Livestock in the surviving Exeter folios for Dorset in 1086

Sheep (*oves*)	22,680
Wethers (*berbices*)	297
Swine (*porci*)	1,567
She-goats (*caprae*)	780
Animals (*animalia*)	541
Cows (*vaccae*)	63
Oxen (*boves*)	9
Rounceys (*runcini* or *roncini*)	126
Mares (*equae*)	13
Unbroken mares (*equae indomitae*)	12
Foals (*pulli*)	12
Donkey (*asinus*)	1

Horses are mentioned in a variety of ways. Rounceys (*runcini* or *roncini*) were most probably pack-horses. The mares at Chelborough (*50*) were described as *equae indomitae*. No reference is made to *equae silvestres* as for other south-western counties. The 12 mares (*equae*) at *Pidela* (Puddle in Bere hundred, *56b*) were said to be there with their foals: *cum suis pullis*. The donkey (*asinus*) was at one of the Frome holdings (*54*). Animals (*animalia*) presumably included all non-ploughing beasts.

---

[1] It is occasionally difficult to know what is the number in each category. Thus the entry for Osmington (*43b*) reads: *c oves et xx et vii*. This could mean one of three things: (*a*) 127 sheep, (*b*) 100 sheep and 27 goats, which normally follow sheep in an entry, (*c*) 120 sheep and 7 goats. We have assumed the total to be 127 sheep. The entry for Bloxworth (*36b*) reads: *x et vii porcos et xx et vi porcos*; here, we have assumed that the second *porcos* is an error for *oves* and have counted 17 pigs and 26 sheep.

Cows (*vaccae*) appear much less frequently. The entries usually record only one, but there were as many as 10 at Cranborne (*29*) and 13 at West Stafford (*55b*), the latter being the highest number of cows recorded for any vill in the Exeter Domesday. The only mention of oxen, apart

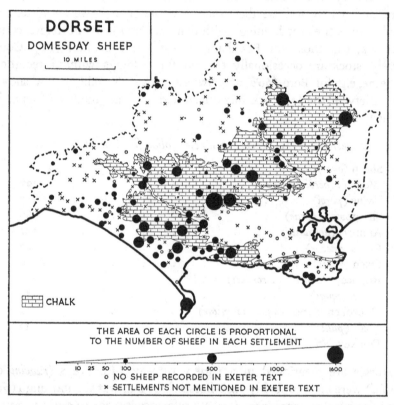

DORSET
DOMESDAY SHEEP
10 MILES

CHALK

THE AREA OF EACH CIRCLE IS PROPORTIONAL
TO THE NUMBER OF SHEEP IN EACH SETTLEMENT

10 25 50 100   500   1000   1600
○ NO SHEEP RECORDED IN EXETER TEXT
× SETTLEMENTS NOT MENTIONED IN EXETER TEXT

Fig. 31. Dorset: Domesday sheep on the demesne in 1086.

Wethers (*berbices*) have been included with sheep.

from plough-teams, is in an entry for Affpuddle (*36b*), where 9 *boves* appear in the enumeration of livestock.

She-goats (*caprae*) were fairly numerous. The number entered on a manor was usually less than 50 and quite often less than 10.

Swine are frequently entered for places without any record of wood; thus 24 swine are entered for Hammoon (*48*), and 37 for West Stafford

(*55b*), but no wood was recorded for either place. The largest number of swine on a single holding was the 60 at Puddletown (*25*).

Sheep were, far and away, the most numerous of the recorded livestock. Flocks of many hundreds could frequently be encountered. It is true that the large figures for some complex manors may be deceptive in the sense that they covered separate unnamed places, e.g. the 1,600 sheep at Puddletown (*25*) and the 600 at Abbotsbury (*39*). Most figures, like the ones just quoted, are in round numbers, but there are occasional entries that suggest precision, e.g. there were 1,037 sheep at Cranborne (*29*), 826 at Ashmore (*29b*), 353 at Long Bredy (*37b*) and 103 at Hurpston (*60b*).[1] Wethers (*berbices*) but not sheep (*oves*) were mentioned for two holdings: there were 47 at Mapperton near Beaminster (*49b*), and as many as 250 at Renscombe (*38b*). Wethers have been included with sheep on Fig. 31.

MISCELLANEOUS INFORMATION

Vineyards

There are references to vineyards in two consecutive entries on fo. 83:

Durweston: *Ibi ii acrae vineae.*
Wootton:[2] *ii arpenz vineae.*

Both were on estates of Aiulf the Chamberlain, who was sheriff of Dorset in 1086. We may note that there was also a vineyard on his estate at Tollard Royal (73) in Wiltshire, near the Dorset border. The arpent was a French unit of measurement.[3]

Waste

Apart from the account of the wasted houses in the boroughs of Bridport, Dorchester, Shaftesbury and Wareham, there is only one other reference to waste in the Exchequer folios for Dorset: a holding of 1 plough-land at Nyland (79) was said to be waste—*Vasta est.* Another holding of half a plough-land at Hurpston rendered only 12s. 6d.; the Domesday Book merely records the fact, but the Exeter text explains

---

[1] Customary dues of sheep and lambs were rendered by Seaborough to the manor of Crewkerne in 1066 (*154*, 87b). See p. 126 below. Seaborough is surveyed in the Somerset folios, and a total of 116 sheep was entered for it in 1086.
[2] Equated here with Abbots Wootton and Wootton Fitzpaine.
[3] For a discussion of the arpent see Sir Henry Ellis, *op. cit.* I, p. 117.

that the holding rendered this amount in spite of the fact that it had been
entirely devastated:

E.D. (*60b*): *haec terra omnino devastata est et tamen valet xii solidos et vi
denarios.*
D.B. (84): *Valet xii solidos et vi denarios.*

Other references

The only reference to a castle occurs in the entry for Kingston (78b):
De manerio Chingestone habet rex i hidam in qua fecit castellum Warham.
This refers to Corfe Castle, which has been included in our total of 319
Domesday settlements.

A sester of honey was included as part of the annual value of Holworth
(*44b*, 78) and four sesters as part of that of Rushton (82).

Gardens (*orti*) are mentioned for Wareham (in two entries: East
Pulham, *48*, and Lychett Matravers, 80b), and, in the Exchequer text,
for Shaftesbury (75). The Exeter text alone also mentions one for Puddle
in Bere hundred (*56b*): *ibi est dimidia hida et quattuor agri et i ortus quae
nunquam gildavit sed celatum est.*

A fruit orchard (*virgultum*) is entered for Orchard in Church Knowle
(*61b*, 84).

A heath is mentioned for Boveridge (77b): *Bruaria ii leuuis longa et
lata.* The phrase occurs in between the statement about pasture and that
about wood; it is the only reference to heath in the Domesday Book.
Boveridge is situated near Cranborne, just outside the main area of heathland in Dorset.

A manor at Puddletown (*25*, 75) received the 'third penny' of the
shire: *Huic etiam manerio Piretonae adjacet tercius denarius de tota scira
Dorsete.*

Abbotsbury Abbey (79) in 1066 had received, from a holding at
Waddon, 6 acres of hay and 3 churchscots by custom: *De eas habebat
ecclesia Abodesberie T.R.E. vi acras messis et iii circsce de consuetudine.*

Finally, in 1066, Seaborough (surveyed in the Somerset folios, *154*,
87b) had rendered customary dues of sheep and lambs and iron to the
manor of Crewkerne: *reddebant mansioni Cruche regis xii oves cum agnis
et de unoquoque homine libero una bloma* [sic] *ferri.*

Dorset has sometimes been characterized as a county of three parts. As John Hutchins wrote in *The history and antiquities of the county of Dorset* (1774), it includes 'the down, the vale, and the heath'. For our purpose, three other regions must also be recognised. Within each of these there are variations, but a division into six will serve to give a broad view of the county in the eleventh century (Fig. 32).

(1) *The Chalk Uplands*

The undulating chalk uplands are bordered along their northern edge by an escarpment overlooking the Vale of Blackmoor and rising to over 800 ft in places. From this northern edge, the dip-slope descends south-eastward until it is covered by the Tertiary deposits of the lowland. The soils of the higher portions are generally thin and shallow, but, especially towards the west, there are stretches of Clay-with-flints. The sloping surface is dissected by streams—the Frome, the Piddle, the Stour and their tributaries—and in the valleys of these streams are to be found fertile loamy soils. The Domesday villages aligned along the valleys were fairly numerous.

The density of plough-teams was about 2 per square mile and that of population about 8, figures that resemble corresponding densities for the Hampshire Downs and for much of southern England in general. The valley villages had moderate amounts of meadow and almost always a mill or mills apiece. Quantities of wood were entered for a number of villages, especially in the Clay-with-flints area to the west, and, surprisingly, in the eastern part of the upland; but the amounts were usually not great and many villages were without wood. Taken as a whole, the chalk upland cannot have been well wooded. Quantities of pasture, on the other hand, were widespread throughout the region.

(2) *The Vale of Blackmoor*

This is a countryside of stiff, heavy soils derived from Kimmeridge Clay, Oxford Clay and Gault. Much of its surface lies between 200 and 300 ft. above sea-level, and it is drained by a network of small streams, the water of which finds its way ultimately into the Stour. The majority of the villages were to be found, not on the clays themselves but on the margins of the region and along the junction between the clay and the Corallian limestone outcrop that runs through the middle of the area. The empty stretches of the chalk upland were here replaced by empty areas of clay. Despite their very different character, the Vale and the chalk uplands had very similar densities of teams and population, i.e. about 2 and about 8

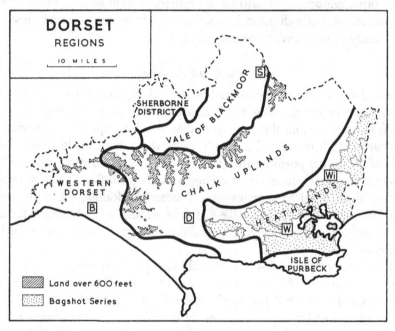

Fig. 32. Dorset: Regional subdivisions.

Domesday boroughs are indicated by initials: B, Bridport; D, Dorchester; S, Shaftesbury; W, Wareham; Wi, Wimborne Minster.

respectively. Substantial amounts of wood and fair amounts of meadow were entered for the villages, which usually also had a mill or mills apiece. In striking contrast to the chalk upland, on the other hand, there were only small quantities of pasture.

(3) *The Heathlands*

Tertiary sands, gravels and clays constitute an area of poor soils that lies below 200 ft O.D. The Bagshot Sands, in particular, cover much of the area, and their light infertile soils were as unrewarding in the eleventh century as in later times. The density of teams per square mile was below one, and that of population below 4, both figures being less than half the corresponding densities for the chalk upland and the Vale of Blackmoor. The figures would be less still but for the fact that the heathland is crossed by the lower valleys of the Frome, the Piddle and the Stour. Set in these valleys, or on the margins of the district, were villages, usually with moderate amounts of meadow and pasture, and sometimes with a mill or mills. There was, on the other hand, very little wood in the district. On the coast, at Ower and Studland, there were salt-works.

(4) The Sherborne District

In the north of the county around Sherborne is a district lying very largely between 200 and 400 ft O.D. It is underlain by Jurassic rocks—clays, marls and shales—that for the most part yield fertile loams. The map of the Domesday settlements does not do justice to the area because the great manor of Sherborne (with nearly 70 teams and nearly 240 people, and with 12 mills) covered a number of nearby settlements that are not mentioned in the Domesday Book. The district was one of the most prosperous in the county, with nearly 3 teams and nearly 10 people per square mile. There were moderate amounts of wood and meadow but very little pasture.

(5) The Isle of Purbeck

The so-called Isle of Purbeck consists very largely of a chalk ridge in the north separated by a clay vale from a southern ridge of Jurassic limestone. Each ridge is about 400 ft above sea-level, rising occasionally to over 600 ft. One feature of the area was the large number of Domesday place-names. The density of teams was below 2 and that of population below 6; but if we could make allowance for the fact that the Purbeck density area includes a stretch of heathland, the densities might well have approximated to those of the chalk upland and the Vale of Blackmoor. In any case, here was a land of small villages. Most of these had some meadow and pasture, and many of them also had a little wood.

(6) Western Dorset

The western part of Dorset is an area of varied relief with steep valleys and exposed ridges that rise to over 500 ft O.D. and, in a few localities, to over 800 ft. The varied geological formations yield soils ranging from sandy loams to very heavy clays. Villages, in a variety of situation, are fairly numerous, except on the tract of very impervious Lower Lias Clay that forms the Vale of Marshwood through which the River Char flows. The abundance of *hay*-names in this Marshwood area, writes Fägersten, 'would seem to point to a comparatively late colonization' in post-Domesday times.[1] Over western Dorset generally, the densities of teams (mostly above 2) and of population (7–9) were similar to those of the chalk uplands and the Vale of Blackmoor. A fair amount of wood covered the area; every village had some meadow and most villages had pasture.

An exception to some of these generalisations is formed by the triangle of land in the east of the area between the chalk escarpment and the coast. Here, the density of teams rose to nearly 3 and that of population to over 10. In view of this, it may be of interest to note that the district contained practically no wood, although almost every village had some meadow and pasture.

Along the coast of western Dorset as a whole there were salt-works at Lyme, at Charmouth, and on one of the Wey manors. Fishermen were entered for Lyme and for Bridge, which apparently was on or near the isthmus that joins the Isle of Portland to the mainland.

BIBLIOGRAPHICAL NOTE

(1) 'A dissertation on Domesday Book' (pp. 1–24), together with the Latin text (pp. 1–xxix), and also a description of the Geld Accounts, together with their text (pp. 1–8), was included in the first volume of John Hutchins, *The history and antiquities of the county of Dorset*, 2 vols. (London, 1774). The second edition was to have been published in 3 volumes, but the third volume was almost completely destroyed by fire, and the edition ultimately appeared in 4 volumes (1796–1815) edited by R. Gough and J. B. Nichols. The fourth volume contained not only the dissertation and an extended Latin text, but also a translation by William Bawdwen, as well as a description of the Geld Accounts and their text (pp. 1–94). The third edition of Hutchins's work was edited by W. Shipp and J. W. Hodson and appeared in 4 volumes (1861–73). Its fourth volume included 'A dissertation on Domesday Book', together with an extension of the text, indexes of places and persons, and a description of the Geld Accounts and their text (pp. i–lix). The translation was discontinued and the relevant Domesday entries were 'almost invariably given in connection

[1] A. Fägersten, *op. cit.* pp. 254, 288.

with the individual parishes' (vol. I, p. xiii) described in the body of the work. A modern translation of the Domesday text by Ann Williams appears in *V.C.H. Dorset*, III (Oxford, 1972). It is preceded by an introduction, and followed by a section dealing with the Dorset Geld Accounts—introduction, text and translation.

(2) R. W. Eyton, *A key to Domesday, showing the method and exactitude of its mensuration, and the precise meaning of its more usual formulae. The subject being specially exemplified by an analysis and digest of the Dorset survey* (London and Dorchester, 1878). It identifies each holding and assigns it to a hundred, but the only statistical information it sets out in tables is that relating to the hidage. Although its ideas of Domesday mensuration and many of its identifications cannot now be accepted, the study is still of interest as an early attempt at Domesday analysis.

(3) The following articles deal with various aspects of the Domesday study of the county:

H. J. MOULE, 'Domesday return for Dorchester', *Notes and Queries, Somerset and Dorset*, II (Taunton, 1891), p. 225.

F. W. MORGAN, 'Domesday woodland in south-west England', *Antiquity*, X (Gloucester, 1936), pp. 306–24. This contains a map of the Domesday woodland of Dorset.

R. WELLDON FINN, 'The making of the Dorset Domesday', *Proc. Dorset Natural History & Archaeological Society*, LXXXI (Dorchester, 1960), pp. 150–7.

(4) For the Exeter Domesday and the Geld Accounts, see p. 393 below.

(5) The English Place-Name Society has begun to produce volumes on Dorset—A. D. Mills, *The Place-Names of Dorset*, pt I (Cambridge, 1977), Pending the completion of the work a useful study is A. Fägersten's *The Place-Names of Dorset* (Uppsala, 1933).

CHAPTER III

SOMERSET

BY R. WELLDON FINN, M.A., AND P. WHEATLEY, D.LIT.

Somerset is one of those south-western counties for which there are two
Domesday descriptions—the Exeter and the Exchequer; but there are
many differences between the two versions. Not only does the Exeter
Domesday enumerate the livestock kept on demesne farms, information
omitted from the Exchequer volume, but it also provides far more detail
about the items common to both versions. The Exeter text, for example,
distinguishes the hidage of the demesne land from that of the villagers,
which the Exchequer version does not always do. It also provides, as we
shall see, much other information not included in the Exchequer version.
The Exeter text has been followed in the present analysis.

Even a cursory reading reveals discrepancies between the two texts,
especially in the arithmetic, and it is usually impossible to decide which
is the more accurate version, though from its provenance the Exeter text
is more likely to be so. In the record of Elworthy (96), we can follow the
workings of a clerk's mind as he lapsed into error. The Exeter Domesday
records 4 virgates as the assessment of this holding (*361*). The Exchequer
scribe wrote *iij virg.*, which was later corrected to *iiji virg.* This he then
decided to express more succinctly as one hide, but he forgot to delete
virg., so that the Exchequer text finally stands as 4 hides. Roman numerals
were prodigal of error, for a carelessly written or miscopied *iiii* was only
too easily mistaken for *iii*, and an indistinct *v* or *x* mistaken for *ii*. This is
seen, possibly, in the entry for Woolley, where the Exeter text (*144b*)
gives the value of the mill as 10*s.*, the Exchequer (88b) as 2*s.* Usually such
discrepancies are relatively small, amounting to no more than half a
plough-team, as at Skilgate (*442*, 94b), or an acre of meadow, as at Stoke
Trister (*277*, 92b), but there are a number of entries in which the texts
diverge more widely. The Exchequer Domesday, for example, credits
Holford St Mary (96) with only 4 acres of woodland, but the Exeter text
allots it 104 (*362*). The Exchequer text, too, omits altogether 25 cottars
who are mentioned in the Exeter Domesday in the entry for Brewham
(*364b*, 96b). It is not difficult to understand how discrepancies such as

these may have arisen through omissions and careless copying.[1] Sometimes the internal evidence of an entry indicates which of the two texts is likely to be the more accurate one. The Exchequer Domesday allots 4 hides to the demesne at Brompton Regis (86b), the Exeter only 3 hides (*103*); but the more detailed Exeter text enables us to see that 3 hides is the figure required to make up the total assessment of 10 hides upon which both texts agree. These are typical of the discrepancies that occur between the texts: others are discussed in the appropriate sections of the following pages, and are set out in Appendix II.

In addition to these two texts, there are auxiliary sources of information. Bound up with the Exeter Domesday in the *Liber Exoniensis* is a series of entries, headed *Terrae Occupatae*, for Devonshire, Cornwall and Somerset, recording for the most part such matters as the additions to and abstractions from manors, unsanctioned occupation of property, and failure to pay customary dues. It is obviously constructed from Inquest material, and from that which produced the Exeter Domesday, which at times it slightly supplements. The Somerset section is of no great assistance for present purposes, but it does tell us of one settlement which is unrecorded in either text—a virgate of land called *Ledforda* which had been added to the royal manor of Williton (*509b*).

Also bound up with the Exeter text are two lists of hundreds (*63 b–64 b*), which do not agree with each other. From bare lists of names no major deductions can be drawn, but the inclusion of Wellow and South Brent as names of hundreds strongly suggests that settlements there were in existence at the time of the Inquest, although neither is mentioned in the Domesday texts,[2] and they have not been counted in our analysis. The Domesday texts, however, mention *Brentamersa* (*170b*, 90b), which was a large 20-hide manor, and which may have covered more than one settlement in the modern Brent Knoll and East Brent.

For the five south-western counties, too, there have survived the so-called Geld Rolls or Geld Accounts, which are the record of a geld levied at the rate of 6s. per hide at a time near that of the Inquest.[3] Eyton leaves us in no doubt as to the value of this record: 'The way in which the two

---

[1] The printed edition of the Exeter text, incidentally, is not infallible: e.g. it gives 55½ tenants' teams at Milborne Port (*91*) where the MS. has 65½, and 24 burgesses at Langport (*89b*) where the MS. has 34.

[2] Kilmersdon also appears as the name of a hundred in the first list; it is entered in the Domesday texts (*198b*, 91 b) as a half-hide holding worth 10s.

[3] See p. 393 below.

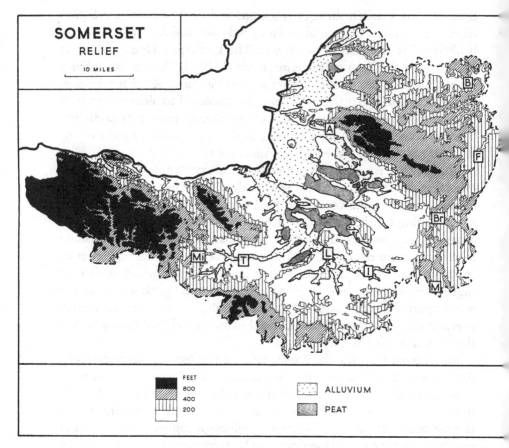

SOMERSET
RELIEF
10 MILES

FEET
800
400
200

ALLUVIUM

PEAT

Fig. 33. Somerset: Relief.

Domesday boroughs are indicated by initials: A, Axbridge; B, Bath; Br, Bruton;
F, Frome; I, Ilchester; L, Langport; M, Milborne Port; Mi, Milverton; T, Taunton.

records, the Gheld-Inquest and Domesday, explain and supplement one
another may be almost said to double the antiquarian value of the greater
record.'[1] By collating the Domesday entries with those of the Geld
Accounts, Eyton was able, though often inaccurately, to reconstruct the
Somerset geld hundreds; a list of these, however, does not agree with
either of the two lists of hundreds on fos. *63b–64b*. Eyton also used the

[1] R. W. Eyton, *Domesday Studies: An Analysis and Digest of the Somerset Survey*
(London, 1880), vol. II, p. 3.

Geld Accounts to identify a number of manors. The anonymous 10-hide manor of the Exchequer (97) and Exeter (*436b*) texts is an example. The statistics of Milborne hundred in the Geld Accounts cannot be harmonised with those of Domesday Book without the inclusion of this large manor, which Eyton showed must be Charlton Horethorne. He was also able to identify the nameless 5-hide holding of the Exchequer (91) and Exeter (*198*) texts with a part of Nunney. The Geld Accounts mention three places which, if they appear in either Domesday text, are there concealed through inclusion as anonymous additions to or subtractions from some manor or manors: *Letfort*, Woodadvent and *Pirtochesworda* in the hundred of Williton (*79b*).[1] Collation with Domesday Book enables us to detect certain omissions from the latter; e.g. Shaftesbury Abbey must have held a manor in Bath hundred which is unrecorded but which probably was Kelston; and royal land at Cranmore and at Stoke near Chew Magna was certainly also omitted.

There is another set of figures that must be mentioned. At the end of the Exeter account of the Glastonbury estates, on fo. *173*, comes a Summary of the abbey lands, repeated, with an error in the number of demesne villeins, on fo. *528*; it is omitted from the Exchequer text. In form and vocabulary it is akin to the Summaries of the *Inquisitio Eliensis* and to that on fo. 382 in the Yorkshire section of the Domesday Book. Though perhaps derived from a feudal rather than a hundredal return to the Inquest, it strengthens a belief that the statistics of the Domesday Book are reasonably accurate. Its figures are arranged in three groups—for the demesne manors (*mansiones dominicae*), for the manors of the *milites*, and for those of the *Angli Taigni*. A comparison of the totals for these groups with the totals for the usual Domesday entries is indicated in the table on p. 137. The Summary treats both cottars and coscets as bordars, and it omits the smiths. But though a collation shows differences between the figures for each group, they are never disturbingly large. There is also a Summary for the Somerset lands of Robert fitzGerold (*530b*) which coincides with the Domesday totals and further shows that the missing value of his unnamed Domesday manor was £17.

The last of these auxiliary texts is a document in the Cartulary of Bath Abbey which R. Lennard has designated the 'Bath Abbey

---

[1] *Letfort* is also mentioned among *Terrae Occupatae*, where it appears as *Ledforda* (see p. 133 above). It is not likely to have been part of *Langefortda* (*104*, 86b), which is Langford Budville, and not in Williton hundred.

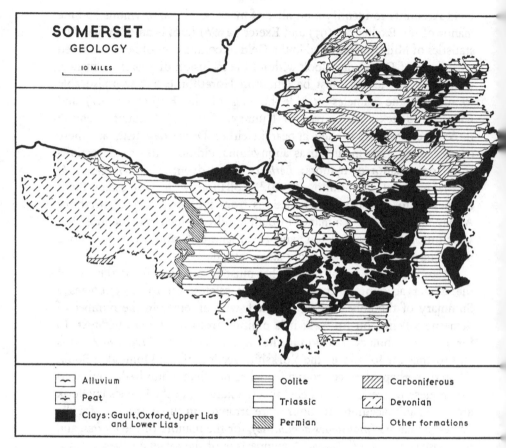

Fig. 34. Somerset: Geology.

Based on: (1) Geological Survey One-Inch Sheets (New Series) 264 and 280, and Quarter-Inch Sheets 18 and 19; (2) Soil Survey One-Inch Sheets 280 and 296.

Domesday'.[1] This is an account of seven of the manors which Domesday Book assigns to the demesne of the Abbey of Bath. In the main, the text is identical with that of the Exeter Domesday, though there are a number of discrepancies, and Mr Lennard believes that it probably represents a stage in the assembling of the returns earlier than that of the Exeter

[1] R. Lennard, 'A neglected Domesday satellite', *Eng. Hist. Rev.* LVIII (1943), pp. 32–41.

Domesday.[1] In addition, the Cartulary contains an abstract of the assessment and demesne hidage of a series of Somerset manors belonging to Bath Abbey, together with those of two Gloucestershire estates.[2]

The Glastonbury Fief

Comparison of the Exeter text entries with those of the Exeter Summary

	Exeter text (*161–73*)	Exeter Summary (*173, 528*)
Assessment	381 ½ hides, 24 exempt hides, 40 exempt carucates	385 hides, 1 virgate, 23 exempt hides, 42 exempt carucates
Plough-lands	507 ½	554 ½
Teams	415 ¾	423
Villeins	567	587
Bordars	529 ⎫	622 bordars
Cottars	54 ⎬	
Coscets	6 ⎭	
Coliberts	19	21
Serfs	218	225
Fishermen	10	10
Smiths	8	—
Value	£454. 8s. 8d.	£454. 19s. 0d.

One of the great difficulties in the geographical interpretation of the Somerset material is that presented by those large composite manors that included a number of places, often widely separated. Among the variety of entries relating to these composite manors we may perhaps discern three broad patterns. In the first place, there is an occasional entry that makes no reference to the constituent parts of a manor, e.g. that for Coker (*107*, 87) with 15 teams and 89 recorded people, or that for Crewkerne (*105b*, 86b) with 25 teams and 125 recorded people. Such large figures

[1] For the text, see W. Hunt, *Two Chartularies of the Priory of Bath* (Somerset Record Society, no. 73, Taunton, 1893, vol. VII), pp. 67–8.

[2] R. Lennard, *op. cit.* pp. 39–41, considers that the figures were derived from Domesday Book itself. But they might be a copy of figures supplied by a Bath official for the Domesday Inquest. The handwriting of the manorial accounts is very like that of one of the clerks who wrote a large portion of the Exeter Domesday.

for any rural holding lead us to suspect the existence of dependent members even if they are not mentioned.

In the second place, there are those entries which mention a large number of sub-tenancies, and such entries form a very characteristic feature of the Somerset record. More usually the Exchequer entry for a large manor is in two parts. The first states the resources of the caput itself, and the second part gives the names of the sub-tenants, together with the combined resources of their holdings in a series of totals, as in the entry for Banwell (*157*, 89b) set out on pp. 148–9 below. Only rarely does the Exchequer version give separate totals for each sub-tenancy as, for example, in the account of Long Ashton (*143b*, 88b) with its details for each of the sub-tenancies of Roger and Guy. The Exeter version, on the other hand, usually sets out the separate details for each sub-tenant, e.g. in the accounts of Chew Magna (*159*, 89b), High Ham (*165*, 90), and Wells (*157b*, 89), and of Banwell, to which we have already referred. Taunton, with its *appenditii* and its groups of sub-tenancies, is a very complicated example of a composite manor (Fig. 35).[1] In all such entries, however, neither the Exchequer nor the Exeter version states the name of the locality of each sub-tenancy. All that can be done is to plot the combined details for the manor as a whole (for the caput and its outliers) at the place of the caput itself.

In the third place, the entries for some large manors specify not only the names of the sub-tenants but also the places where they—or at any rate some of them—held their lands. The Exchequer version only rarely gives separate details for each sub-tenancy; e.g. for the five dependencies (three of them named) of Keynsham (*113b*, 87)[2] or for the two dependencies (one of them named) of Brompton Regis (*91b*, 86b). The Exeter text, on the other hand, normally does separately describe each sub-tenancy, so enabling us to obtain a more realistic picture of each. Thus the Exchequer text describes four members of the manor of Shapwick (Sutton Mallet, Edington, Chilton upon Polden, and Catcott) by means of a single set of totals (90). If this version alone existed, we could but divide the totals equally among the four places, whereas the Exeter version (*162*), with its separate account of each locality, shows that there was nothing like equality between them. Or again, the Exchequer account of Pilton

[1] See p. 202 below.
[2] The Exeter text gives the name of one of the anonymous holdings—Burnett, not mentioned anywhere in the Exchequer text.

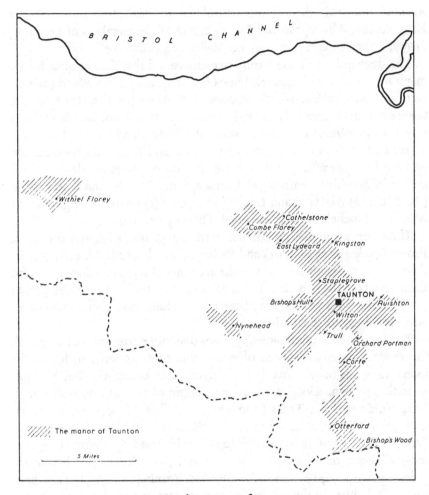

Fig. 35. The manor of Taunton.

The shading covers the area of the modern civil parishes.

(90–90b) describes its five sub-tenancies in two groups: a combined statement for Shepton Mallet and Croscombe, and another for North Wootton, Pylle and an unnamed holding; the Exeter version (*166*), however, describes each of the five holdings separately, and so enables us to assign correct figures to each of the four named villages. The same is true of the three members of Walton (*163b*, 90) at Compton Dundon,

Ashcott and Pedwell; of three members of Ditcheat (*169b*, 90b) at Hornblotton, Alhampton and Lamyatt; and of the members of Doulting (*167b*, 90) at Charlton in Shepton Mallet and elsewhere (*alibi*).

The descriptions of some sub-tenancies reveal that the composition of many manors had changed. Holdings had frequently been added (*addita*) or taken away (*ablata*); both processes had taken place in, for example, the manor of Martock (*113*, 87). The only time a holding is called a member (*membrum*) is in the account of Cheddar and its member Wedmore; and various constituents of Cheddar and Somerton together are described as *appenditii* (*90*, 86). Three groups of places are described together in combined entries (*a*) Carhampton, Williton and Cannington (*89*, 86b); (*b*) Alfoxton and Dyche (*373b*, 97); (*c*) the three neighbouring islands of Muchelney, Midelney and Thorney (*189*, 91).

There are a number of parallel entries, e.g. those for the church at Frome (*90b*, *197b*; 86b, 91) and Belluton (*113b*, *282b*; 87, 91b); sometimes the same details are given in the accounts of separate manors, e.g. for Eastham under its own name (*272*, 92) and again under that of Crewkerne (*105b*, 86b) from which it had been taken. These have been remembered when making our calculations.

The Somerset of the Domesday folios does not correspond exactly with the modern county in terms of which this study is written, for in the intervening centuries boundary changes have brought about a slight reduction in area. Along the southern margin of the county, Seaborough (*154 bis*, *513*; 87b); Trent (*279*, 93) and Adber (*279*, *355b*, *493b*; 93, 95b, 99) within it, Sandford Orcas (*466b*, *521b*; 99) and Weathergrove (*152*, *522b*; 89) within it, Goathill (*278b*, 92b), and Poyntington (*279*, 93) have all been transferred to Dorset, while, on the eastern boundary, Kilmington (*91*, *453*, *520*; 86b, 98) and Yarnfield (*447*, 95) have been lost to Wiltshire. In the north-east, two manors surveyed in the Somerset folios, those of Bedminster (*90b*, *525*; 86b) and Knowle (*447*, 98), are now included in the new county of Bristol. Brislington and part of Whitchurch (both unmentioned in the Domesday texts) have also been transferred to Bristol. So has part of Bishopsworth, but the main portion with its Domesday settlement still remains in Somerset. The gain to be set against these losses is small. Churchstanton (*382*, 115b), surveyed in the Devonshire folios, has been transferred to Somerset.

SETTLEMENTS AND THEIR DISTRIBUTION

The total number of separate places mentioned in the Domesday Book, and its associated documents, for the area now within Somerset seems to be at least 611, including the eight places which appear to have been boroughs: Axbridge, Bath, Bruton, Ilchester, Langport, Milborne Port, Milverton and Taunton. The total includes four places not mentioned in the Exchequer version: Burnett appears only in the Exeter text (*114*); *Ledforda* (*509b*) appears as an addition to the manor of Williton in *Terrae Occupatae* and also in the Geld Accounts (*Letfort, 79b*); and Woodadvent (*Oda*) and *Pirtochesworda* appear in the Geld Accounts alone (*79b*).[1] The total of 611, for a variety of reasons, cannot accurately reflect the total number of settlements in 1086. In the first place, there are some entries in which no place-names appear.[2] They speak only of *terra* and *mansio* and *alibi*, and it is possible that some of these refer to holdings at places not named elsewhere in the text. With these may perhaps be included many of those unnamed holdings added to, or taken away from, named manors, e.g. Wellington: *hic addita est i mansio unius hidae* (*156b*, 89), and Littleton in Compton Dundon: *additae ii mansiones* (*433b*, 94). In the second place, as we have seen, the entries for some large manors cover not only the caput itself but also several unnamed dependencies, and again we have no means of telling whether these are named elsewhere in the text.

In the third place, when two or more adjoining villages bear the same basic name today, it is not always clear whether more than one existed in the eleventh century.[3] There is no indication in the Domesday text that,

[1] The Hundred-Lists for Somerset, but not the Geld Accounts, include the hundreds of Wellow (*63b*, 64b) and South Brent (*64b*), which suggests the existence of settlements of those names not mentioned in either Domesday text. The Hundred-Lists, but not the Geld Accounts, also mention Kilmersdon, but this is a Domesday name (*198b*, 91b).

[2] The relevant entries are: Edith the nun, 12 acres of land (*196*, 91b); St Mary of Montebourg, 5 hides (*198*, 91); *Terra Alwini* (*424b*, 93b), 1 virgate and 1 ferling; *Terra Colgrini* (*423b*, 93), ½ virgate; *Terra Olta* (*426*, 93b), 1 virgate; *Terra Teodrici* (*425b*, 93b), 1 virgate; Roger de Courseulles's *mansio* (*167b*) which the Exchequer text describes as *alibi* and combines with Charlton (90b). The Exeter text also mentions *i mansio* (*116*) which the Exchequer version (87) combines with Pitney. Finally, a blank space appears in both versions of the account of the holdings of Robert son of Gerold (*436b*, 97); the missing name is Charlton Horethorne.— R. Eyton, *op. cit.* I, p. 176.

[3] For the place-names in this section, see E. Ekwall, *The Concise Oxford Dictionary of English Place-Names* (Oxford, 4th ed. 1960).

say, the East Coker, West Coker and North Coker of today existed as separate villages; the Domesday information about them is entered under one name only (*107*, 87) though they may well have been separate in the eleventh century, as they certainly were by the middle of the thirteenth.[1] When the holdings under such names were in different hundreds (as deduced from the Geld Accounts), we have counted them as separate villages. Thus there are four entries for *Harpetreu*, and we have counted East Harptree in Winterstoke hundred (*139b*, *272b*; 88, 92), and West Harptree in Chewton Mendip hundred (*150*, *354b*; 89, 95 b). This is not satisfactory, because if two or more such places lay in the same hundred, they have been counted as one in the present analysis. Moreover, the criterion of separate hundreds may not be infallible.

For some counties the Domesday text occasionally differentiates between the related units of a group by such designations as *alia* or *parva*. The only example of this is the distinction between *Seveberge* and *alia Seveberge*, but there is no trace of two villages here in later times, and in any case Seaborough is now in Dorset.[2] The Somerset text does occasionally indicate separately a village belonging to a group with the same basic name. It distinguishes North Curry (*Nortchori*, *105*; *Nortcuri*, 86b) from Curry Rivel (*Churi*, *89*, 86; *Chori*, *197*, 91 b) and Curry Mallet (*Curi*, *429*, 93). South Cadbury is *Sut Cadeberia* but North Cadbury is merely *Cadeberia* (*383*, *383b*; 97b). Upper Cheddon is *Opecedra*, but Cheddon Fitzpaine is merely *Cedra* (*444*, 94b), and in the list of contributories to Taunton they are *Ubcedena* and *Succedena* (*174*, 87b) respectively. We likewise find Babcary (*Babekari*, *277b*, 92b; *Babecari*, *466*, 94) distinguished from Lyte's Cary and Cary Fitzpaine (*Cari*, *443b*, *bis*; 94b, 98b *bis*).

Quite different are those groups of places bearing the same Domesday name but not lying adjacent. Thus the references to *Nortona* cover four widely separated places—Norton Malreward in Chew hundred (*140b*, 88), Norton Fitzwarren in Taunton hundred (*174*, *273b*; 87b, 92b), Norton-sub-Hamdon in Houndsbarrow hundred (*275*, 92b), and Norton St Philip in Frome hundred (*473*, 98). Again, *Alra* is represented by Aller Farm in Sampford Brett (*286b*, 91 b), Aller in Carhampton (*430b*, 94),

[1] Thus Trebles Holford and Riches Holford have been counted as one place, although a document of 968 refers to 'the two Holfords': A. J. Robertson, *Anglo-Saxon Charters* (Cambridge, 1939), pp. 238, 487. The two Domesday entries are consecutive (*433b bis*, 94 *bis*).

[2] See p. 73 above.

and Aller near Somerton (*464b*, 97); while *Bera* is Beere Manor Farm in Cannington (*196*, 91 b) and Beer Crocombe in Abdick hundred (*217*, 92). On one occasion the Domesday text itself makes the distinction, and speaks of *Nortperret* or North Petherton, which is some thirteen or so miles away from *Sudperet* or South Petherton (*89*, 86 *et al.*).

The total of 611 includes 20 or so places about which very little information is given. The record may be incomplete or the details for some places may have been included within those for parent manors. Thus all we hear of Maidenbrook, East Nynehead and Shopnoller is their names in a list of lands (*terrae*) which paid customary dues to Taunton (*174*, 87b). We hear of Wemberham merely as the name of a pasture in the manor of Yatton (*189b*, 89b). Athelney appears only as a Domesday land-holder.[1] For the three places mentioned only in the Geld Accounts or there and in *Terrae Occupatae*, we have no details—Woodadvent (*79b*), *Pirtochesworda* (*79b*) and *Ledforda* (*509b*, *79b*). A number of places are mentioned only incidentally, usually as former portions of a complex manor, and the most we are told about them is the name of their tenant and the figures for their assessments, plough-lands and values, the other details presumably having been included in the totals for the caput. Thus all we hear of Over Stratton (*88b*, 86) is that it had been part of the manor of South Petherton in 1066, the name of its tenant at that time, together with its assessment and its contribution to the manorial *firma*. Or, again, we are told of half a hide which used to belong (*pertinuit*) to the king's demesne manor of Barrington, and which was worth 10*s.* a year in 1086 as in 1066 (*435b*, 94b); but a royal manor of Barrington is described neither here nor elsewhere by name. The only details we are given for Kilmersdon are the presence of a church and half a hide worth 10*s.* (*198b*, 91 b). Here also might be mentioned those other villages which have no population or no teams or neither mentioned for them.[2]

Not all the Domesday names appear on the present-day map of Somerset parishes. Over one third of the total of 611 are represented by the names of hamlets or, more usually, by those of individual houses and farms or those of topographical features. Thus *Cilela* (*452b*, 97b) is now Chillyhill Farm in Chew Stoke; *Waltuna* (*480*, 98b) is Walton Farm in Kilmersdon; *Alduica* (*452b*, 97b) is Aldwick Court in Blagdon; *Sideham* (*443*, 94b) is Sydenham Manor in Bridgwater; *Illega* (*426b*, 93b) is

---

[1] Athelney is a pre-Domesday name, but no manor of Athelney is recorded, the home-farm being at Lyng nearby (*191b*, 91).   [2] See pp. 160–1 below.

Eleigh Water in Combe St Nicholas; *Hascecomba* (*137*, 87b) or *Hetsecoma* (*172b*, 91) is Hiscombe Mead in West Coker. *Turnietta* (*153b*, *467*) or *Tornie* (87b) is represented by Thorent Hill and Field now in Milborne Port. In the east of the county about three quarters of the Domesday names appear as those of modern parishes, but in the west a single parish frequently contains three or more Domesday names in addition to its own. Thus Spaxton[1] and Stogumber[2] each contains six, Cannington[3] and North Petherton[4] as many as seven. The name of Dodo's holding of *Stawe* (*491*, 99) has changed to that of himself, and appears from the thirteenth century onwards as Dodington. The name of *Biscopestone* (*280*, 93) has disappeared and the place is now known as Montacute, which is the Domesday name of a castle nearby. The Domesday *Torre* (*359*, 95b) is represented by the modern Dunster and may appear to have changed its name, but the present form results merely from the addition of Dunna, possibly an early owner. Some names have disappeared entirely from the map. Thus *Wdewica* (*186b*, 89b) was among the fields now comprising Peipard's Farm in Freshford, among which stood a lost 'Woodwick'; *Holcumbe* (*424*, 93b) was a lost place in Aisholt; *Pantesheda* (*433*, 94) was a lost place in Banwell; and *Evestia* (*186b*, 89b) was a lost place in Dunkerton. There are some names which remain unidentified or which cannot be identified with certainty.[5] It is unlikely that any two scholars will agree about all of them. Whether they will yet be located, or whether the places they represent have completely disappeared leaving no trace or record behind, we cannot say. Finally, it must be remembered that many earlier identifications, including some of those of Eyton and of the *V.C.H.* translation, are now known to be incorrect.

[1] Spaxton (*372*, 97): Currypool Farm (*423*, 93), Pightley (*423b*, 93), Merridge (*374b*, 97b), Radlet (*372b*, *425b*; 97, 93b), Swang Farm (*425b*, 93b), Tuxwell (*441b*, *490b*; 94b, 99).

[2] Stogumber (*Warverdinestoc*, *197*, 91): Capton (*104*, 86b), Coleford (*361b*, *427*; 93b, 96), Combe Sydenham (*362*, 96), Emble Farm (*428*, 93b), Hartrow (*362*, 96), Vexford (*427b bis*, 93b *bis*).

[3] Cannington (*89*, *196b*, *478*; 86b, 91b, 98b): Beere Manor Farm (*196*, 91b), Blackmoor Farm (*426*, 93b), Chilton Trivett (*424b*, *425*; 93b *bis*), Clayhill House (*422b*, 93), Edstock (*426*, 93b), Pillock's Orchard (*425*, 93b), Withiel Farm (*426*, 93b).

[4] North Petherton (*88b*, *196*; 86, 91b): Hadworthy Farm (*422*, 93), Huntworth (*371b*, 97), Melcombe (*477b*, 98b), Newton (*282*, *422*, *441*, *477*, *477b*; 91b, 93, 94, 98b *bis*), Shearston (*422b*, 93), Shovel (*477b*, 98b), Woolmersdon (371b, 97).

[5] The names unidentified in this analysis are as follows: *Dudesham* (*424b*, *509b*; 93b), *Ledforda*, *Letfort* (*79b*, *509b*, not in Exch. Dom.), *Pirtochesworda* (*79b*, not in Exch. Dom.), *Rima* (*423*, 93), *Shepwurda* (*493*, 99), *Wlftuna* (*382b*, *520b*; 97b).

On the other hand, about 80 or so parishes on the modern map are not mentioned in the Domesday text. Their names do not appear until the twelfth or thirteenth century or even later, and, presumably, if they existed in 1086, they are accounted for under the statistics relating to other place-names. Thus, so far as record goes, Priddy was first mentioned in 1180 and Nempnett Thrubwell in *c*. 1200. Some places not mentioned in the Domesday Book must have existed, or at any rate borne names, in Domesday times, because they are named in pre-Domesday documents and again in the twelfth or thirteenth century. Thus the account of the manor of Wells (*157b*, 89) may have covered the nearby villages of Binegar, Dinder and Wookey, all named at least as early as 1065.[1] But the most notable example of omission is in the account of the great manor of Taunton (*173b*, 87b), 'that classical example of colossal manors'.[2] It covers, wrote Maitland, 'far more than the borough which bears that name; it covers many places which have names of their own and had names of their own when the survey was made'.[3] They were identified by R. W. Eyton and they most of them lie not far from Taunton itself (Fig. 35).[4] Occasionally, although a modern parish is not named in the Domesday Book, it contains a name or names that are mentioned. Thus Weston-super-Mare contains Ashcombe (*140*, 88) and Milton (*350b*, *479*; 95, 98b). Or, again, neither Domesday text describes a manor of Wellow, although it appears in the Hundred-Lists (*63b*, *64b*), but the modern parish contains Eckweek (*276b*, 92b), Stony Littleton (*149*, 89), Woodborough (*447*, 98), and White Ox Mead (*434*, 94). The history of *Littelaneia* or *Litelande* (*156b*, 89) is of interest in this connection. Its Domesday entry seems to have included details for Huish Episcopi, named in pre-Domesday times but not in either Domesday text. Its name is now represented only by that of Litnes Field and the name of the parish as a whole has become Huish Episcopi. From this account it is clear that there have been many changes in the village geography of the county, and that the list of Domesday names differs considerably from that of present-day parishes.

Fig. 36 shows that two types of country were very sparsely settled—the uplands above 600 ft O.D. or so and the marshy lowlands of the

[1] See H. P. R. Finberg, *The Early Charters of Wessex* (Leicester, 1964), p. 152.
[2] F. W. Maitland, *Domesday Book and Beyond* (Cambridge, 1897), p. 276.
[3] *Ibid.* p. 113.
[4] R. W. Eyton, *op. cit.* I, p. 193; II, p. 34.

Somerset Levels. In the west, the bleak Exmoor upland was uninhabited but there were settlements in the valleys of its northern margin and along the interior valleys of the Exe and its tributaries. Such were Old Stowey (*359b*, 96), at about 800 ft, with 2 recorded people and 1 team, and

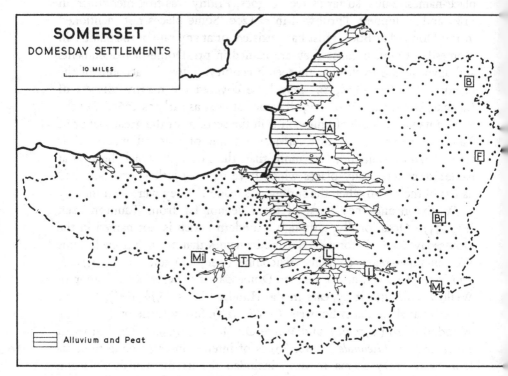

Fig. 36. Somerset: Domesday place-names.

Domesday boroughs are indicated by initials: A, Axbridge; B, Bath; Br, Bruton; F, Frome; I, Ilchester; L, Langport; M, Milborne Port; Mi, Milverton; T, Taunton.

Almsworthy (*430*, 94), at about 1,000 ft, with as many as 17 recorded people and 4 teams; but the latter entry may well have covered several scattered settlements. The empty area continued into that of the Brendon Hills, and, further east still, there were the smaller empty areas of the Quantock Hills and the Blackdown Hills. In the other half of the county, in the north-east, settlements also avoided the dry limestone upland of the Mendips, but the outcrops of conglomerate along its northern and southern flanks were marked by lines of vills which included some of the largest and wealthiest manors in the county, e.g. Banwell, Cheddar, Chewton Mendip and Wells. Between the uplands of east and west stretches a lowland which has always been liable to flooding by the rivers that cross it, the Parrett, the Brue, and their associated streams. It would seem that most of this area was marshland in the eleventh century, but the surface of the marsh was broken by hillocks and knolls of higher ground such as Brent and Meare, and it was here that settlements were to be found, and also along the peninsula of the Polden Hills, and on the large island of Wedmore. But on the silt and clay coastal strip fronting the Bristol Channel there were frequent and prosperous manors.

The closely settled areas of the county show a contrast between east and west—there are many more names to the west of the lower Parrett than to the east. But this fact is somewhat misleading, especially for western Somerset, because we can hardly think of each Domesday name as representing a true vill. One of the reasons why vills seem to be more numerous in the west than in the east is that in the former the character of the terrain does not favour the growth of nucleated villages. The tendency in the west is to form hamlets and isolated farms; the number of separate settlements is accordingly large, but each supports only a small population. We can hardly think of, say, Downscombe (*430*, 94), where a solitary inhabitant was recorded, or Pixton (*429*, 93b), where neither inhabitants nor teams were mentioned, as vills, in spite of the fact that they were called 'manors'. On the other hand Exford may perhaps, in some sense, be considered a vill (*359b bis, 430 bis, 431b*; 94 *ter*, 95b *bis*), a small vill, for the total of recorded inhabitants was only six. But the five holdings that comprise it may well have represented, not a single settlement, but five distinct, though perhaps neighbouring, settlements in and about the area of modern Exford. A study of the modern settlement pattern of Somerset notes the difference between the two halves of the county: 'Judged purely on a basis of proportion of dispersion and agglomeration, Somerset may be divided into two sharply contrasted zones separated by a line running southward from the Parrett estuary[1]— agglomeration in the east, dispersion in the west.

[1] B. M. Swainson, 'Dispersion and agglomeration of rural settlement in Somerset', *Geography*, XXIX (1944), pp. 1–8.

THE DISTRIBUTION OF PROSPERITY AND POPULATION

Some idea of the nature of the information in the Domesday folios for Somerset, and of the form in which it is presented, may be obtained from the account of the manor of Banwell, situated below the northern border of the Mendip Hills (see below). The contrast between the Exeter and Exchequer versions has already been noted. The description of Banwell does not include all the kinds of information that appear elsewhere in the folios for the county. There is no mention, for example, of cottars or coliberts, not is there any reference to underwood or to a fishery or to a vineyard. But it is a fairly representative and straightforward entry, and it sets out the recurring standard items that are entered for most villages. These are five in number: (1) assessment, (2) plough-lands, (3) plough-teams, (4) population, and (5) values. The bearing of these five items of information upon regional variations in the prosperity of the county must now be considered.

BANWELL

Exeter Domesday (*157*)

The bishop of Wells has one manor which is called Banwell which Earl Harold held on the day on which King Edward was alive and dead, and it rendered geld for 30 hides. 40 teams can plough these [hides]. Thence the bishop has 6 hides and 3 teams in demesne and the villeins have 7 hides and 18 teams. There the bishop has 23 villeins and 12 bordars and 5 serfs and 15 'animals' (*animalia*) and 15 swine and 30 sheep and 20 she-goats and 2½ leagues of woodland in length and in breadth and 100 acres of meadow and 1 league of pasture in length and breadth, and it is worth to the bishop's use £15 a year.

Of the aforesaid 30 hides Serlo de Burcy holds 3 hides of the bishop and has land for (*ad*) 8 teams[1] and has 1 team there and the villeins 3 teams. There S. has 5 villeins and 3 bordars and 2 serfs, and it is worth 60s., and when Serlo received it, it was worth £6.

Of these 30 [hides] Ralph Twistedhands (*Radulfus tortesmanus*) holds 5½ hides of the bishop and has there 3 teams in demesne, and his villeins 5, and he has 6 villeins and 2 bordars, and it is worth 100s.

Of these 30 hides Rohard holds 5½ hides of the bishop and has there 2 hides and 2 teams in demesne and his villeins 3½ hides and 4 teams. There R. has 9 villeins and 10 bordars and 3 serfs and 6 unbroken mares and 1 rouncey and

---

[1] This is a variant of the usual Exeter formula—see p. 156 below.

20 'animals' and 30 swine and 100 sheep and 2 mills which render 10s., and it is worth 100s., and when R. received it £4.

Also, of these aforesaid hides Fastrad holds 1 hide of the bishop and has there 1 team and 1 villein, and it is worth 20s.

Moreover, of these Bono holds 1 hide of the bishop and has there 1 team and 1 villein who has 1 team, and it is worth 20s.

Also, of these 30 hides Ælfwig 'Haussonna' has 1 hide and has there 1 team and 1 villein who has half a team, and it is worth 10s., and Ordulf has there 1 mill which renders 40d.

Exchequer Domesday (89 b)

The same bishop [of Wells] holds Banwell. Earl Harold held in it the time of King Edward and it gelded for 30 hides. There is land for 40 teams. Of it 6 hides are in demesne, and there are there 3 teams and 5 serfs, and 24[1] villeins and 12 bordars with 18 teams. There are 100 acres of meadow, pasture 1 league long and broad, woodland 2½ leagues in length and breadth.

Of the land of this manor there hold of the bishop: Serlo 3 hides, Ralph 5½ hides, Rohard 5½ hides, Fastrad 1 hide, Bono 1 hide, Ælfwig 1 hide. There are 9 teams in demesne and 5 serfs, and 25[2] villeins and 15 bordars having 13½ teams.

Two mills there belonging to Rohard are rented for 10s.; Ordulf has a mill which is rented for 40d.

The whole manor is worth £15 for the use of the bishop, and similarly £15 for the use of his men.

(1) *Assessment*

The Somerset assessment is stated in terms of hides, virgates and ferlings (or ferdings or fertings), and sometimes of what appear to be geld acres. Fractions of hides, virgates and ferlings are encountered, e.g. Hawkwell (*49ι*, 99) was assessed at 1 virgate 1¼ ferlings, which implies estimation to the sixty-fourth part of a hide. The usual Exchequer formula is *geldabat pro n hidis*,[3] but there are occasional variations, as the following examples show:

Draycott in Rodney Stoke (*492*, 99): *defendebat se pro una virgata.*
Crewkerne (*197*, 91): *Ibi sunt x hidae.*
Carhampton church (*196b*, 91 b): *In ecclesia Carentone jacet i hida et dimidia.*

---

[1] Note that the Exeter text says 23 villeins.
[2] Note that the Exeter total amounts to 23 villeins.
[3] The corresponding Exeter formula runs: *reddidit gildum pro n hidis.*

Some people have thought that the hides of the south-west were small and that they comprised not the usual 120 acres but only 40 or 48.[1] The fact that the quantities of acres are often 5, 8, 10, 12 or 15 suggests that the Somerset hide comprised 120 acres,[2] but the evidence cannot be regarded as conclusive.

The Exchequer text sometimes states also the assessment of the demesne portion of an estate; the Exeter text not only almost invariably does this, but in addition it often states explicitly the assessment of the portion in the occupation of the peasantry. It is interesting to compare the sum of the demesne and the other hidage with the total given for a holding as a whole. In most entries there is agreement, even though the figures may be complicated ones. At Weston near Bath (*448 b*), for example, 4 h less ½ v and 3 ac in demesne appears with 1 h ½ v and 3 ac to make a correct total of 5 h. But in nearly fifty entries there are discrepancies between the stated total and the actual total. Some of these discrepancies may be due to simple clerical errors; others may be the result of unknown changes in the composition of manors. In occasional entries, the greater detail of the Exeter text allows us to decide between the variant readings in that and the Exchequer version. Thus the Exchequer assessment for Ashbrittle (92) is 5 hides, whereas the Exeter assessment (269) amounts to 5½ hides, but this latter figure is divided between 2½ hides on the demesne and 3 with the peasants; the extra circumstantial detail would seem to imply that the Exeter figure is the correct one. The account of Hatch Beauchamp (271, 92) is interesting in that the texts agree on an assessment of 5 hides, but the figures of the expanded version add up to 6 hides.

The account of twelve manors comprising the Ancient Demesne of the Crown says they never paid geld.[3] The statement or some variant of it runs: *nunquam reddidit gildum quia nescitur quot hidae ibi sint* in the Exeter version, and *nunquam geldavit nec scitur quot hidae sint ibi* in the Exchequer version.[4] Each of the manors concerned was responsible for the ancient render of one night's *firma* or part of a night's *firma*.[5]

---

[1] See pp. 14 and 80 above.

[2] The Exeter entry for Pillock (*425*) shows that 'ten acres' is less than half a ferling. Even with a 120-acre hide, half a ferling would amount to only 3¾ acres, so that these 'ten acres' may have been arable acres, not geld acres.

[3] One of the twelve manors (Bedminster) is now in the county of Bristol.

[4] The Exeter version of the phrase for the combined entry for Cannington, Williton and Carhampton specifically says that they had never been hidated—*nunquam reddiderunt gildum et nunquam fuerunt hidatae et nescitur quot hidae in ea sint* (*89*).

[5] The manors were those of North Petherton, South Petherton, Curry Rivel, Frome,

There were also eight other holdings that did not pay geld. One of these was the royal manor of Crewkerne, which had belonged to the Godwine family in 1066 and which is described in words similar to those for the other royal holdings, but, of course, it had not been responsible for a night's *firma*. The number of hides at each of the other seven places is stated. The relevant entries are set out below. Fourteen holdings had their assessments reduced. The full list is set out on p. 152. This geld-free land did not necessarily correspond either with the manorial demesne of the Exeter Domesday or with the fiscal demesne of the Geld Accounts. For example, at Queen Camel (*106b*, 87) there were 15 hides of which only 8½ paid geld, yet 5 (not 6½) were in manorial demesne. Or, again, in Chewton Mendip (*114b*, 87) there were 29 hides of which only 14 paid geld, yet 18 (not 15) represented manorial demesne, and the Geld Accounts show that none of these had paid geld (*78*).

Eight Somerset holdings not paying geld

(*Exchequer version*)

Andersey (*172*, 90): *In qua sunt ii hidae quae nunquam geldaverunt.*

Crewkerne (*105b*, 86b): *non geldabat nec scitur quot ibi hidae habentur.*

Glastonbury (*172*, 90): *habet in ipsa villa xii hidas quae nunquam geldaverunt.*

Lyng (*191b*, 91): *Ibi est i hida sed non geldabat T.R.E.*

Stoke-sub-Hamdon (*267b*, 92): *Superest ibi una virgata terrae quae non geldabat T.R.E.*

Tuxwell (*490b*, 99): *Ibi est dimidia virgata terrae et non geldabat T.R.E.*

Wearne (*479b*, 98b): *ii virgatae terrae et dimidia quae nunquam geldaverunt.*

Wells (*158*, 89): *Praeter has l hidas habet episcopus ii hidas quae nunquam geldaverunt T.R.E.*

Note: (1) In addition, in the Geld Accounts (Milborne hundred, *80*) the church of Milborne had a hide in fiscal demesne 'which never rendered geld'.

(2) The Glastonbury island of Andersey is mentioned above. The other two Glastonbury islands of Meare and Panborough (*172*, 90) had respectively 60 and 6 acres of land, but nothing is said about their geld.

There are seven other entries relating to holdings which did not pay geld, and they are set out on p. 153. Three of these entries speak not of hides

Bruton, Cannington, Carhampton, Williton, Somerton, Cheddar, Milborne Port and probably Bedminster (now in Bristol). The last-named was said never to have paid geld, but no *firma* is mentioned in connection with it—see p. 170 below.

but only of *carucatae terrae*. The other four entries have orthodox assessments in terms of hides but mention also additional *terra ad n carucas* which did not pay geld. A reasonable assumption seems to be that they represent land brought into cultivation, or added to a manor, since the time when its assessment had been made or last revised. An alternative explanation is that the owners were trying to account for land the hidage

Fourteen Somerset holdings paying reduced geld
(*Exchequer version apart from Wincanton*)

Batheaston (*114*, 87): *Ibi sunt ii hidae et geld[at] pro una hida.**

Chewton Mendip (*114b*, 87): *Ibi sunt xxix hidae. T.R.E. geldabat pro xiiii hidis.*

Coker (*107*, 87): *Ibi sunt xv hidae et geldabat pro v hidis.*

Crowcombe (*266*, 91 b): *Ibi sunt x hidae sed non geldat nisi pro iiii hidis.*

East Pennard (*166b*, 90b): *T.R.E. geldabat pro x hidis. Ibi sunt tamen xx hidae.*

Hardington Mandeville (*107*, 87): *Ibi sunt x hidae et geld[at] pro v hidis.*

Kingston Seymour (*151*, 89): First entry: *geldabat pro i hida.* Second entry: *geldabat pro iiii hidis et dimidia . . . Hoc manerium T.R.E. non geldabat nisi pro una hida.*†

Leigh in Milverton (*363*, 96): *geldabat pro dimidia hida. Tamen ibi est i hida.*

Martock (*113*, 87): *Ibi sunt xxxviii hidae. T.R.E. geldabat pro xiii hidis.*

Puriton (*197b*, 91): *Ibi sunt vi hidae sed non geldat nisi pro v hidis.*

Queen Camel (*106b*, 87): *geldabat pro viii hidis et dimidia. Ibi sunt tamen xv hidae.*

Rodney Stoke (*137*, 87b): *Ibi sunt v hidae et una virga terrae et pro iii hidis geldabat.*

Wedmore (*159b*, 89b): *geldabat pro x hidis. Sunt tamen ibi xi hidae.*

Wincanton (*352*, 95): *reddidit gildum pro iii hidis et dimidia . . . huic mansione est addita dimidia hida terrae. . . . Istae supradictae iiii hidae reddiderunt gildum pro iii hidis.*††

---

* But note that this reduction for Batheaston cannot be presumed with certainty, for the Bath Geld Account (*76*) says that the king's villeins had failed to pay on one hide, and the other hide was exempt as fiscal demesne. The Exchequer text has, most unusually, *geld'* not *geldb'*.

† The Exeter text has *Istae v predictae hidae et dimidia non reddiderunt gildum nisi pro i hida tempore E. regis.*

†† This is the Exeter version because the Exchequer entry omits the last sentence.

of which did not tally with that of their charters.[1] The 20 carucates at Taunton are listed in the Geld Account of the hundreds of Taunton and Pitminster (*75*) and are said never to have gelded. None of the other *carucatae terrae*, however, appears in the relevant entries of the Geld Accounts.

Somerset: *Carucatae terrae nunquam geldantes*

(*Exchequer version*)

A. *Carucatae terrae* mentioned

Frome (*198*, 91): *Reinbaldus tenet ecclesiam de Frome cum viii carucatis terrae.*

Frome (*90 b*, 86b): *De hoc manerio* [Frome] *tenet ecclesia S. Johannis de Froma viii carucatas terrae et similiter tenuit T.R.E.*

Muchelney, Midelney, Thorney (*189*, 91): *Ecclesia Sancti Petri de Micelenye habet iiii carucatas terrae quae nunquam geldaverunt.*

B. *Carucatae terrae* implied

Henstridge (*107*, 87): *Terra est xvi carucis. Praeter has x hidas est terra ad viii carucas quae nunquam geldavit.*

Pilton (*165b*, 90): *Terra est xxx carucis. Praeter hanc habet abbas ibi terram xx carucis quae nunquam geldavit.*

Shapwick (*161b*, 90): *Terra est xl carucis. Praeter hanc habet abbas terram xx carucis quae nunquam geldavit.* The Exeter text mentions that in these the villeins had respectively 12 and 10 teams *supra illam terram quae non reddit gildum.*

Taunton (*173b*, 87b): *Terra est c carucis. Praeter hanc habet episcopus in dominio terram ad xx carucas quae nunquam geldavit.*

Note: (1) The list does not include the strange reference to carucates at Norton St Philip (*437*, 98): *De his x hidis dedit Rex Edwardus predicto Luing ii carucatas terrae.*

(2) The carucates at Frome are not specifically described as non-gelding, but they would seem to have been so.

(3) The two entries for Frome refer to the same holding.

A glance through the Somerset folios reveals many examples of the 5-hide unit. Thus Baltonsborough (*167*, 90b) was assessed at 5 hides, Compton Dando (*144b*, 88b) at 10 hides, and Pitminster (*173b*, 87b) at 15 hides. One Somerset holding was called *Fifhida* or *Fihide*, represented today by Fivehead, but its recorded assessment was for only 1½ hides

---

[1] See p. 82 above.

(*431 b*, 94). When a manor or village included a number of holdings, the same feature can occasionally be demonstrated. Thus Barton St David (*434b*, *480*; 94, 98 b) with 2 holdings was assessed at 5 hides (2½ + 2½); so was Foddington (*278*, *466*, *466b*; 92 b, 99 *bis*) with 3 holdings (1⅜ + 2¼ + 1⅜ hides). Sixty-eight vills in the modern county were assessed at 5 hides, 41 at 10 hides, and 9 at 15 hides. There were also a number of other places assessed at large multiples of 5, e.g. 20, 30 and 50 hides. We cannot be far wrong in saying that the 5-hide principle is readily apparent in about two out of nine Somerset villages. But there were very few 5-hide settlements west of the Parrett. A count restricted to the area east of the Parrett shows the 5-hide principle to be readily apparent in as many as about one out of three places. It is also possible that some villages were grouped together for the purpose of decimal assessment. J. H. Round, with the help of Eyton's analyses, gave as an example the hundred of Crewkerne with 40 hides and that of Whitstone with 120 hides.[1] The first detailed attempt to discover how far the 5-hide principle was applicable to Somerset was that of the Rev. E. H. Bates, who claimed that it was possible to discover many groups which suggested that a 20-hide unit had played a definite part in the scheme of assessment.[2] Both Bates and Round admitted that beyond the Parrett, in a region of small settlements which are often rated in virgates and ferlings, the 5-hide unit was less prominent, and that 'in the western half at least of this district we lose sight altogether of the "five-hide unit"'.[3] But Bates's theories received stern criticism. The Rev. T. W. Whale asserted that 'as regards Somerset the five-hide unit is a myth'.[4] This view is somewhat exaggerated, but his stigmatisation of Bates as 'manipulating manors *ad libitum*, transgressing the bounds of Hundreds, and utterly disregarding Geld Rolls and Domesday principles' may be justifiable, and so is his further contention that mistakes in the identification of place-names (now known to be many) invalidate a number of Bates's tables.

Whatever may have been the incidence of the 5-hide unit, it is clear that the assessment was largely artificial in character, and bore no constant

---

[1] J. H. Round, *Feudal England* (London, 1895), pp. 61–2. But note that the Geld Accounts give only 39 hides for Crewkerne.

[2] E. H. Bates, 'The Five-Hide Unit in the Somerset Domesday', *Proc. Som. Arch. & Nat. Hist. Soc.* XLV (Taunton, 1899), pp. 51–107.

[3] J. H. Round, *V.C.H. Somerset*, I, p. 387.

[4] T. W. Whale, 'Principles of the Somerset Domesday', *Proc. Bath Nat. Hist. and Antiq. Field Club* (Bath, 1902), p. 39.

relation to the agricultural resources of a vill. The variation among a representative selection of 5-hide vills speaks for itself:

	Plough-lands	Plough-teams	Population	Valuit	Valet
Chelwood (447b, 97)	5	2	11	100s.	100s.
Wanstrow (384, 97b)	5	3	12	£6	£3
Farrington Gurney (149b, 89)	7	7	18	100s.	£4
Odcombe (274, 92b)	5	5	20	100s.	100s.
Minehead (358, 95b)	12	13	61	100s.	£6

An extreme example of low rating was the royal manor of Milverton (113, 87), which had belonged to Queen Edith in 1066. It answered for only half a virgate, yet there were 29 recorded people with 10 teams at work, and the annual value of the whole holding amounted to £25.[1] Or, again, Nettlecombe (104, 86b) was rated at but 2¾ hides, yet it had 22 recorded people with 9 teams.

The total assessment including non-gelding hides amounted to 2,900 hides 1 virgate 1¼ ferlings and 92 acres. This is for the area included within the modern county, and is not strictly comparable with the totals of Eyton[2] or Maitland,[3] which were for the Domesday county. The nature of some entries and the possibility of unrecognised parallel entries make exact calculation very difficult. All that any estimate can do is to indicate the order of magnitude involved. We must furthermore remember those thirteen manors of which it could be said, *nunquam geldavit nec scitur quot hidae sint ibi,*[4] and also those 80 carucates of land that had never paid geld or never been assessed.

(2) Plough-lands

Plough-lands are systematically entered for the Somerset holdings, and the Exchequer formula normally runs: 'There is land for *n* plough-teams'

[1] Milverton seems to have included a borough, which may explain the low assessment—see p. 204 below.

[2] Eyton counted 2,952 hides in the county. In a second estimate he made allowance for *carucatae terrae* and other land which never paid geld and brought the estimate up to 3,488 83/192 hides.—R. W. Eyton, *op. cit.* I, p. 23; II, pp. 7, 43.

[3] F. W. Maitland, *op. cit.* p. 400, reckoned 2,936 hides.

[4] The figure of thirteen includes the twelve manors of the Ancient Demesne of the Crown together with Crewkerne.

(*Terra est n carucis*). The Exeter text usually says: *hanc terram possunt arare n carucae*, with variants.[1] A reckoning in half-teams is not unusual; but there are only very occasional references to oxen, such as that for Exford (*359b*, 95b): *Terra est ii bobus*. In two Exchequer entries the phrase *Terra est* is followed by a blank space in which a figure was never inserted: Lilstock (*478b*, 98b) and Stawley (*373b*, 97); nor does the Exeter text supply the missing information. There are also ten entries which make no reference to plough-lands, although they state the number of teams at work, e.g. those for Bagley in Stoke Pero (*430b*, 94) and Farleigh Hungerford (*434*, 94). The twelve entries without record of plough-lands account for a total of 24½ teams. Not included in this total are the teams of those other holdings whose plough-lands seem to be recorded under those of parent manors, e.g. High Ham (*165*, 90). Many plough-land figures look suspiciously artificial, e.g. Bruton (*90b*, 86b) with 50 for 21 teams.

The relation between plough-lands and teams varies a great deal in individual entries. They are equal in number in about 35 per cent of the entries which record both. Occasionally we are explicitly told that this is so, as in the entry for Huish in Burnham-on-Sea (95b): *Terra est i caruca quae ibi est cum i servo et i cotario*.[2] A deficiency of teams is more usual, and is encountered in 51 per cent of the entries which record both plough-lands and teams. The values of some deficient holdings had fallen, but on others they had remained constant,[3] and on yet others they had even increased. No general correlation between plough-team deficiency and decrease in value is therefore possible, as the following table shows:

		Ploughlands	Teams	*Valuit*	*Valet*
Decrease in value	Brushford (*268*, 92)	12	3	£8	£4
	Hatch Beauchamp (*271*, 92)	6	5	£8	£4
Same value	Donyatt (*270*, 92)	5	3	100s.	100s.
	Wootton Courtenay (*369*, 96b)	10	6	£5	£5
Increase in value	Broomfield (*363b*, 96)	10	4	40s.	£3
	Whitestaunton (*265b*, 91)	8	5	30s.	60s.

[1] One variant is *terra ad n carucas* as for Banwell on p. 148 above.
[2] The Exeter version (*355*) reads: *hanc potest arare i carruca et est ibi*.
[3] See p. 350 below for F. W. Maitland's explanation of deficient teams.

These figures are merely indications of unexplained changes in individual holdings.

An excess of teams is a good deal less frequently encountered, being found in only 14 per cent of the entries which record both plough-lands and teams. The values on some of these holdings had risen, e.g. at Emborough (*150*, 89) from 20*s.* to 70*s.* and Lullington (*147b*, 88b) from £4 to 100*s.*, but very frequently this was not so. Other holdings with excess teams were valued at the same sum at both dates, e.g. at Crowcombe (*266*, 91b) and Orchardleigh (*147b*, 88b). The values of yet other of these holdings had even dropped, e.g. at Ilminster (*188*, 91) from £26 to £20 and at Standerwick (*434*, 94) from 50*s.* to 20*s.* Our attention is sometimes specifically drawn to excess teams. An entry for Coleford (*427*, 93b), for example, reads: *Terra est dimidia caruca. In dominio tamen est i caruca.* This phrase, or something like it, occurs altogether in six entries, but no entry hints at the reason for overstocking.[1]

The total number of plough-lands for the area within the modern county is 4,750¼. If, however, we assume that the plough-lands equalled the teams on those holdings where a figure for the former is not given, the total becomes 4,774¾. The presence of 80 non-gelding carucates must also be remembered.[2] For various reasons the figures given by Eyton[3] and Maitland[4] are not strictly comparable with the present total. In view of the nature of some entries and of the possibility of unrecognised parallel entries, no estimate can be anything other than approximation.

(3) *Plough-teams*

The Somerset entries for plough-teams are fairly straightforward, and, like those for other counties, they normally draw a distinction between teams on the demesne and those held by the peasantry, the Exeter text being far more regular than that of the Exchequer in this respect. Thus the Exeter account of West Lydford (*493*) tells of 3 teams on the demesne and of 4 with the peasantry, but the Exchequer version (99) merely says:

---

[1] The six entries are: Tarnock (second entry, 95), Watchet (96), Weston Bampfylde (97b), Pillock (93b), Coleford (93b), Babcary (99).

[2] See p. 153 above.

[3] R. W. Eyton, *op. cit.* II, p. 7. The total of 4,812¾ was for the Domesday county. As far as we can tell, no allowance was made for missing figures.

[4] F. W. Maitland, *op. cit.* pp. 401, 410. Maitland's total of 4,858 was for the Domesday county, but, as in the present estimate, allowance was made for the missing plough-land figures.

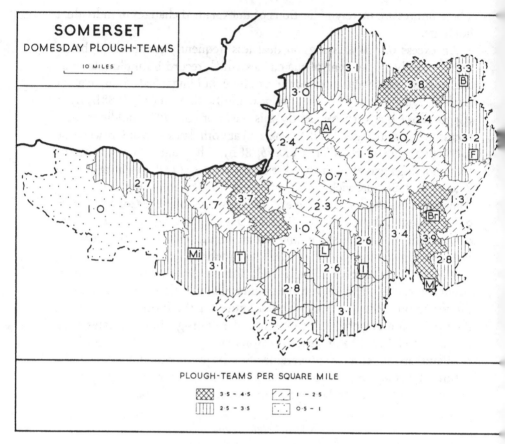

Fig. 37. Somerset: Domesday plough-teams in 1086 (by densities).

Domesday boroughs are indicated by initials: A, Axbridge; B, Bath; Br, Bruton; F, Frome; I, Ilchester; L, Langport; M, Milborne Port; Mi, Milverton; T, Taunton.

Ibi sunt vii carucae. The teams of the peasantry are usually described in the Exeter folios as belonging to the *villani* even when the recorded population includes no villeins, e.g. in the entry for Stawell (*163*, 90).

Half-teams and oxen are not uncommon. Occasionally, when the Exeter text mentions 4 oxen, the Exchequer version speaks of half a team, e.g. in entries for Compton Pauncefoot (*383b*, 97b) and Eastrip (*493b*, 99). The two versions for Street (*357*, 95b) likewise indicate that 4 oxen + 4 oxen = 1 team; an unusual Exeter formula runs: *iiii animalia in carrucam,*

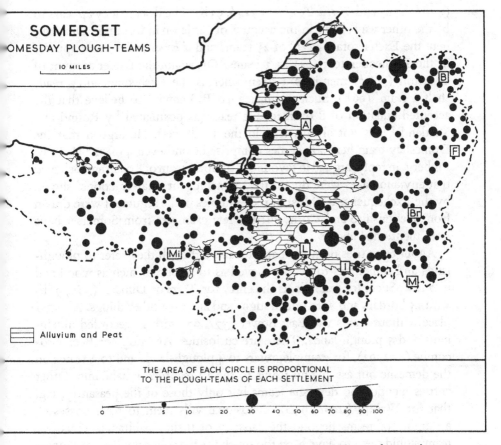

Fig. 38. Somerset: Domesday plough-teams in 1086 (by settlements).

Domesday boroughs are indicated by initials: A, Axbridge; B, Bath; Br, Bruton; F, Frome; I, Ilchester; L, Langport; M, Milborne Port; Mi, Milverton; T, Taunton.

as in the entry for Combe Sydenham (*362b*, 96). There are, however, a number of discrepancies between the Exeter and the Exchequer texts. Oxen entered in the former go unmentioned in the latter. Thus the Exeter account of Cannington (*478*) mentions 2 plough oxen (*ii boves in carrucam*) in-addition to a demesne team, but the Exchequer version records only the team (98b). There are also other holdings for which the Exeter text mentions a pair of oxen that do not appear in the Exchequer text: at Draycott in Rodney Stoke (*492*, 99), Monksilver (*427*, 93b), Puckington

(*429b*, 93 b) and *Terra Teodrici* (*425b*, 93 b). The discrepancy appears to be the other way round in the account of Littleton in Compton Dundon; here the Exeter total (*433b*) of 2½ teams and *ii boves in carrucam* appears in the Exchequer text (94) as 3 teams. Or, again, the Exeter account of Cheriton (*364b*) mentions 3 oxen whereas the Exchequer entry reads 'half a team' (96b). Such anomalies led R. Lennard to believe that the accepted equation of 8 oxen to one team (as postulated by Round and Maitland)[1] was not always true in the south-west. He argued that the Domesday team here sometimes comprised 6 and even 4 oxen.[2] H. P. R. Finberg, in reply, argued in favour of a uniform team of 8 oxen and believed that any anomalies were due to the 'contempt for small fractions' shown by the Exchequer text.[3] Most of the examples in the argument were drawn from Devonshire and Cornwall; the few examples from Somerset have been mentioned above.

About 60 entries do not mention teams although they refer to plough-lands and sometimes to people as well as to resources such as wood and meadow. Such, for example, is the entry for *Torre* or Dunster (*359*, 95 b) with 15 bordars and with one plough-land amongst other things. Another place without teams was East Bower (*371*, 97) with 17 recorded inhabitants and 5 plough-lands. There are curiosities. An entry for Beer Crocombe (*271*, 92), for example, refers to 4 plough-lands and to 3 teams on the demesne but assigns no teams to the 17 recorded inhabitants. Other entries mention no demesne teams but only those of the peasantry, e.g. that for Wreath (*426b*, 93 b). Occasional vills seem to have possessed a number of teams beyond the capacity of their inhabitants. Thus one team would seem to have been too much for the solitary villein of Goose-bradon (*491b*, 98b) to manage without help.

The total number of plough-teams amounted to 3,885⅝, but this refers to the area included in the modern county, and, in any case, a definitive total is hardly possible.[4] This count has been made on the assumption of an 8-ox team.

[1] J. H. Round, *Feudal England*, p. 35; F. W. Maitland, *op. cit.* p. 417.

[2] R. Lennard, 'Domesday plough-teams: the south-western evidence', *Eng. Hist. Rev.* LX (1945), pp. 217–33. See also R. Lennard, 'The composition of the Domesday caruca', *Eng. Hist. Rev.* LXXXI (1966), pp. 770–5.

[3] H. P. R. Finberg, 'The Domesday plough-team', *Eng. Hist. Rev.* LXVI (1951), pp. 67–71.

[4] F. W. Maitland's total for the Domesday county was 3,804 teams (*op. cit.* p. 401). R. W. Eyton's analysis did not cover plough-teams.

(4) *Population*

The bulk of the population was comprised in the five main categories of villeins, bordars, serfs, cottars and coliberts. In addition to these were the burgesses and a miscellaneous group that included swineherds, coscets, fishermen and others. The details of the groups are summarized on p. 162. There are three other estimates, by Sir Henry Ellis,[1] W. Phelps[2] and R. W. Eyton,[3] but the present estimate is not strictly comparable with these because it is in terms of the modern county. In any case, an estimate of Domesday population can rarely be definitive, and all that can be claimed for the present figures is that they indicate the order of magnitude involved. These figures are those of recorded population, and must be multiplied by some factor, say 4 or 5, in order to obtain the actual population; but this does not affect the relative density as between one area and another.[4] That is all that a map such as Fig. 39 can roughly indicate.

It is impossible to say how complete were these Domesday figures. We should have expected to hear more, for example, of priests. Then there are some 40 entries which do not mention inhabitants although they sometimes refer to plough-lands and even teams and to such other resources as wood or meadow. Thus the Exchequer text records no peasantry on one of the holdings of Buckland St Mary (98b) and the Exeter version (*492*) refers only to an unspecified number of villeins; there was no team on the 4 plough-lands but there were animals and pasture and wood. A holding at Coleford (*427*, 93b) had land for half a team; we are told, however, that there was a full team there, but no mention is made of the people who might have worked it. Or, again, no inhabitants are recorded for Emble (*428*, 93b). Can we assume that its half-team was worked by Ælfric, the sub-tenant, and his family? It is certainly unlikely that the two nuns who held land at Huntscott (*196b*, 91b) worked the team there. Or, yet again, there were two plough-lands at Pixton (*429*, 93b) but no people. It is

---

[1] Sir Henry Ellis, A *General Introduction to Domesday Book*, II (London, 1833), pp. 483–5. His total for the county (excluding tenants and sub-tenants) came to 13,316.

[2] W. Phelps, *The History and Antiquities of Somersetshire with an Historical Introduction*, II (London, 1839), p. 29. His total (excluding tenants and sub-tenants) came to 11,467.

[3] R. W. Eyton, *op. cit.* I, p. 223; II, p. 43. His total (excluding tenants and sub-tenants) came to 13,307.

[4] But see p. 368 below for the complication of serfs.

true that no team is mentioned, but there was pasture and wood. Could they
have been used by the men and teams of another holding? Then there
are those other place-names for which hardly any information is given.[1]
We cannot be certain about the significance of these omissions, but
they (or some of them) may imply unrecorded inhabitants.

Unnamed thegns are mentioned quite frequently as holders of estates
in 1066 and they are sometimes entered in small groups, e.g. the 3 at

Recorded Population of Somerset in 1086

A. Rural Population

Villeins	5,239
Bordars	4,743
Serfs	2,106
Cottars	390
Coliberts	208
Miscellaneous	172
Total	12,858

Details of Miscellaneous Rural Population

Swineherds	84	Rent-payers (*gablatores*)		7
Coscets	56	Men (*homines*)	.	4
Fishermen	10	Priests	.	3
Smiths (*fabri*)	8	Total	.	172

B. Urban Population

AXBRIDGE	32 burgesses (see p. 202 below).
BATH	192 burgesses, 7 *domus*, 6 *domus vastae*, 1 *mansura vacua* (see p. 199 below).
BRUTON	17 burgesses (see p. 201 below).
FROME	not enumerated (see p. 204 below.)
ILCHESTER	108 burgesses (see p. 200 below).
LANGPORT	39 burgesses (see p. 202 below).
MILBORNE PORT	67 burgesses, 2 *masurae* (see p. 200 below).
MILVERTON	1 *domus* (see p. 204 below).
TAUNTON	64 burgesses (see p. 202 below).

[1] See p. 143 above.

Congresbury (*106*, 87), the 7 at Rode (*148*, 88b), and the 9 at Newton St Loe (*149*, 89). Altogether, some 295 unnamed thegns are enumerated for 1066. *Terrae Occupatae* adds another dozen or so.[1] Unnamed thegns are not mentioned for 1086 except for one each on holdings at Chard (*156*, 89) and Marksbury (*169b*, 90b). Both these (and the named thegns), being landholders, have been excluded from our totals. So have the various *milites* and *Angli* mentioned for Somerset only when they held land. We have also excluded those priests who were landholders.[2]

Villeins constituted the most important element numerically in the population, and amounted to 41 per cent of the total. A common Exeter formula states the hidage and plough-lands occupied by the villeins before going on to enumerate the categories of people on a holding; and the word *villani* is sometimes used in this connection even when no villeins appear in the subsequent enumeration. The 8 villeins on the land added to Dulverton (*103b*, 86b) are said, in the Exeter text, to dwell (*manent*) there. Two villeins at Ashwick (*187*, 89b) paid 42*d.* a year (*reddunt per annum xlii denarios*), and, as this sum is also given as the value of the holding, they were apparently renting the whole. One out of 7 villeins at Alford (*277b*, 92b) paid 8 blooms of iron (*reddit viii blumas ferri*).[3] Next in numerical order came bordars, who amounted to 37 per cent. They call for no special comment. The Exeter account of Weacombe (*428b*, 93b) gives us the additional information that the one bordar (the only recorded inhabitant) paid 7*s.* 6*d.* to a sub-tenant. Serfs amounted to nearly 16 per cent of the recorded population, and their incidence over the county varied considerably. It was about 25 per cent in the north-east scarplands around Bath, and only about 12 per cent in the southern lowland. It was 24 per cent on the demesne lands of the bishop of Winchester, 28 per cent on those of Bath Abbey, but only about 10 per cent in the royal manors, although the inclusion of coliberts would bring this last figure up to about 18 per cent. At Moortown (*429b*, 94) six serfs are said, in the Exeter text, to hold (*tenent*) a virgate. At Kenn (*143*, 88) the serf who alone in-habited it is said to 'dwell' (*manet*) there; but the entry says nothing

---

[1] Sometimes a man named in the Domesday text appears in that of *Terrae Occupatae* as an anonymous thegn, and vice versa. In arriving at our total we have included all those specifically described as thegns whether or not they are named elsewhere.

[2] See p. 194 below.

[3] The Exeter text occasionally states the amount of land held by the villeins, e.g. at East Pennard 4 villeins held 1 hide (*166b*, 90b); at Avill 1 villein held 12 acres (*359*, 95b); and at Isle Abbots the 3 bordars held 15 acres (*188b*, 91).

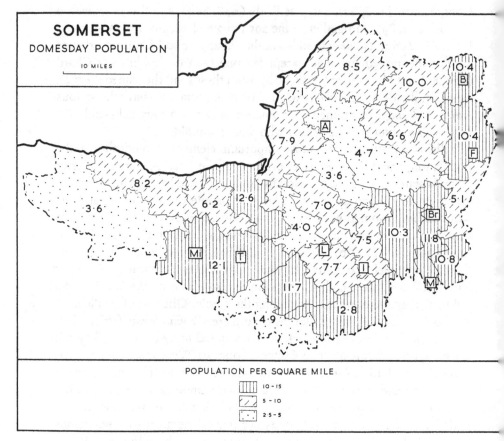

Fig. 39. Somerset: Domesday population in 1086 (by densities).

Domesday boroughs are indicated by initials: A, Axbridge; B, Bath; Br, Bruton; F, Frome; I, Ilchester; L, Langport; M, Milborne Port; Mi, Milverton; T, Taunton.

about plough-lands, teams, meadow, pasture, or other resources. Cottars accounted for 3 per cent of the total, and were to be found mainly in the south-east of the county. Another group (1·6 per cent) was that of coliberts, whom Maitland considered to be 'distinctly superior to the *servi*, but distinctly inferior to the villeins, bordiers and cottars';[1] four fifths of these coliberts were on royal manors where they were frequently grouped with serfs as connected with the demesne. They amounted to nearly one

[1] F. W. Maitland, *op. cit.* p. 28.

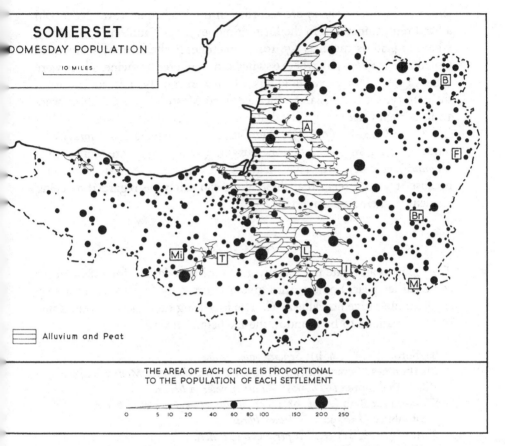

SOMERSET
DOMESDAY POPULATION
10 MILES

Alluvium and Peat

THE AREA OF EACH CIRCLE IS PROPORTIONAL
TO THE POPULATION OF EACH SETTLEMENT

0   5 10 20   40   60 80 100   150   200   250

Fig. 40. Somerset: Domesday population in 1086 (by settlements).

Domesday boroughs are indicated by initials: A, Axbridge; B, Bath; Br, Bruton;
F, Frome; I, Ilchester; L, Langport; M, Milborne Port; Mi, Milverton; T, Taunton.

quarter of the total number of coliberts recorded in the Domesday Book
as a whole. At Frome (*90b*, 86b), where there were no serfs, Round believed that they were replaced by the six coliberts in the normal proportion of 'two to each demesne team'.[1] But in other entries, notably
that for Taunton (*173b*, 87b), coliberts were sometimes grouped with
villeins.

In the miscellaneous category were 84 swineherds (*porcarii*) in 14

[1] J. H. Round in *V.C.H. Somerset*, I, p. 426.

entries, of whom only 57 are recorded in the Exchequer text.[1] Most paid
a fixed rent, sometimes in the form of money, as at Taunton (*173b*, 87b),
where 17 paid as much as £7. 10s.; sometimes in the form of swine, as at
Cutcombe (*357b*, 95 b), where 6 swineherds rendered 31 swine. There were
some 56 coscets[2] on eleven holdings, which were mostly in the south-east of
the county. On 'the island which is called Meare' (*172*, 90) there were
10 fishermen (*piscatores*); at Glastonbury (*172*, 90), there were 8 smiths
(*fabri*); at Cheddar (*90*, 86), there were 7 rent-payers (*gablatores*) who
rendered 17s.; and at Compton Durville (*113b*, 87), there were 4 men
(*homines*). Finally, we have included only three priests in our count
because it would seem that these alone can safely be regarded as being
resident in their respective vills.[3]

(5) *Values*

The value of an estate is generally given for two dates—for 1086 and for
some earlier year. Some entries give a figure only for 1086, and a very
small number give no value at all. The following examples illustrate some
of the variations to be found in the Exchequer text:

> Bagborough (*464*, 96b): *Semper valet l solidos.*
> Charlton near Somerton (*443b*, 94b): *Valuit vi libras. Modo c solidos.*
> Milton Podimore (*161b*, 90): *Valuit et valet vi libras.*
> Weston near Bath (*448b*, 98): *Totum olim et modo valet vi libras.*
> Glastonbury (*172*, 90): *Valet xx libras.*
> Puriton (*197b*, 91): *Reddit per annum xii libras.*
> Stoke near Chew Magna (*492*, 99): Value omitted (25s. in Exeter text).

Occasionally, when the Exchequer text uses *valet*, the Exeter text uses
reddit, as in the entries for Rimpton (*173b*, 87b) and Wiveliscombe (*156b*,
89). The use of the words *valuit* and *olim* in the Exchequer folios leaves
the exact date of the earlier valuation in doubt, but a few entries are more
explicit and refer either to the time when an existing holder received

---

[1] The Exchequer entry (94b) for Whitelackington mentions 7 swineherds, whereas
the Exeter version (*443*) records only 6. The total of 84 is tied to the Exeter
version.

[2] The Bath document sometimes calls *coceti* what the Domesday Book calls *bordarii*
(R. Lennard, 'A neglected Domesday satellite', *op. cit.* p. 37). See p. 137 above.

[3] See p. 194 below.

an estate or to some other definite date or period. Here are some examples:

Stogursey (*369*, 96b): *Quando recepit valebat xxv libras. Modo xx libras.*
Ilminster (*188*, 91): *Totum valet xx libras. Quando abbas obiit valebat xxvi libras.*
Chewton Mendip (*114b*, 87): *Reddit l libras ad numerum. T.E. reginae reddebat xxx libras.*

In contrast to the Exchequer version, that of the Exeter text almost invariably gives the earlier date as *quando recepit*. Here are the two versions for the valuation of Clapton in Gordano.

E.D. (*142 b*): *Valet per annum iii libras et x solidos et quando recepit valebat xl solidos.*
D.B. (88): *Valuit xl solidos. Modo lxx solidos.*

Very often when the Domesday Book states a value for only one date, the Exeter text gives two values, sometimes the same, sometimes different, as the following examples show:

Chaffcombe
E.D. (*136b*): *Valet xl solidos per annum et quando episcopus recepit valebat tantundem.*
D.B. (87b): *Valet xl solidos.*

Swell
E.D. (*268*): *Valet lx solidos et quando recepit valebat xl solidos.*
D.B. (92): *Valet lx solidos.*

Downscombe
E.D. (*430*): *Valet ii solidos et quando recepit erat vastata.*
D.B. (94): *Valet ii solidos.*

These and other discrepancies are noted in Appendix II.

The amounts are usually stated in round numbers of pounds and shillings, e.g. £6 or 100s. Occasionally, we hear of 20 pence to the ounce, e.g. at Somerton (*89b*, 86). Occasional precise figures, however, suggest either a sum of rents or careful appraisal, e.g. the £154 and 13d. at Taunton (*173b*, 87b) and the 13s. 2d. at Dinnington (*172*, 90b). In some entries the gold and the silver mark and even a render in kind (cheese, bacon-pigs, cows) appear:

Tintinhull (*266b*, 91 b): *Valet xvi libras. Drogo tenet de comite unam virgatam de ipsa terra et valet i markam argenti.*

Bath and Batheaston (*114b*, 87): *Istud burgum cum predicta Estone reddit lx libras ad numerum et unam markam auri.*

Anonymous holding (*436b*, 97): *Quando recepit valebat xviii libras. Modo reddit c caseos et x bacons.*[1]

Stogumber (*197*, 91): *Valet iii libras et iiii vaccas.*[2]

The account of Martock (*113*, 87) gives an unusual piece of information. The manor paid £70, and 100s. more if Bishop Walchelin had borne witness (*Reddit lxx libras ad numerum et c solidos plus si episcopus Walchelinus testatus fuerit*).

The Exeter text frequently gives more detail about the values of individual constituents of a manor than does the Exchequer version, e.g. in the account of the manor of Wells (*157b*, 89). Frequently, the two texts are at variance, but the discrepancies are usually small, e.g. 7s. 6d. for 8s. at Aller in Carhampton (*430b*, 94) and at Doverhay (*431*, 94), or 7s. 6d. plus 25s. (=32s. 6d.) for 32s. at Weacombe (*428b*, 93b). Occasionally, as in the entry for Laverton (*438b*, 96b), the two texts reverse the values of a holding for the dates for which they are given. In an entry for Broford (*429*, 93b) we seem to see the result of careless copying, for the 2s. 6d. of the Exeter text becomes 26d. in that of the Exchequer. Lastly, to the 60s. paid by Over Stratton (86), the Exeter text (*88b*) adds 24 sheep, presumably a customary render.

The 'comital manors' (*103–7*, 86b–87) deserve special mention here for three reasons.[3] In the first place, their values are entered as 'renders' (*reddere*, not *valere*, is the verb used). In the second place (with the exception of that for Langford Budville), the amounts were tested by assay; hence they are said to be 'of white silver'. The entry for Brompton

[1] This is a most unusual statement. The Exeter entry in which the statement also occurs is notable as being the work of an Exchequer clerk (see R. Welldon Finn, 'The evolution of successive versions of Domesday Book', *Eng. Hist. Rev.* LXVI, 1951, p. 563). The place-name is omitted in both versions and R. W. Eyton showed that Charlton Horethorne was implied—*op. cit.* I, p. 176. The Summary (*530b*) shows that the missing value for 1086 was £17—see p. 135 above.

[2] This may well be a scribal error. Possibly the rent was not paid in cows, but owing to an omission the demesne stock has appeared in the wrong place.

[3] The term 'comital manors' was first proposed by R. W. Eyton (*op. cit.* I, p. 78) and adopted by J. H. Round (*V.C.H. Somerset*, I, p. 397) and F. W. Maitland (*op. cit.* p. 167) to denote the fifteen manors which had passed into King William's hands by forfeiture from the House of Godwine.

Regis (*103*, 86b) is typical: *Reddit per annum xxvii libras et xii solidos et i denarium de albo argento.* The third feature is more strange. As Round noticed as long ago as 1895, ten out of the fifteen entries state values that are multiples of a unit of 23*s*.[1]

Somerset: The 'comital manors'

	£	s.	d.	Multiple
Crewkerne	46	0	0	40
Congresbury	28	15	0	25
Old Cleeve	23	0	0	20
North Curry	23	0	0	20
Henstridge	23	0	0	20
Queen Camel	23	0	0	20
Dulverton	11	10	0	10
Creech St Michael	9	4	0	8
Langford Budville	4	12	0	4
Capton	2	6	0	2

What Round did not note is the fact that the renders of at least three of the other five manors were very close to multiples of 23*s*. It is even conceivable that the values of the remaining two were multiples. Here are the details:

Winsford: If the value of the half-hide added to the manor is included, the total is £11. 10*s*. or 10 units.

Brompton Regis: The render was £27. 12*s*. 1*d*., which is only 1*d*. more than 24 units.

Hardington Mandeville: The render was £12. 14*s*. 0*d*.; only one shilling more than eleven units.

Nettlecombe: The entry is incomplete, the number of pounds being omitted; but the number of shillings is given as 12. If (as is very possible) the render was £4. 12*s*. 0*d*., that would make 4 units.

Coker: The render of £19. 1*s*. 0*d*. is wholly incompatible with a 23*s*. unit, but if one shilling was a mistake for 11*s*., the result would be 17 units.

This strange unit has not been explained, but it clearly formed the basis of a reckoning quite independent of the assessment, and bearing no constant relationship to the agricultural resources of these manors.

The renders of the royal manors which had been held by Queen Edith (*113–14*, 87) may owe their origins to a system which is not an obvious

[1] J. H. Round, *Feudal England*, pp. 114–15.

one. They are all said to be counted by tale (*ad numerum*), and the three for which we are given comparative figures show marked increases—Milverton from £12 to £25, Keynsham from £80 to £108, Chewton Mendip from £30 to £50.

We obtain a glimpse of an archaic system, which had almost disappeared, in the valuations of the twelve manors which comprised the Ancient Demesne of the Crown (*88b–91*, 86). In the time of King Edward, all but one of them had contributed to the *firma unius noctis* and had been grouped into units for this purpose, each unit contributing one *firma*. By 1086, these obligations had been commuted for a money payment at 20*d*. to the ounce (*de xx in ora*), a whole *firma* amounting to something over £100. The details are summarised on p. 171. No manor is specifically linked with Milborne Port as the complement of its three quarters of a *firma*. Bedminster, however, is the only manor of the ancient demesne for which no *firma* was recorded, and its valuation was £21. 0*s*. 2½*d*., the same amount as was paid by Cheddar. It is tempting to regard Bedminster and Milborne Port as constituting one unit in the same way that Somerton and Cheddar were combined, but whereas the other units comprised neighbouring manors, Bedminster (now in Bristol) and Milborne Port were separated by the whole width of a shire.[1] All twelve manors (including Bedminster) had never paid geld nor apparently had they ever been assessed in hides.[2]

Values remained the same or increased; under a fifth had decreased. Occasionally the variations are striking, e.g. the rise from £1 to £4 at Combe Hay (*492b*, 99) and from £10 to £24 at Westonzoyland (*162b*, 90), or the fall from £5 to £2 at Timberscombe (*442b*, 94b) and from 100*s*. to 30*s*. at Sutton Bingham (*444*, 94b). Some decreases may have been due to a variety of causes—crop failure, animal murrain or bad management. Thus the value of Porlock had fallen from £4 to 25*s*., and no teams were recorded for its 12 plough-lands (*315*, 93). The decrease in the value of some manors in the prosperous district to the south of the River Brue may have been connected with the rising against Montacute Castle in 1069. Some changes, however, may be more apparent than real, and may be explained by the fact that the composition of some manors had been altered between 1066 and 1086. Thus the values of two holdings at Barton St David had fallen from £6 to £3 and from 40*s*. to 30*s*.; both had lost land, one probably and the other certainly to Keinton Mandeville (*480*, 98b; *434b*, 94).

---

[1] Similar groups in Dorset were also widely separated.　　[2] See p. 150 above.

Commutation of the firma unius noctis (*88b–91b*, 86b)

	Render in 1086			Fraction	Firma commuted		
	£	s.	d.		£	s.	d.
North Petherton	42	8	4	$\frac{2}{5}$			
South Petherton	42	8	4	$\frac{2}{5}$	106	0	10
Curry Rivel	21	4	2	$\frac{1}{5}$			
Frome	53	0	5	$\frac{1}{2}$			
Bruton	53	0	5	$\frac{1}{2}$	106	0	10
Cannington Williton Carhampton	105	16	6¼	—	105	16	6¼
Somerton	79	10	7	?$\frac{4}{5}$			
Cheddar	21	0	2½	?$\frac{1}{5}$	100	10	9½
Milborne Port	79	10	7	?$\frac{4}{5}$			
Bedminster	21	0	2½	?$\frac{1}{5}$	See text		

Generally speaking, the greater the number of teams and men on an estate, the higher its value; but there is much variation, and it is impossible to discern any consistent relationship, as the following figures for five holdings, each rendering £2 in 1086, show:

	Teams	Population	Other resources
Adsborough (*356*, 95 b)	4½	21	Meadow, pasture, wood
Clatworthy (*357*, 95 b)	7	23	Mill, meadow, pasture, wood
Enmore (*432*, 94)	4	8	Wood
Preston Bermondsey (*467*, 99)	1	9	Meadow
Redlynch (*276*, 92 b)	2	14	Meadow, wood

It is true that the variations in the arable, as between one estate and another, did not necessarily reflect variations in total resources, but, even taking the other resources into account, the figures are not easy to explain.

[1] The Exeter Text records 3 villeins, 3 bordars, 2 serfs and 2 serfs. This is obviously an error and we have counted not 4 but 2 serfs.

Conclusion

For the purpose of calculating densities, Somerset has been divided into twenty-six units, based as far as possible upon the main physical and geological features. The boundaries of the density units have been drawn to coincide with those of civil parishes, and this gives the units an artificial character because many parishes extend across more than one kind of soil and country. The upland of the Mendips, for example, is divided between two units, one of which includes part of the low-lying Brue valley. Another difficulty arises from the fact that the details of composite entries have been divided equally among their constituent holdings. But even this rough solution is not possible for those manors covering a number of anonymous settlements that may not be contained within the same unit. The division into twenty-six units is therefore not a perfect one, but it may provide a basis for distinguishing broad variations over the face of the countryside.

Of the five standard formulae, those relating to plough-teams and the population are most likely to reflect something of the distribution of wealth and prosperity throughout the county. Taken together, certain common features stand out on Figs. 37 and 39. The most densely occupied areas of Somerset seem to have fallen into two groups. In the first place, there was the Vale of Taunton with its extension northwards around the Quantocks to the coast, a region of varied soils comprising Keuper marls and sandstones and Lias clays, all reasonably fertile. Secondly, there were the oolitic scarplands and adjoining areas along the eastern and northern borders of the county, extending from the neighbourhood of Crewkerne in the south towards Bath in the north. Over both these groups of districts there were between 10 and 12 people per square mile, and between 3 and 4 teams. The areas with densities below the average for the county, on the other hand, were the uplands and the lowlands; on the one hand, Exmoor, the Quantocks, the Mendips and the southern hills that are an extension of the Blackdown Hills of Devonshire; on the other hand, the low-lying levels of the Yeo, the Brue and the Parrett rivers. In these districts there were under 7 recorded people per square mile (sometimes very much under) and less than 2 teams.

Figs. 38 and 40 are supplementary to the density maps, but it is necessary to make reservations about them. As we have seen on p. 141, some Domesday names covered two or more settlements, and several of the symbols should thus appear as two or more smaller symbols. This limitation, although important locally, does not affect the main impression conveyed by the maps. Generally speaking, they confirm and amplify the information on the density maps.

WOODLAND

Types of entries

The Somerset folios give information both for wood itself and for underwood. The amount of each is normally expressed in one of two ways—in terms of linear dimensions or in terms of acres. There are also a number of variant entries.

Woodland is called *silva* in the Exchequer folios and *nemus* in the Exeter folios. The linear dimensions are stated in terms of leagues and furlongs and, very rarely, of perches.[1] The exact wording varies slightly, but the following examples indicate the kind of entry that is encountered:

Curry Rivel
    E.D. (*89*): *ii leugae nemoris in longitudine et i in latitudine.*
    D.B. (86): *Silva ii leuuis longa et una leuua lata.*

Frome
    E.D. (*90b*): *i leuga nemoris in longitudine et alia in latitudine.*
    D.B. (86b): *Silva i leuua longa et tantundem lata.*

Ilton
    E.D. (*191*): *i leuca nemoris in longitudine et alia in latitudine.*
    D.B. (91): *Silva i leuua longa et alia in latitudine.*

Swell
    E.D. (*268*): *v quadragenariae et x perticae nemoris in longitudine et ii quadragenariae in latitudine.*
    D.B. (92): *Silva v quarentenis et x perticis longa et ii quarentenis lata.*

Where the quantities are in furlongs, the number is usually less than the twelve commonly supposed to make a Domesday league, but Ashill (*268b*, 92), for example, had wood 40 furlongs by 20; and amounts of 18 furlongs, as well as 12, are also recorded. The largest amount entered under a single name is the 5 leagues by 1 league for Bruton (*90b*, 86b).[2]

---

[1] The two entries which mention perches are for Penselwood (*445*, 94b) and Swell (*268*, 92).
[2] Bruton was a royal manor, and the wood may possibly have been part of Selwood Forest.

The exact significance of these linear dimensions is far from clear, and we cannot hope to convert them into modern acreages.[1] All we can do is to plot them diagrammatically as on Fig. 41, and we have assumed that a league comprised 12 furlongs. One set of measurements (4 leagues by 2½) is found in the composite entry for the three royal manors of Cannington, Williton and Carhampton (*89*, 86b), the first-named being over a dozen miles to the east of the other two. This may imply, though not necessarily, some process of addition whereby the dimensions of separate tracts of wood have been consolidated into one sum. On Fig. 41 the amount has been plotted at Cannington.[2]

The Exeter text sometimes gives separate quantities for manorial components that are combined into a single amount in the Exchequer version. With one exception, the amounts involved are in acres, so that only simple addition is involved. The exception is the entry for Hornblotton, Alhampton and Lamyatt, where wood is measured in terms of linear dimensions as well as in acres. Here are the details:

E.D. (*169b*):
Hornblotton	Wood 4 furlongs by 1 furlong
Alhampton	Wood 5 acres
Lamyatt	Wood 3 furlongs by ½ furlong

D.B. (90b)   Wood 9 furlongs by 1½ furlongs

It seems impossible to reconcile the total with the sum of the components, and this impossibility defeats any attempt to find a clue to what arithmetic lay behind the ubiquitous formula '*m* furlongs by *n* furlongs'. The Exeter dimensions have been plotted on Fig. 41.

Fifteen entries state the amount of wood to be 'in length and breadth' or 'between length and breadth'. Eyton thought that these variant formulae indicated 'areal leagues' or 'areal furlongs'.[3] On the other hand, when the Exchequer version gives one form, the Exeter text occasionally gives another:

Banwell
E.D. (*157*): *ii leugae et dimidia nemoris in longitudine et in latitudine.*
D.B. (89b): *Silva ii leuua et dimidia in longitudine et latitudine.*

[1] See p. 372 below.
[2] Williton and Carhampton also had separate amounts of wood, which appear on Fig. 41.
[3] R. W. Eyton, *A key to Domesday...the Dorset survey* (London, 1878), pp. 31–5.

Eastrip

E.D. (*382b*): *i quadragenaria nemoris in longitudine et latitudine.*

D.B. (97b): *Silva i quarentena longitudine et latitudine.*

Curry Mallet

E.D. (*429*): *dimidia leuga nemoris in longitudine et latitudine.*

D.B. (93): *dimidia leuua silvae inter longitudinem et latitudinem.*

One is tempted to assume that the variant formulae imply the same facts as the more usual formula, and that they occur occasionally when the length and breadth of woodland are the same. We have followed this assumption in drawing Fig. 41.

Three entries give only single dimensions for wood. At Newhall (*464*, 96b) there was half a league, and at Nether Stowey the Exeter text (*373*) enters 1 league of wood that is omitted from the Exchequer version (97). That these entries also imply that the length and breadth were the same may be indicated by the Exeter version for Pitcott:

E.D. (*146*): *i quadragenaria nemoris inter longitudinem et latitudinem.*

D.B. (88b): *Una quarentena silvae.*

On Fig. 41, however, such entries have been plotted as single lines, except that for Pitcott itself, for which intersecting lines have been drawn.

The amount of wood is very frequently stated in acres. The figures range from one acre at, for example, Tadwick (*465*, 99) to as much as 300 acres at Porlock (*315*, 93) and 400 acres at Pitminster (*173b*, 87b), but the great majority do not rise above 30 acres. The figures are often in round numbers that suggest estimates, e.g. the 100 acres at Cranmore (*170b*, 90b); but many entries seem to be precise and the two holdings at Tellisford (*148*, 88b), for example, had 1½ and 4½ acres of wood. No attempt has been made to convert these figures into modern acres, and they have been plotted merely as conventional units on Fig. 41. Occasionally both acres and linear dimensions appear in entries for one place-name, e.g. for Tickenham, with 110 acres of wood (*438b*, 96b) and wood 3 furlongs by 1 furlong (*448b*, 98). The two types of measurement occasionally appear together in the same entry. At Lilstock (*478b*, 98b) there were in one place (*in uno loco*) 20 acres of wood, and in another place (*in alio loco*) woodland a league long and half a league broad. An entry for Dowlish (*136b*, 87b) speaks of a wood of 8 by 3 furlongs and 20 acres in addition (*insuper*); the Exeter text makes it clear that these were in different manors that had been added to Dowlish.

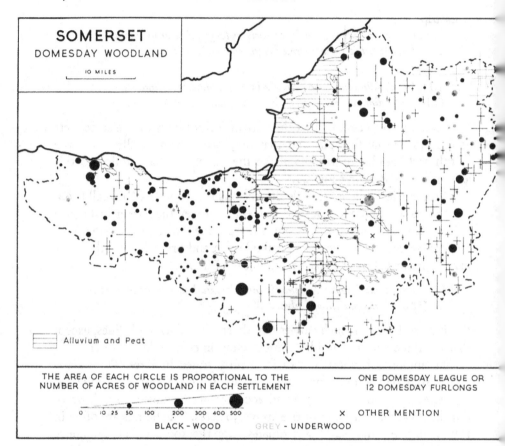

Fig. 41. Somerset: Domesday woodland in 1086.

When the wood of a settlement is entered partly in linear dimensions and partly in some other way, both are shown. When wood and underwood are entered for the same settlement, both are shown.

There are occasional entries in which the acre seems to have been used as a linear measure, e.g. at Kingweston (*283*, 91 b) for which we read: *Silva iii quarentenis longa et una acra lata.*[1] On Fig. 41 these have been indicated under the category of 'other mention'. There are also a few other unusual entries. The nameless holding of Edith the nun (*196*, 91 b) is said to have 80 acres of wood and pasture (*quater xx acrae silvae et*

[1] For the Dorset evidence, see p. 101 above.

pasturae), and at Bathwick and Swainswick (*492b*, 99) there were 10 acres of spinney (*spinetum*). Finally, we hear of a park (*parcus*) at Donyatt (*270*, 92), which has not been plotted on Fig. 41.

A number of entries mention underwood—*nemusculus* in the Exeter text and *silva minuta* in that of the Exchequer. In three entries the Exchequer version speaks of *silva modica*, which also appears in the Exeter text as *nemusculus*.[1] It would seem that *silva modica* and *silva minuta* were interchangeable. Occasionally what is called *nemusculus* in the Exeter text appears as *silva* in that of the Exchequer, e.g. in the entry for Seavington (*265b*, 91b), and such entries have been plotted as underwood on Fig. 41. The amounts are usually in acres, and range from a single acre at Bleadon (*173b*, 87b) to as much as 300 acres on the manor of Glastonbury (*172*, 90). Two entries mention underwood and wood together:

> Glastonbury (*172*, 90): *xx acrae silvae et ccc acrae silvae minutae.*
> Norton Malreward (*140b*, 88): *vi acrae silvae minutae et una leuua silvae in longitudine et tantundem in latitudine.*

There were other vills with record of both wood and underwood, but only in separate entries. The Exchequer text occasionally adds the wood and underwood of the Exeter text and records the total as entirely wood. Thus the Exchequer entry (88b) for Rode speaks of 33 acres of wood which in the Exeter version (*148*) appear as 2 + 6 + 13 acres of wood and 12 acres of underwood.[2] One version occasionally records as wood what the other calls underwood, e.g. for Halswell (*443*, 94b), the Exchequer 14 acres of wood appear as 14 acres of underwood in the Exeter version. The underwood on eleven holdings was measured in linear terms. Three give two dimensions as normally; seven speak of 'in length and breadth' or 'between length and breadth'; the amount entered for Batheaston (*114*, 87) was as much as 2 leagues in length and breadth (*in longitudine et latitudine*). There are occasional variants, as may be seen in two versions of an account of a holding at Bathford:

> E.D. (*185b*): *i leuga nemusculi in longitudine et latitudine.*
> D.B. (89b): *una leuua silvae minutae inter longitudinem et latitudinem.*

[1] Claverham (*141*, 88), Milverton (*113*, 87) and Winterhead (*140*, 88). It is interesting to note that *nemusculus* is sometimes rendered as *pascua* in the Bath document (R. Lennard, 'A neglected Domesday satellite', *op. cit.* p. 37).

[2] Other examples of such addition can be seen in the entries for Taunton (*174b*, 87b) and for North Wootton and Pylle (*166b*, 90b).

Such entries, like the corresponding ones for wood itself, have been plotted as intersecting lines. Finally, the account of Odcombe (*274*, 92b) speaks only of one furlong of underwood, and this has been plotted as a single line.

Distribution of woodland

As Fig. 41 shows, much of Somerset was still well wooded in 1086. Most of the villages of the oolitic scarplands, for example, had considerable amounts of wood entered for them, although some of this may well have been actually situated on the heavy clays to the east and the west. The northern part of the scarplands towards Bath, however, was poor in woodland proper but well provided with underwood. Wood was plentiful over most of the Lias Clay plain, particularly in the west where there were such extensive tracts as the 3 by 1½ leagues at Ilminster (*188*, 91) and as much at Isle Abbots (*188*, 91); but the middle part of the plain, the district around Ilchester, seems to have been largely treeless.

There was fairly extensive wood among the hills and valleys of north-eastern Somerset, and, although the Mendip upland itself was apparently an empty area, the villages along its flanks seem to have been well endowed with wood, e.g. Wells with 2 leagues by 2 furlongs (*157b*, 89) and Winscombe with 2 by 1 leagues (*161*, 90), but we have to remember that these are complex manors and that individual holdings may not have had much wood. In the west of the county, beyond the Parrett, about three quarters of the villages were credited with some wood, though the quantities were small, frequently not more than 4, 6, 10 or 12 acres, and rarely above 30 acres. These were to be found in the northern valleys between the Exmoor upland and the coast, in the depression between Exmoor and the Quantocks, and on the north-east flanks of the Quantocks reaching down to the lower Parrett.

The exceptions to the wooded nature of the landscape were the lowland marshes and the upland moors. The most extensive of these were the marshes of the Parrett and the Brue in central Somerset, with an extension to the north-east into the valleys of the Yeo and the Kenn. Apart from a few isolated woods on the islands formed by the higher ground, the whole area seems to have been devoid of woodland proper, but a tongue of underwood ran out along the length of the Polden Hills. The Exmoor upland itself, as might be expected, was without wood except in the sheltered valleys of the Exe and its tributary the Barle.

No Somerset forest is mentioned by name nor are any foresters or huntsmen, except that three foresters had held Withypool in Exmoor in 1066 (*479*, 98b), and Ailward, who had held Pendomer (not far from Selwood Forest) in 1066, is described in the Exeter text as a huntsman (*275*, 92b).

MEADOW

Types of entries

The entries for meadow in the Somerset folios are comparatively straightforward. For holding after holding, the same phrase is repeated, '*n* acres of meadow' (*n prati acrae*). The amount of meadow entered under each place-name ranges from only half an acre each at Chipstable (*188*, 91), Hone (*431*, 94) and *Pantesheda* (*433*, 94) to 300 acres at Wells (*157b*, 89) and 250 acres at Congresbury (*106*, 87)—but these were giant complex manors—and 150 acres at each of the two Burnham manors (*354*, 95). Not many entries record more than 100 acres, and some of these, as we have seen, are for complex manors or in joint entries such as that for Cannington, Williton and Carhampton (*89*, 86b) with its 104 acres. By far the majority of settlements had under 30 acres apiece. Round figures such as 20 or 50 acres suggest that they are estimates, and it is noticeable that the figures for meadow are often the same as those for pasture or wood or both. At Queen Camel (*106b*, 87), for example, meadow, pasture and wood were estimated at 100 acres each; and at Keynsham (*113*, 87) there were 100 acres each of meadow and pasture. But there are also detailed figures that suggest precision, e.g. the 4½ acres at Langridge (*144*, 88b), the 26¼ acres at Draycott in Limington (*267*, 92) and the 107 acres at Clutton (*140*, 88). Comparison of the Exchequer and Exeter texts reveals occasional discrepancies, but usually only of a few acres, e.g. between 15 and 16 acres for Stoke Trister (*277*, 92b) and between 10 and 15 acres for Chard (*156*, 89). As for other counties, no attempt has been made to translate these figures into modern acreages, and they have been plotted as conventional units on Fig. 42. The meadow in a composite entry has been divided equally among the places involved.

There are two unusual entries. One, for Abbas and Temple Combe (*193b*, 91), is the sole entry for Somerset in which meadow is recorded in terms of linear dimensions, the amount being 4 by 2 furlongs. The other entry is the '25 acres of moor and meadow' at Seavington which had been transferred to the nearby manor of South Petherton: *De hoc manerio*

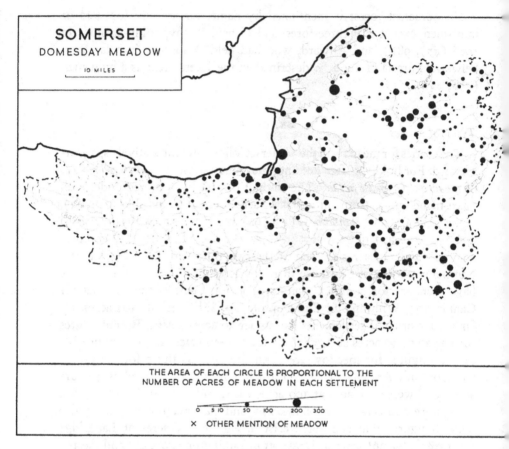

Fig. 42. Somerset: Domesday meadow in 1086.

The meadow of Abbas and Temple Combe, in the south-east, is measured partly in acres and partly in some other way; both amounts are shown. Areas of alluvium and lowland peat are shown in grey; rivers passing through these areas are not marked.

sunt ablatae x acrae silvae[1] et xxv acrae morae et prati et sunt in Sudperet manerium regis (265b, 91b).[2] Finally, of the 130 acres of meadow in the manor of Sherborne in Dorset (77), three were said to be in Somerset near Milborne Port, just across the border (iii acrae sunt in Sumersete juxta Meleburne), and they have been included in the total for Milborne Port.

[1] The Exeter text refers to x agrae nemusculi.
[2] For the purpose of plotting on Fig. 42, only 12½ acres of meadow have been counted.

Distribution of meadowland

Meadow was widely distributed throughout most of lowland Somerset, especially on the Lias Clay plain with its high water-table (Fig. 42). The extension of Lias Clay along the length of the Polden Hills was likewise marked by villages with fair amounts of meadow—often 50 acres or over. Meadow was also abundant in the valleys of the oolitic scarplands and in those of northern Somerset. It was not so abundant in the Taunton region, but to the north between the Quantocks and the lower Parrett most villages had some meadow, often small in amount, e.g. 1 acre at Petherham (*424b*, 93 b) and 3 at Clayhill (*422b*, 93), but occasionally over 50 acres, e.g. at Combwich (*282*, *462b*; 91 b, 97).

Comparatively little meadow was entered for the villages in the Exmoor region. Villages here had either none at all or only small quantities. Thus in the valleys of the Tone, the Exe and the Barle there were half-acres at Chipstable (*188*, 91) and Hone (*431*, 94) and a single acre at Ashway (*428b*, 93 b). Amounts were also very frequently below ten acres in the northern valleys between the Exmoor upland and the coast and in the depression between Exmoor and the Quantocks. The villages of the southern hills likewise had little or none.

PASTURE

Types of entries

Pasture is recorded for just over one half of the settlements of Somerset. It is entered in the Exeter text as *pascua*, and in the Exchequer version as *pastura*. Its amount is usually recorded in one of two ways—either in linear dimensions or, much more frequently, in acres. The linear dimensions are given in terms of leagues and furlongs. The following examples from the Exchequer text indicate the kind of entry that is encountered:

Cheddar (*90*, 86): *Pastura i leuua longa et tantundem lata.*
Clapton in Gordano (*142*, 88): *Pastura xviii quarentenis longa et iii quarentenis lata.*
Somerton (*89*, 86): *una leuua pasturae in longitudine et dimidia leuua latitudine.*
Staple Fitzpaine (*270*, 92): *Pastura dimidia leuua longa et una quarentena lata.*

The exact significance of these linear dimensions is far from clear, and we cannot hope to convert them into modern acreages.[1] All we can do is to

---

[1] See p. 372 below.

plot them diagrammatically as on Fig. 43, and we have assumed that a league comprised 12 furlongs. Pasture 5 leagues by 3 appears in the combined entry for the three royal manors of Cannington, Williton and Carhampton (*89*, 86b) and it has been plotted at the first-named place.[1] There is likewise a total of 3 furlongs by 1 in the Exchequer combined entry for Shepton Mallet and Croscombe (90), but the Exeter version ascribes the wood to Shepton Mallet alone (*166*). The measurements (3 by 1 leagues) in an entry for a large manor such as Wells (*157b*, 89) may possibly cover a number of separate places and may imply, although not necessarily, some process of addition whereby the dimensions of separate tracts have been consolidated into one sum.

Fifteen entries state the amount of pasture to be 'in length and breadth' or 'between length and breadth'. Judging from occasional variations between the Exeter and the Exchequer texts, the formulae seem to be interchangeable and, moreover, to imply the same facts as the more usual statement. At any rate that is what the following examples seem to indicate:

Pendomer
E.D. (*275*): *iiii quadragenariae pascuae inter longum et latum.*
D.B. (92b): *iiii quarentenae pasturae in longitudine et in latitudine.*
Winscombe
E.D. (*161*): *i leuga pascuae in longitudine et in latitudine.*[2]
D.B. (90): *una leuua pasturae in longitudine et latitudine.*

There is thus no need to invoke the 'areal leagues and furlongs' of Eyton.[3] All these entries have been plotted in the normal manner as intersecting lines.

Seven entries give only one dimension.[4] The Exchequer and Exeter versions agree, with the exception of those for Pitcott:

E.D. (*146*): *ii quadragenariae pascuae inter longitudinem et latitudinem.*
D.B. (88b): *ii quarentenae pasturae.*

[1] There was pasture on another holding at Carhampton which has been indicated on Fig. 43. Williton had no other pasture and is shown under the category of 'other mention'.
[2] A similar variation is encountered in entries for Banwell (*157*, 89b), Merriott (*271b*, 92), Newton in Bicknoller (*361*, 96), and Haselbury Plucknett (*492*, 99).
[3] See p. 174 above.
[4] The seven entries are for Chilcompton (*154*, 87b), Pitcott (*146*, 88b), Stoke-sub-Hamdon (*267b*,92), Brompton Ralph (*357*,95b), Alcombe (*358*,95b), Merriott (*491b*, 98b) and North Petherton (*88b*, 86).

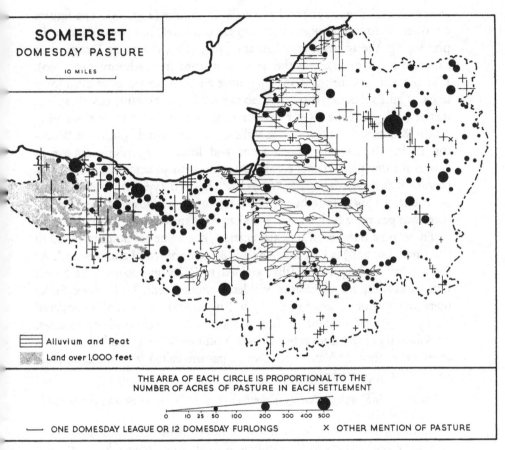

SOMERSET
DOMESDAY PASTURE

10 MILES

Alluvium and Peat
Land over 1,000 feet

THE AREA OF EACH CIRCLE IS PROPORTIONAL TO THE
NUMBER OF ACRES OF PASTURE IN EACH SETTLEMENT

0   10 25 50   100   200   300 400 500

—— ONE DOMESDAY LEAGUE OR 12 DOMESDAY FURLONGS   ✕ OTHER MENTION OF PASTURE

Fig. 43. Somerset: Domesday pasture in 1086.

When the pasture of a settlement is entered partly in acres
and partly in some other way, both are shown.

Eyton thought that these one-dimension entries also indicated areal
quantities, but one is tempted to assume that they imply the same facts
as the more usual formula when length and breadth are equal.[1] On Fig. 43,
however, they have been plotted as single lines, that is with the exception
of that for Pitcott.

Measurements in acres are much more frequent. Amounts are generally
below 30 acres, and range from 2 acres at Moreton (*453b*, 98) to 1,000

[1] See p. 101. See also p. 175 for the Pitcott wood measurements.

acres at Litton (*160*, 89b) and 550 at Withycombe (*139*, 88). The figures are often in round numbers that suggest estimates, but some seem to be precise, e.g. the 19 acres at Stogursey (*369*, 96b), the 31 at Draycott in Limington (*267*, 92), and the 43 at Blackford in Wedmore (*163*, 90). Like the acreages for meadow, they have all been treated as conventional units. The Exchequer text gives 20 acres for the constituents of Pilton manor (90b), but the Exeter version (*166b*) shows that these were all at North Wootton and none was at Pylle or on the second holding at Wootton itself. Occasionally, both acres and linear dimensions appear in entries for one place-name, e.g. at Quarme, with 50 acres (*473b*, 98) and pasture 5 furlongs by 5 (*358b*, 95b). The two types of measurement occasionally appear together in the same entry. At Dowlish (*136b*, 87b) there was pasture 4 furlongs by 4, and 20 acres in addition (*xx acrae plus*).[1]

There are a number of unusual entries. The nameless holding of Edith the nun (*196*, 91b) is said to have had '80 acres of wood and pasture'. At Exford half a plough-land (that yielded 12*d.*) lay in pasture.[2] At North Petherton 2 leagues of pasture yielded a money rent. At Rodney Stoke there was pasture 2 leagues by 1, and in addition as much as rendered 2*s.* a year. At Whitestaunton four blooms of iron were paid for 50 acres. At Kilton there was a virgate of pasture on one holding in addition to 60 acres on another. At Yatton there was a pasture called Wemberham. Here are the Exchequer entries:

> Exford (*359b*, 95b): *Terra est dimidia caruca. Sed jacet in pastura et reddit xii denarios.*
>
> North Petherton (*88b*, 86): *ii leuuae pasturae reddunt xx solidos per annum.*
>
> Rodney Stoke (*137*, 87b): *Pasturae ii leuuis longa et una leuua lata et ii solidi desuper plus.*[3]
>
> Whitestaunton (*265b*, 91b): *l acrae pasturae reddunt iiii blomas ferri.*
>
> Kilton (*360b*, 96): *lx acrae pasturae...una virgata pasturae.*[4]
>
> Yatton (*159b*, 89b): *Una pastura Waimora dicta ibi est quod T.R.E. pertinebat ad Congresberie manerium regis.*[5]

[1] The Exeter version gives the breadth as 3 furlongs (plotted on Fig. 43), and shows that the pasture was at separate manors combined by 1086 with that of Dowlish.

[2] One of the other four holdings at Exford had but 1 team, 1 bordar and 1 serf, and it was worth only 3*s.* It had been received by Roger de Courseulles in a completely waste condition—see p. 189 below.

[3] The Exeter text reads: *tantum pascuae quod reddit ii solidos per annum.*

[4] The contraction in both texts is *virg'*. Could the entry imply one perch of pasture?

[5] The Exeter text reads: *i pascua quae vocatur Weimorham quae iacebat in mansione regis de Congresberia.* Yatton and Congresbury are adjacent parishes.

Two other entries, for Hardington and for Hemington, have unusual interest for they take us behind the bare Domesday figures and show us an arrangement for intercommoning between two adjacent villages and, moreover, two villages in different fiefs. Here are the Exchequer entries:

Hardington (*147*, 88b): *In hoc manerio est una hida pertinens ad Hamintone. Balduuinus tenet et habet communem pasturam huic manerio.*
Hemington (*315*, 93): *De hac terra i hida est in communi pastura in Hardintone.*

The entry for Hemington also records other pasture half a league by half a league, but evidently the men of the village also had the right to send their livestock to graze with those of the neighbouring village of Hardington.

Distribution of pasture

Fig. 43 shows that those villages most plentifully endowed with pasture were situated close to upland areas unsuitable for tillage. Thus most of the villages in the valleys, and on the seaward slopes of Exmoor, had some pasture, and amounts were often large. At Winsford (*104b*, 86b), for example, there were 4 by 2 leagues; at Brompton Regis (*103*, 86b) there were 3 leagues by 1 league, and at Almsworthy (*430*, 94) there were 2 by 2 leagues. Near the coast, Minehead (*358*, 95b) had 4 by 2 leagues. The Quantock Hills, too, seem to have provided abundant pasture for the villages on their slopes. The third group of villages with large amounts lay along the northern and southern slopes of the Mendips. Here were such manors as Wells (*157b*, 89) with 3 leagues by 1 league of pasture, Chewton Mendip (*114b*, 87) with 2 leagues by 1 league, Cheddar (*90*, 86) with 1 league by 1 league, and many more with only slightly smaller amounts.

Apart from one area, pasture was moderately plentiful over the rest of the county. The exception was the low-lying country around the middle and lower courses of the Parrett and the Brue where, in any case, there were but four villages.

MARSH

There are eleven references to marsh or moor (*mora*), of which six are measured in acres and four in linear dimensions. Moor is linked with

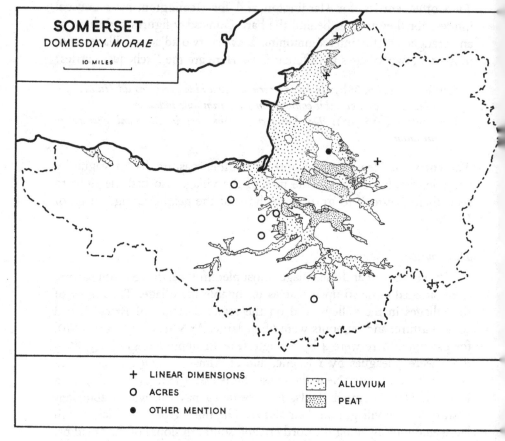

Fig. 44. Somerset: Domesday *morae* in 1086.

meadow in the entry for Seavington. The Exchequer text omits that at
Wedmore. The full list is as follows:

Adsborough (*356*, 95 b)	10 acres
Fiddington (*441 b*, 94 b)	43 acres
Huntworth (*371 b*, 97)	10 acres
North Newton (*477 b*, 98 b)	20 acres
Seavington (*265 b*, 91 b)	25 acres of moor and meadow[1]
Tuxwell (*441 b*, 94 b)	41 acres

[1] See p. 179 above.

Milborne Port (*91*, 86b)	1 league
Weston in Gordano (*142b*, 88)	6 furlongs
Wells (*157b*, 89)	3 leagues
Yatton (*159b*, 89b)	1 league in length and breadth
Wedmore (*159b*, 89b)	Besides these (wood, meadow, pasture) are marshes there which render nothing (*praeter haec sunt ibi morae quae nichil reddunt*)

It is surprising to find so few references to the Somerset marshes. It has been suggested, however, that the increase in the values of some marshland manors may have been due to reclamation.[1]

With one exception, the villages with recorded moor were not far from the central alluvial and peat area with its extension north-eastwards (Fig. 44). The exception was Milborne Port in the south-east of the county. Its league of moor may have been at Yeovil Marsh along the headwaters of the river Yeo, or in the Somerset Levels.

FISHERIES

Fisheries are mentioned less frequently than might have been expected in view of the extensive marshes of the shire. They appear in connection with only ten places; some rendered money and one rendered eels (Fig. 45). Holdings at six places were on royal manors; the rest belonged to ecclesiastical foundations. The complete list, in the Exchequer version, is as follows:

Cheddar (*90*, 86): *iii piscariae reddunt x solidos.*
Creech St Michael (*104b*, 86b): *Ibi est piscaria sed non pertinet ad firmam.*
Martock (*113*, 87): *Piscaria reddit v solidos.*
Meare (*172*, 90): *x piscatores et iii piscariae reddunt xx denarios.*
North Curry (*105*, 86b): *Ibi est piscaria sed non pertinet ad firmam.*
Somerton (*89b*, 86): *ii piscariae reddunt x solidos.*
Wedmore (*159b*, 89b): *ii piscariae reddunt x solidos.*
Muchelney, Midelney and Thorney (*189*, 91): *Ibi ii piscariae reddunt vi millia anguillarum.*

They appear to have been inland river fisheries, a conclusion at any rate suggested by the one reference to an eel render. In addition to these specific references to fisheries, we hear that to the manor of Staple Fitzpaine (*270*, 92) belonged a garden in Langport which rendered 50 eels in

[1] W. G. Hoskins, *Fieldwork in Local History* (London, 1967), pp. 166–7.

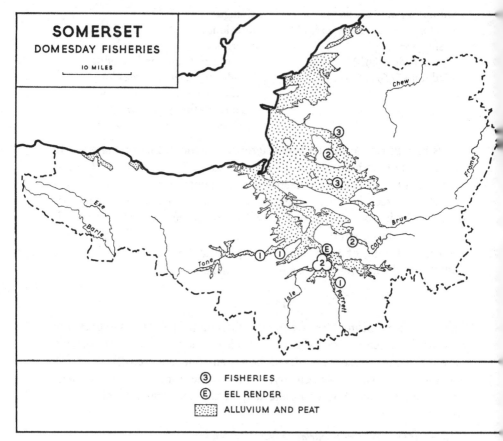

Fig. 45. Somerset: Domesday fisheries in 1086.

The figure in each circle indicates the number of fisheries. The three islands of Muchelney, Midelney and Thorney had two fisheries between them. Rivers flowing across the alluvium and peat are not marked.

rent (*Huic manerio pertinet unus ortus in Langport reddens l anguillas*). Fig. 45 shows how the fisheries were located in and around the Somerset Levels.

WASTE

A minor but interesting feature of the Somerset folios is the occasional reference to waste. Two holdings seem to have been waste, or almost so, in 1086—at Stone and Treborough, but another five had previously been waste. For three of the holdings, waste is mentioned in both texts; but for the other four, a reference to waste occurs only in the Exeter text, and for Capland there is a reference both in the main text and among *Terrae Occupatae*.[1] The entries are set out below. We might also remember those other holdings without teams or people, yet for which no waste was specifically entered (Fig. 82).

Domesday Waste in Somerset

Stone in Exford

    E.D. (*431 b*): *Hanc possunt arare ii carrucae sed semper fuit vasta postquam Rogerus recepit.*

    D.B. (94): *Terra est ii carucis sed vasta est.*

Treborough

    E.D. (*463b*): *de eo quod devastata non valet nisi vii solidos et quando recepit tantundem.*

    D.B. (97): *Valet vii solidos. Nam vastata est.*

Wearne

    E.D. (*479 b*): *valet xv solidos per annum et quando Robertus recepit fuit vasta.*

    D.B. (98 b): *Valet xv solidos. Vastam accepit.*

Almsworthy (*430*): *valet xxv solidos et quando recepit erat penitus vastata.*

Downscombe (*430*): *valet ii solidos et quando recepit erat vastata.*

Exford (*430*): *valet iii solidos et quando recepit erat penitus vastata.*

Capland (*491 b*): *valet vi solidos et quando Hardinc recepit fuit vastata.*

Capland (*517b*): *valet per annum vi solidos et quando Hardingus eam accepit erat vastata.*

The wasted holdings had recovered, or at any rate recovered sufficiently to be of value in 1086. Almsworthy had as many as 6 villeins, 9 bordars and 2 serfs with 4 teams, but the other holdings were small. Treborough had a villein with a team at work. Downscombe had a bordar with only half a team. Wearne had a bordar and a serf but no teams. Capland had a bordar and a serf with one team. The holding at Exford had 1 serf and a bordar with half a team; but three other Exford holdings also had a villein, 3 bordars and a team between them (*430, 431 b, 359 b*; 94 *bis*, 95 b); and there was a fifth holding of 1½ ferlings which lay in pasture: *jacet in pastura* (*359 b*, 95 b).[2]

---

[1] The Geld Accounts (*78 b*) include a section dealing with 'part of the land of Giso which belongs to the honour of his bishopric', and it contains a reference to 4½ hides which the bishop's villeins hold uncultivated (*vacuas*). This suggests land which had gone out of cultivation, but the site cannot be identified.   [2] See p. 184 above.

The waste of these entries was presumably not the waste of moor and heath but land that had gone out of cultivation. It may be significant that Almsworthy, Downscombe, Exford and Stone were all on Exmoor (in the upper valley of the Exe). It seems as though there had been some local tragedy or as if marginal land had gone out of cultivation. A fifth place—Treborough—was also in an upland area, in the Brendon Hills some 10 miles to the east.

Finally, to complete the picture of Somerset waste, we must note that there were 7 waste houses in Bath (*114b*, 87).[1]

MILLS

Mills are mentioned in connection with 250 out of the 611 Domesday settlements in Somerset; they also appear in entries for two anonymous holdings.[2] The only mills allotted to a group of villages are the 2 in the joint entry for Carhampton, Williton and Cannington (*89*, 86b), and 1 mill each has been allotted to the last two, as providing the most probable sites. But there are other entries that cover two or more unnamed places, and we cannot apportion the mills among these places, e.g. the 9 mills in the entry for the large manor of Wells (*157b*, 89). The number of mills on a holding is stated, and normally their annual value, ranging from a mill at Swang (*425b*, 93b) that rendered only 3*d.* to one at Bathwick and Swainswick (*144*, 88b) which was worth as much as 35*s.* a year. This was a very high render, and more than one half of the Somerset mills paid only 5*s.* or under. Such sums as 20*d.* (i.e. the *ora*), 30*d.*, and 40*d.* were frequent, so were 5*s.*, 10*s.*, 20*s.* Renders of uneven amounts were rare, e.g. the mill at Bishop's Lydeard, which paid 31*d.* (*157*, 89), and the two at Englishcombe which paid 11*s.* 7*d.* (*146*, 88b). The 4 mills on three holdings at Lexworthy (*282*, *432*, *432b*; 91b, 94 *bis*) paid a total of 6 blooms of iron (*plumbae ferri*).[3] We are specifically told in the Exchequer text that eight mills did not pay rent (*sine censu*);[4] and the Exeter entries describe them

[1] See p. 199 above.

[2] They were: (1) *Terra Alwini* (*424b*, 93b), and (2) Roger de Courseulles's *mansio* (*167b*), which appears in the Exchequer text merely as *alibi* (90b).

[3] This is the only example in the Domesday Book of mills rendering blooms of iron. It has been suggested that this entry implies the use of water-power at forges (L. White, *Medieval Technology and Social Change*, Oxford, 1962, p. 84).

[4] They were at Combe Sydenham (*362b*, 96), Donyatt (*270*, 92), Shepton Montague (*276*, 92b), Nether Stowey (*491*, 99), Sutton Montis (*276*, 92b), Torweston (*361*, 96), Whitestaunton (*265b*, 91b), and Willett (*361b*, 96).

as mills which ground their own grain (*qui molit annonam suam*). The phrase is found also in the Exeter account of Raddington (*442*) and it is rendered in the Exchequer version (94b) as 'grinding for the hall' (*ad aulam molens*). In each of the villages of Nether Stowey (*373, 491*; 97, 99) and at Shepton Montague (*276*, 92b) there were pairs of mills, one of which worked solely for the manor while the other paid a rent. At Shepton Montague both were on the same holding and are described in the Exchequer version thus: *Ibi ii molini, unum sine censu, alterum* [sic] *reddit vii solidos et vi denarios*. The Exchequer version of Sampford Brett (91b) leaves a gap where the render of its mill should have been inserted, and the text of the Exeter account (*286b*) seems confused at this point.

Fractions of mills are sometimes encountered, and it is often possible to combine these fractions to show that neighbouring estates shared a common mill. Thus in each of two of the four holdings at Moreton the Exeter text records half a mill paying 30*d.* (*453b*) and this appears as 1 mill in the Exchequer text (98). It seem probable that the half-mill at Woodwick (*186b*, 89b) was matched by the half at Freshford nearby (*144*, 88b), for each yielded 5*s.* But the halves of mills did not always pay equal sums. Thus of those at Tellisford (*148*, 88b), one half paid 7*s.* 6*d.*, and the other paid 9*s.*[1] The figures for the half-mills at Weston Bampfylde (*278, 383*; 92b, 97b) are 30*d.* and 45*d.* The mills at Rode provide an interesting example of division. The Exchequer version (88b) says merely that an unspecified number paid 27*s.* (*de molinis exeunt xxvii solidi*). The Exeter account (*148*) ascribes the following fractions to five of the six holdings:

half a mill rendering	6*s.*
quarter of a mill rendering	3*s.*
one sixth of a mill rendering	30*d.*
'two parts' of 2 mills rendering	8*s.*
half a mill rendering	7*s.* 6*d.*

We cannot therefore say for certain how many mills there were, and we have taken the number as 2.[2] There are some fractions which, apparently, cannot conveniently be combined with others, e.g. the half-mill yielding 10*d.* at Gothelney (*433b*, 93). There are also occasional discrepancies

[1] Could the original information have said 90*d.*, not 9*s.*?
[2] This is discussed in R. Welldon Finn, *An Introduction to Domesday Book* (London, 1963), p. 190.

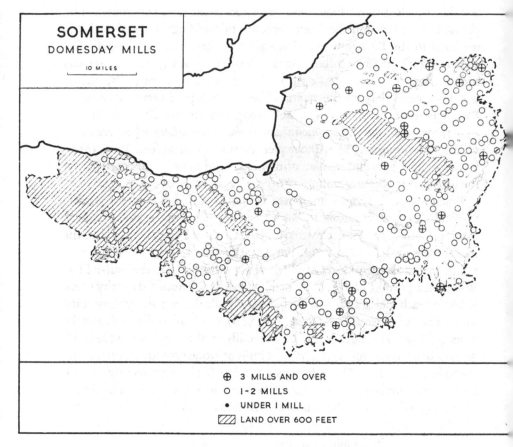

Fig. 46. Somerset: Domesday mills in 1086.

There were also mills on two anonymous holdings—see p. 190. Areas of alluvium and lowland peat are shown in grey; rivers passing through these areas are not marked.

between the two texts. The mill yielding 20s. at Chilton Trivett in the Exeter text (*425*) appears only as half a mill yielding 20s. in the Exchequer version (93b); and the mill which the Exeter text records for Pitminster (*173b*) goes completely unmentioned in the Exchequer version (87b).

There are a number of anomalies, or apparent anomalies, about the incidence of these mills. The mill at Sutton Bingham (*444*, 94b), without recorded teams, paid as much as 16s. Crandon (*485b*, 99), too, had a small mill yielding 30d. although it had no team at work. Conversely, there were

large and seemingly prosperous manors for which no mills were recorded,
e.g. North Curry (*105*, 86b) with 39 teams or Somerton (*89*, 86) with as
many as 49.

Most Somerset settlements with mills had 1 or 2 apiece, but the table
below cannot be very accurate, because a group of mills in a large manor
may have been distributed among its members in a way unknown to us:

Domesday Mills in Somerset in 1086

Under 1 mill	3 settlements	5 mills	4 settlements
1 mill	176 settlements	6 mills	1 settlement
2 mills	52 settlements	7 mills	2 settlements
3 mills	6 settlements	8 mills	1 settlement
4 mills	4 settlements	9 mills	1 settlement

The group of 9 mills was entered for the large manor of Wells (*157b*,
89), which included a number of unnamed places. The group of 8 mills
belonged to Taunton (*173b*, 87b), another large manor covering many
places. The two groups of 7 were in the manors of Bruton (*90b*, *434b*;
86b, 94) and Milborne Port (*91*, *278b*; 86b, 93). The group of 6 was at
Keynsham (*113b*, 87), and the four groups of 5 were at Batheaston (*114*,
186, *465*; 87, 89b, 99), Chew Magna (*159*, 89b), Chewton Mendip (*114b*,
87), and Crewkerne (*105b*, *197*; 86b, 91). These examples illustrate the
unsatisfactory nature of the table above. We might go so far as to say that
a large number of mills entered under one name creates a suspicion that
more than one settlement was involved. This is not to say that the table is
without value, because, at any rate, it suggests that there were some
villages with more than one mill at work.

Fig. 46 shows how the mills were associated with the rivers, and they
were most frequent along those rivers flowing through the most extensive
areas of arable land—along the Tone and its tributaries in the Taunton
area, along the Chew, the Frome, and other tributaries of the Bristol
Avon in the north-east of the county, along the Brue, the Parrett, the
Isle and their tributaries in the oolitic belt. The Quantocks and the
Mendips each had a line of mills in the spring-line villages at its foot.
Among the western hills, there were few mills, owing partly to the nature
of the moorland rivers and partly to the small amount of cultivated land.
Such mills as were found here were very often very small, e.g. those at
Clatworthy (*357*, 95 b) and at Winsford (*104b*, 86b) each yielded only 6*d*.

The central marshlands with their arable land limited to the Polden Hills
and to a few knolls and islands also had very few mills, but some situated
on the edge of the marsh were unusually profitable, e.g. two mills at
Weare (*350b*, 95), to the south of Axbridge, yielded as much as 42*s*.

CHURCHES

Churches are not regularly enumerated in the folios for Somerset, and
they are mentioned in connection with but 17 out of the 611 places recorded for the county. The relevant entries are set out on p. 195. The
churches at Horsey Pignes and Kilmersdon are mentioned only in the
Exeter text. The total number of churches in Somerset in 1086 must have
been very much more than 17. We can assume, for example, that there was
a church at Taunton, for 'the priests of Taunton' are mentioned in the
Geld Accounts (*75*).

Eight churches are named in a special section of the Exchequer text
(*Clerici tenentes de rege*) sandwiched between the accounts of the ecclesiastical and lay tenants-in-chief (91–91b). The analogous section in the
Exeter text (*Terrae quae datae sunt Sanctis in elemosina in Sumerseta*) also
records a church at Kilmersdon (*196–8b*). The endowments of the Somerset churches ranged from those of 1 hide or under, yielding 10*s.* to 40*s.*
a year, up to that at Crewkerne assessed at 10 hides and yielding £7, and
to that with eight *carucatae terrae* at Frome yielding £6.

Priests were entered for three of these churches—at Carhampton,
Curry Rivel and Horsey Pignes. The Exchequer entry (86b) for Frome
says 'Reinbald is priest here', but the Exeter version (*90b*) shows him to
be a landholder and he was probably the Reinbald who also held churches
elsewhere.[1] Other priests who were sub-tenants have also been excluded
from our count: the priest at Evercreech (*158*, 89b), Ralf the priest at
Thorne (*279b*, 93), Alviet the priest at South Petherton (*196b*, 91b), Guy
and another priest at Long Ashton (*143b*, 88b), Reinbald the priest at
Milborne Port (*91b*, 86b), Stephen, *capellanus*, who held the church at
Milverton (*197*, 91b) but who was not necessarily resident. Some people
named in the special sections of the Exchequer text (91–91b) and Exeter
text (*196–8*) mentioned above may also have been priests, e.g. the Godwine
who held 'in the manor called Ridgehill' (*198*, 91b) may have been
Godwine the priest who was a sub-tenant in Brent not far away (*170b*,90b).

[1] J. H. Round in *V.C.H. Somerset*, I, p. 406.

Domesday churches in Somerset in 1086

The extracts are from the Exchequer text with the exception of: (1) those for Horsey Pignes and Kilmersdon, for which the Exeter version alone mentions churches; (2) that for Milborne Port, which appears as a place-name only in the Exeter version.

Cannington (*196b*, 91 b): *Erchenger tenet de rege in ecclesia de Cantetone ii virgatas terrae et dimidiam.*

Carhampton (*196b*, 91 b): *In ecclesia Carentone jacet i hida et dimidia. Ibi est in dominio i caruca et dimidia cum presbytero et i villano et viii bordariis.*

Chewton Mendip (*114b*, 87): *Ecclesiam hujus manerii tenet abbas de Gemetico cum dimidia hida terrae.*

Congresbury (*106*, 87): *Hujus manerii ecclesiam tenet Mauricius episcopus cum dimidia hida.*

Crewkerne (*197*, 91): *Ecclesia S. Stephani tenet de rege ecclesiam Cruche. Ibi sunt x hidae.*

Curry Rivel (*197b*, 91 b): *In ecclesia de Curi est dimidia hida. Ibi habet presbyter i carucam.*

Frome (*90b*, 86b): *De hoc manerio tenet ecclesia S. Johannis de Frome viii carucatas terrae et similiter tenuit T.R.E. Reinbaldus ibi est presbyter.*

Frome (*198*, 91): *Reinbaldus tenet ecclesiam de Frome cum viii carucatis terrae.*

Horsey Pignes (*477*): *Sacerdos de ecclesia istius villae habet dimidiam hidam.* The Exchequer entry (98b) mentions the priest but not the church.

Ilchester (*197b*, 91): *Episcopus Mauricius tenet de rege ecclesiam S. Andreae. Brictric tenuit T.R.E. et geldabat pro iii hidis.*

Ilchester (*171 b*, 91): *Mauricius episcopus tenet ecclesiam S. Andreae de Givelcestrie cum iii hidis terrae de rege. Hanc tenebat Brictric T.R.E. de ecclesia Glastingberiae nec ab ea poterat separari.*

Kilmersdon (*198b*): *In ecclesia Chinemeresdonae est dimidia hida terrae.* The Exchequer entry (91 b) does not mention a church.

Long Ashton (*143b*, 88b): *Ad ecclesiam huius manerii pertinet una virgata de eadem terra.*

Milborne Port (*91 b*, 86b): *Reinbaldus presbyter tenet ecclesiam Sancti Johannis de Meleborna qui servit ecclesiam.*

Milverton (*197*, 91 b): *Stefanus Capellanus tenet ecclesiam de Milvertone cum una virgata terrae et uno ferding.*

North Curry (*105b*, 86b): *Ecclesiam hujus manerii tenet Episcopus Mauricius cum iii hidis de eadem terra.*

North Petherton (*196b*, 91 b): *In ecclesia de Peretune jacet iii virgatae terrae.*

Stogumber (*197*, 91): *Richerius tenet ecclesiam de Warverdinestoch de rege. T.R.E. geldabat pro ii hidis.*

Yatton (*159b*, 89b): *Ecclesiam hujus manerii cum i hida tenet Benthelinus de episcopo.*

The Geld Accounts also tells us of the priest of Keynsham (*75b*), of six parochial priests (*presbiteri parrochiani*) on the land of the bishop of Wells (*78b*), of the *clericus* (? priest) in South Petherton (*81*), of the priest at Ilminster (*81b*), and of Bristward the priest in Frome hundred (*52*), who may have been at Writhlington (*493, 99*). All these would seem to imply the existence of unmentioned priests. Then, again, the cathedral at Wells and the abbey churches of Athelney, Bath, Glastonbury and Muchelney are mentioned only as landholders. Furthermore, such place-names as Pitminster (*173b*, 87b), Ilminster (*188*, 91), *Pennarministra* (*166b*, 90b, Pennard) and *Michelescerca* (*477b*, 98b, St Michael Church) suggest the existence of former or surviving churches.[1]

URBAN LIFE

The information about urban life is the least satisfactory part of the Somerset record. At the beginning of the description of most counties there is an account of the principal borough of the county, but this space in the Somerset text is left blank. There has, moreover, been some difference of opinion as to how many settlements should be called boroughs. Round and Eyton were in agreement in producing a basic list of eight—Bath, Ilchester, Milborne Port, Bruton, Langport, Axbridge, Taunton and Frome.[2] Ballard omitted Frome from his list.[3] It is true that no burgesses are recorded for Frome, but the payment of the 'third penny' suggests burghal status, and Tait included it. Tait also included Milverton, which likewise paid the 'third penny' and which, moreover, was incidentally styled a *burgus* in one Domesday entry. But he added that both 'had practically ceased to be boroughs by the date of Domesday, though Milverton retained some burghal features'.[4]

There remains the possibility that in 1086 there were two other claimants to burghal status in Somerset. One was Yeovil. Round thought he saw in Yeovil 'the beginning of a Somerset town', although it had no burgesses, nor did it pay the 'third penny'.[5] The case rests upon the

---

[1] Glastonbury documents, confirmed by William of Malmesbury's chronicle, tell of numerous churches and chapels in the abbey's ownership about which the Domesday Book is silent.

[2] J. H. Round in *V.C.H. Somerset*, I, pp. 420–1; R. W. Eyton, *op. cit.* I, pp. 49–53.

[3] A. Ballard, *The Domesday Boroughs* (Oxford, 1904), p. 9.

[4] J. Tait, *The Medieval English Borough* (Manchester, 1936), p. 55. Eyton, too, had noted the third penny at Milverton (*op. cit.* I, pp. 52, 81).

[5] *V.C.H. Somerset*, I, p. 422.

The Boroughs of Somerset

	Recorded burgesses, *domus* or *masurae*	Third penny	Market	Domes-day Mint	Pre-Conquest Mint	No recorded rural popula-tion
Bath	/	/		/	/	/
Ilchester	/	/	/		/	/
Milborne Port	/	/	/		/	
Bruton	/	/			/	
Langport	/	/			/	/
Axbridge	/	/			/	/
Taunton	/		/	/	/	
Frome		/	/		?	
Milverton	/	/	/			
Yeovil	?					
Watchet		?			/	

interpretation of a sentence in the account of the manor of Yeovil, which was part of William of Eu's fief: *Huic manerio additae sunt xxii masurae quas tenebant xxii homines in paragio T.R.E. Reddunt xii solidos* (439, 96b).[1] Round pointed out that its 22 households compared 'not unworthily' with the numbers recorded for Langport, Axbridge and Bruton. On the other hand, the manor was held *T.R.E.* by a man of no particular importance, and in 1086 it was in a sub-tenant's hands. Moreover, holding *in paragio* usually implies manorial, not urban, status. Still, the phrase *xxii masurae* suggests that here was a possible quasi-borough, although it has not been regarded as one by Eyton or Tait. Nor has it been so regarded in the present analysis.

[1] The Exeter text uses the phrase *mansurae terrae* (439).

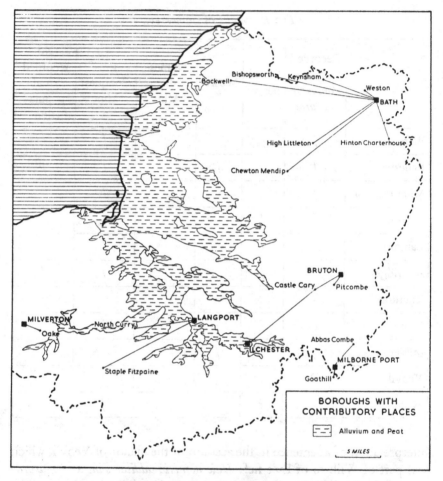

Fig. 47. Places contributory to six Somerset boroughs.

Castle Cary had appurtenant holdings in both Bruton and
Ilchester; Goathill is now in Dorset.

The other possible place with burghal status is Watchet. It is recorded
only as a small manor with half a team for its single plough-land and with
but two recorded inhabitants and a mill (*361 b*, 96). This seems very exig-
uous, but Watchet had been a *burh* at the time the Burghal Hidage
was drawn up, and it had had a mint in pre-Conquest times.[1] The adjacent

[1] R. H. M. Dolley (ed.), *Anglo-Saxon coins* (London, 1960), p. 146.

manor of Old Cleeve (*103b*, 86b) is said to have the third penny of the *burhriht* or *burgherist* from four hundreds for which no borough is recorded but in which Watchet lay. *Burgerist* is recorded as one of the dues payable to the manor of Taunton (*173b*, 87b) which was certainly a borough, and it is possible that Watchet was still a kind of borough in 1086.[1]

Some of the main features of the evidence relating to these eleven settlements is set out in the table on p. 197. For the purpose of this analysis, we have recognised only the first nine as being certain boroughs, and each of these is discussed separately below.

Places contributory to Bath

Keynsham (*113b*, 87): *viii burgenses in Bade reddentes v solidos per annum.* The Exeter text says 7 burgesses.

Chewton Mendip (*114*, 87): *In Bade iiii burgenses reddunt xl denarios.*

Bishopsworth (*141b*, 88): *In Bade ii domus reddunt x denarios.* Ten houses in Bristol are also mentioned.

High Littleton (*151b*, 89): *In Bada i burgensis reddit xv denarios.*

Hinton Charterhouse (*437*, 98): *In Bade ii domus, una reddit vii denarios et obolum.* The Exeter text runs: *i domus in Bada quae reddit per annum vii denarios et i obolum et alia mansura in eodem burgo vacua.*

Weston (*448b*, 98): *In Bade iii domus reddunt xxvii denarios.*

Backwell (*143*): *i borgisum qui manet ad Badam et reddit xxxii denarios per annum* (omitted from the Exchequer entry on fo. 88).

Bath

The two main entries dealing with Bath appear in the section dealing with the king's land (*114b*, 87)[2] and in that dealing with the land of Bath Abbey (*185*, 89b). The royal entry refers to the 'third penny' of the borough,[3] and both entries together mention a total of 178 burgesses, together with 7 houses (*domus*), of which 6 were waste (*vastae*). But Bath shares one interesting feature with a number of other Domesday boroughs; we read of burgesses and properties in Bath but belonging to rural manors elsewhere, and these are set out above (Fig. 47). We cannot say whether these other 13 or 14 burgesses and 7 houses (including 1 *mansura vacua*) were included in the details of the two main entries. If not, the grand total would be 192 burgesses, 7 occupied houses and 7 waste houses.

[1] J. Tait. *op. cit.* p. 61 n. See p. 210 below.
[2] Some of the information for the royal holding appears also in the Exeter folios among the account of *Terrae Occupatae* in Somerset (*518b*).
[3] The 'third penny' from Bath is also mentioned in the Wiltshire folios (64b).

In any case, the figures may imply an urban population of about a thousand or so.

We hear nothing of the commercial life of the city beyond the existence of a mint (*moneta*) which rendered 100s. The Exeter version of the Bath entry says that the abbey had a manor called Bath which was the seat of that abbey (*caput ipsius abbatiae*), but there is no reference to any specifically agricultural population or activity beyond a mention of 12 acres of meadow and a mill rendering 20s.

Ilchester

The main entry (*91b*, 86b) states merely that the king has 107 burgesses who rendered 20s. There was also a single burgess appurtenant to Castle Cary (*352b*, 95) who, with another at Bruton, rendered 1s. 4½d. (Fig. 47). Thus there may have been a total population of about 500. There was also a market which, together with its appendages (*cum suis appenditiis*), rendered £11 *in firma regis*. On another folio (*107b*, 87) we hear of receipts from the 'third penny'. There is no reference to a mint, but we know that there had been one here in pre-Conquest times.[1] In addition, Ilchester had a church dedicated to St Andrew (*171b*, 91), and its 3-hide holding is described in a separate entry (*197b*, 91).

Milborne Port

Milborne Port is described in two entries. The more important relates to a substantial royal manor that had never paid geld, that included 50 plough-lands, and that was responsible for three quarters of a night's *firma* (*91b*, 86b). Among the details for the manor there is reference to 56 burgesses, with a market, paying 60s. The second entry relates to a holding of only 1 hide and 1 plough-land in the count of Mortain's fief (*278b*, 93) and it includes a reference to 5 burgesses paying 3s. (3s. 9d. in the Exeter text). Two other entries refer to appurtenant holdings in Milborne Port (Fig. 47). Attached to Abbas Combe there were 6 burgesses rendering 50d. (*193b*, 91), and attached to Goathill but mentioned only in the Exeter text were 2 *masurae* (*278b*).[2] On the assumption that the appurtenant holdings were not included in the other entries, this makes

---

[1] R. H. M. Dolley (ed.), *op. cit.* pp. 131, 145.
[2] Goathill is now in Dorset. The omission of the 2 *masurae* from the Exchequer text may be not unconnected with the fact that the entry for Goathill is written across the foot of fo. 92b in the manner of an afterthought.

a total of 67 burgesses and 2 *masurae*, which may imply an urban element of, say, some 350 people. There was a church dedicated to St John. Receipts from the 'third penny' are mentioned on another folio (*107b*, 87).

The king's manor and the count of Mortain's minor holding also had between them 70 villeins, 20 bordars and 9 serfs, but we cannot say how many of these 99 people were in Milborne Port itself. There were altogether 71½ teams (71 in the Exchequer version) and 7 mills besides livestock and tracts of wood, meadow, pasture and moor. The large figures for the royal manor suggest the existence of unspecified constituent holdings. We must, apparently, envisage a small town with a number of scattered holdings in other places, most of which constituted the royal manor. The entry for the manor speaks of 'appendages' (*appenditii*), and the order of the text in its Exchequer version seems to suggest that the market and the burgesses were among these. There had been a pre-Conquest mint here.[1]

Bruton

Bruton is described in two entries. The more important relates to a royal manor that had never paid geld, that included 50 plough-lands, and that, together with Frome, was responsible for one night's *firma* (*90b*, 86b). Among the details for the manor there is reference to 5 burgesses. The second entry relates to a holding of only 1 hide 1 virgate, and mentions no burgesses (*434b*, 94). Two other entries refer to appurtenant holdings in Bruton (Fig. 47). Attached to the neighbouring settlement of Pitcombe there were eleven burgesses who rendered 23s. (*382b*, 97b), and attached to Castle Cary nearby there was one burgess who, together with another at Ilchester, rendered 1s. 4½d. (*352b*, 95). This total of 17 burgesses may imply an urban element of, say, some 85 people. Receipts from the 'third penny' are mentioned on another folio (*107b*, 87). There is no reference to a market or to a mint, but we know that there had been a pre-Conquest mint here.[2]

The king's manor together with that of Roger de Courseulles also had between them 29 bordars, 28 villeins, 4 coliberts, 5 serfs and a swineherd, but we cannot say how many of these 67 people were in Bruton itself. It may well be that the royal manor comprised unspecified constituent holdings. There were, altogether, 22 teams and 7 mills besides livestock and tracts of wood, meadow and pasture. The evidence suggests a small urban development in a rural setting.

[1] R. H. M. Dolley (ed.), *op. cit.* p. 135.   [2] *Ibid.* p. 131.

Langport

The account of the royal manor of Somerton includes a brief reference to
the borough of Langport, some 5 miles to the west: *Ibi burgus quod vocatur
Lanporth in quo manent xxxiiii burgenses reddentes xv solidos* (*89 b*, 86).
In addition to these 34 burgesses, there were also 5 burgesses, rendering
38*d.*, who were appurtenant to North Curry (*105*, 86b). There was a
garden, rendering 50 eels, belonging to the manor of Staple Fitzpaine (*270*,
92): *Huic manerio pertinet unus ortus in Langeport reddens l anguillas*
(Fig. 47).[1] On the assumption that the 5 burgesses were additional, the
total of 39 implies an urban element of, say, some 200 people. On another
folio there is mention of receipts from the 'third penny' (*107b*, 87).
There is no mention of a market or of a mint, but we know that there had
been a pre-Conquest mint here.[2]

Axbridge

The account of the royal manor of Cheddar includes a brief reference
to the borough of Axbridge nearby: *In Alsebruge xxxii burgenses reddunt
xx solidos* (*90*, 86). We also hear on another folio of receipts from the
'third penny' (*107b*, 87). That is all we are told about Axbridge, and we
can only assume that this meagre information implies an urban element of,
say, some 160 people. We know, from other sources, that there had been
a pre-Conquest mint here.[3]

Taunton

The account of the bishop of Winchester's great manor of Taunton
includes a reference to 64 burgesses rendering 32*s.* (*173b*, 87b). There
were also a market and a mint, each rendering 50*s.*; the Exchequer text
alone mentions the mint. There is no other reference to commercial life.
We are, however, given much detail about customary dues payable to
Taunton from some nineteen places (Fig. 48).[4] The manor comprised 100
plough-lands and 20 non-gelding carucates, and we know that it included
holdings in many villages unnamed in the Domesday folios.[5] The main
part of the entry mentions 73 teams and what is apparently a rural popula-

---

[1] The statement about Langport in the Somerton entry (*89b*, 86) is followed by
a reference to 2 fisheries rendering 10*s.*, but it is not clear whether these were at
Langport or at Somerton.
[2] R. H. M. Dolley (ed.), *op. cit.* p. 146.   [3] *Ibid.*
[4] The customary dues are set out on p. 210.
[5] For the constituents of the manor, see Fig. 35 and p. 138 above.

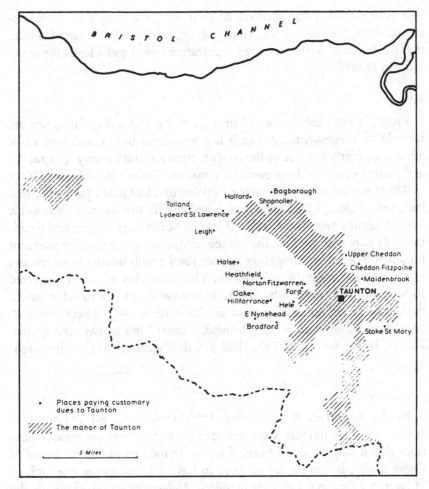

Fig. 48. Places paying customary dues to Taunton.

For the manor of Taunton, see Fig. 35.

tion totalling 265 people: 82 bordars, 80 villeins, 70 serfs, 17 swineherds
and 16 coliberts; there were also wood, pasture, meadow, mills, and livestock. But some of these people and resources must have been dispersed
among appendages (*appenditii*), and in addition there were over a dozen
other unnamed sub-tenancies with another 36½ teams and another 150
people and also with wood, pasture, meadow, mills and livestock. How
many people lived within the town, we cannot say. All we can infer is

that here was an urban element of over, say, 300 people set in a substantial rural community with outlying dependencies. We can also infer that the borough itself was an administrative centre and a focus for many villages around.

Frome

Frome is not described as a borough in the Domesday folios nor are we told of burgesses there. But it has sometimes been considered to be a borough partly because of the receipts from its 'third penny' (*107b*, 87) and partly because of the presence of a market yielding 46s. 8d. (*90b*, 86b).

There was also a royal manor of Frome that had never paid geld, that included 50 plough-lands, and that, together with Bruton, was responsible for one night's *firma* (*90b*, 86b). It had 36 bordars, 31 villeins and 6 coliberts. It may well be that the manor comprised unspecified constituent holdings. There were altogether 43 teams and 3 mills besides livestock and tracts of wood, meadow and pasture. There was also a church dedicated to St John (*90b*, *198*; 86b, 91). It is impossible to say to what extent there was an urban development in this rural setting. There seems to have been a mint here in pre-Domesday times,[1] but it may well be that Frome, in the words of Tait, 'had practically ceased to be a borough' by 1086.[2]

Milverton

Neither Round nor Ballard included Milverton in their lists of boroughs. It is true that no burgesses are recorded for it, yet there are some indications that it may have had burghal status. In the first place, the manor of Oake is said, in the Exeter version, to have a house in the borough of Milverton (Fig. 47), *i domum in burgo Meluertone* (*433*), although the Exchequer text merely says *in Miluertone* (94). Then, secondly, we hear of the 'third penny' of Milverton in the account of the royal manor of Brompton Regis (*103*, 86b). And, thirdly, there is mention of a market rendering 10s. (*113*, 87), although this in itself does not necessarily indicate a borough.

There was also a rural manor of Milverton (*113*, 87). It had 16 villeins, 7 bordars, 3 cottars and 3 serfs. These 29 people may have been in

[1] R. H. M. Dolley (ed.), *op. cit.* p. 146: 'There are other coins from a puzzling mint at FRO, which could be Frome in Somerset.'
[2] J. Tait, *op. cit.* p. 55.

Milverton itself, but we cannot be sure that the Domesday entry did not cover another place or places. Then there were 16 plough-lands, 10 teams and a mill together with livestock and tracts of meadow, pasture and underwood. It is impossible to say how the borough fitted into this rural setting. It may well be that Milverton, in the words of Tait, 'had practically ceased to be a borough' by 1086.[1]

LIVESTOCK

The entries of the Exeter text record information about livestock—not about total stock but only about that on the demesne. We cannot say how complete even these restricted figures are, and it may well be that some demesne animals were omitted. A number of entries for holdings with demesne land make no reference to livestock, e.g. those for Bradney (353b) and for Sutton Bingham (444).[2] Conversely, stock is occasionally entered for holdings without recorded demesne, e.g. for East Myne (360). Other entries mention customary renders of sheep and lambs, which would seem to imply the existence of unrecorded stock.[3] The Taunton district ranked relatively low in livestock as compared with its population and its resources in teams. There is no obvious aspect of the physical environment which might be held to account for this, but the explanation is possibly to be found in the form of the entries. Although the manors of midland England could be fitted into the framework devised by King William's commissioners, other social and economic systems on the periphery of England were accommodated only with difficulty. Perhaps among those small settlements to the west of the Parrett, demesne implied something different from what it meant in the eastern parts of the county, and so a smaller proportion of livestock was recorded.

There are some curiosities. The entry for Stoke Pero (430b) gives the numbers of swine, of sheep and of goats, but refers merely to *animalia* without specifying a number. The two Exeter entries that describe a disputed portion of Long Sutton disagree with one another: Dodeman's

---

[1] J. Tait, *op. cit.* p. 55.
[2] A special case of an Exeter entry that omits reference to livestock is that for a holding the name of which was never inserted in either the Exeter or the Exchequer text. It probably refers to Charlton Horethorne (436b, 97), and it seems to have been inserted in the Exeter text by an Exchequer clerk (R. Welldon Finn, 'The evolution of successive versions of Domesday Book', *Eng. Hist. Rev.* LXVI, 1951, p. 563).
[3] See p. 212 below.

holding in the fief of Athelney Abbey had 8 animals, 2 swine and 80 sheep
(*191*), but the account of what is apparently the same two-hide holding,
in the fief of Roger de Courseulles, speaks of 1 cow, 4 swine and 50 sheep
(*435b*). The table below shows the variety of livestock mentioned, and
also the number of animals in each category.

Demesne Livestock in Somerset in 1086

Sheep (*oves*)	46,033
Wethers (*berbices*)	948
Swine (*porci*)	6,847
She-goats (*caprae*)	4,505
Animals (*animalia*)	4,289
'Otiose' animals (*animalia ociosa*)	16
Cows (*vaccae*)	117
Rounceys (*roncini* or *runcini*)	443
Unbroken mares (*equae indomitae*)	318
Forest mares (*equae silvaticae/silvestres*)	38
Mares (*equae*)	35
Donkeys (*asini*)	3

Horses are mentioned in a variety of ways. Rounceys (*roncini* or
runcini) were widely distributed, and were most probably pack-horses.
We hear of *caballi* (riding horses) at Woolavington (*162*), but *roncini* has
been interlined above. Apart from 35 mares (*equae*), a distinction is made
between *equae indomitae* and *equae silvaticae* or *silvestres*, but we may
wonder whether there was any difference between the two categories.
They appear very frequently on the estates of the bishop of Coutances
(who kept about one third of the total), the count of Mortain, and William
de Mohun. Thus there were 60 *equae indomitae* in the manor of Long
Ashton (*144*), 38 at Cloford (*275b*), and 36 at Cutcombe together with 3
equae silvestres on a sub-tenancy (*357b*). Finally, 3 donkeys are mentioned,
two at Hardington (*147*) and one at White Ox Mead (*434*).

Animals (*animalia*) presumably included all non-ploughing beasts.[1]
One entry for a dependency of Taunton speaks of 16 *animalia ociosa* (*175*).
Generally speaking, the number on a holding was small, usually less than

---

[1] An exception to the normal use of the word *animalia* occurs in the entry for
Combe Sydenham. Where the Exchequer text speaks of 'half a plough' (*96*), the
Exeter text reads *iiii animalia in carrucam* (*362b*), and records no other *animalia*
on the holding.

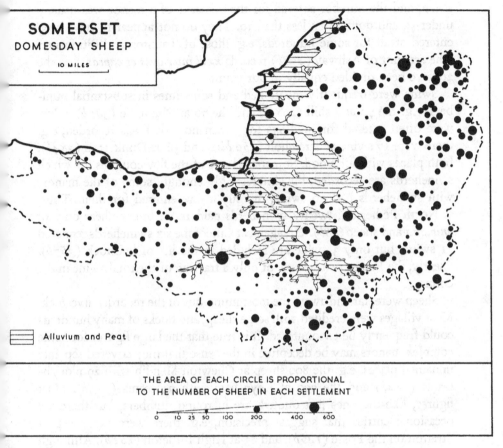

SOMERSET
DOMESDAY SHEEP

|⌐ 10 MILES ⌐|

≣ Alluvium and Peat

THE AREA OF EACH CIRCLE IS PROPORTIONAL
TO THE NUMBER OF SHEEP IN EACH SETTLEMENT

0   10 25 50   100   200   400   600

Fig. 49. Somerset: Domesday sheep on the demesne in 1086.

Wethers (*berbices*) have been included with sheep.

forty, and often only two or three. Cows (*vaccae*) appear only infrequently.
Out of the 42 entries that mention a total of 117 cows, nineteen entries
record only one apiece, and Round suggested that these solitary animals
were kept to supply milk for the lords of manors and their families.[1]

Goats were surprisingly numerous until we remember the important
part they still played as milk-producers in the eleventh century; only she-
goats (*caprae*) are recorded. The flocks were much smaller than those of

[1] *V.C.H. Somerset*, I, p. 424.

sheep, and the number entered for the demesne of a manor was usually under 50 and quite often less than 10. They do not appear to have been entered at all for some hundreds, e.g. those of Taunton and Pitminster. The account of Ashway (*428b*) records *xxvi inter oves et capras*, and the 26 have been divided equally in our count.

Swine were frequently mentioned and sometimes in substantial numbers, e.g. the 70 at Ashill (*268b*), and the 70 at Pitcombe (*382b*). They were often entered for holdings for which no wood was recorded, e.g. there were 57 swine at Stretcholt (*350 bis*), and 36 at Dunkerton (*384b*), both places without wood. Somerset is one of the few counties for which swineherds are recorded, but, interestingly enough, some of the manors with swineherds had no swine, or only a few, entered for them. Thus, at North Petherton, twenty swineherds paid 100s., but we hear only of *animalia* and sheep (*88b*). Or, again, at Cutcombe six swineherds rendered 31 swine, but only 3 swine are included in the list of livestock (*357b*). Clearly, the Exeter text tells us of only a fraction of the total swine in the county.

Sheep were, far and away, the most numerous of the recorded livestock. Most villages were credited with some sheep, and flocks of many hundreds could frequently be encountered. It is true that the large figures for some complex manors may be deceptive in the sense that they covered separate unnamed places, e.g. the 800 sheep at Chewton Mendip with 29 plough-teams (*114b*) and the 700 at Keynsham with 73 teams (*113b*). Most figures, like the ones just quoted, are in round numbers, but there are occasional entries that suggest precision, e.g. there were 317 sheep at Stratton on the Fosse (*145b*), and 83 at High Littleton (*151b*). Although Fig. 49 is a map only of demesne sheep, it may give some hint about the general distribution of sheep throughout the county. The most heavily stocked lands seem to have been in the north-east, especially among the hills and valleys of the oolitic scarplands. But sheep were also fairly abundant elsewhere, except in the marshland area of central Somerset and in the uplands of the Exmoor region to the west. Wethers (*berbices*) are recorded only for five holdings.[1]

[1] There were 10 at Cannington (*196b*), 158 at Emborough (*150*), 438 at Henstridge (*107*), 300 at Old Cleeve (*103b*) and 42 at Quarme (*358b*).

Markets

MISCELLANEOUS INFORMATION

Markets are recorded for seven places, and the Exchequer entries read as follows:

Frome (*90b*, 86b): *Mercatum reddit xlvi solidos et viii denarios.*
Ilchester (*91b*, 86b): *Mercatum cum suis appenditiis reddit xi libras.*[1]
Milborne Port (*91*, 86b): *In hoc manerio sunt lvi burgenses et* [*cum* interlined] *mercato reddentes lx solidos.*[2]
Crewkerne (*105b*, 86b): *Mercatum reddit iiii libras.*
Milverton (*113*, 87): *Ibi mercatum reddit x solidos.*
Taunton (*174*, 87b): *Mercatum reddit l solidos.*
Ilminster (*188*, 91): *Ibi mercatum reddit xx solidos.*

The first five places were royal manors. Taunton was in the fief of the bishop of Winchester, and Ilminster in that of Muchelney Abbey. Burgesses were recorded for only three of these places—Ilchester, Milborne Port and Taunton, although Frome and Milverton also seem to have had some claims to be considered boroughs.

Vineyards

The vine seems to have been grown on a small scale in the lowlands of central Somerset. The references to vineyards number five. One was measured in acres and the others in terms of the French arpent which appears as *agripenna* in the Exeter text.[3] Here are the Exchequer phrases:

Glastonbury (*172*, 90): *iii arpenᶎ vineae.*
Meare (*172*, 90): *ii arpenᶎ vineae.*
Panborough (*172*, 90): *iii arpenᶎ vineae.*
Muchelney, Midelney and Thorney (*189*, 91): *unum arpent vineae.*
North Curry (*105*, 86b): *vii acrae vineae.*

The first three holdings were in the fief of the abbey of Glastonbury; the three island sites were in that of the abbey of Muchelney. North Curry was a royal manor.

[1] The Exeter version reads: *i mercatum in Guilecestra quod cum suis appenditiis reddit in firma regis xi libras.*
[2] The Exeter version reads: *In hac eadem Meleborna habet Rex lvi borgenses et i mercatum et inter borgenses et mercatum reddunt per annum lx solidos in firma regis.*
[3] For a discussion of the arpent see Sir Henry Ellis, *A General Introduction to Domesday Book*, I, p. 117.

Customary dues

A feature of some royal manors was the customary due (*consuetudo*) received from other holdings. But by 1086 all these payments seem either to have lapsed or to be withheld (usually by the count of Mortain) with the exception of those from Monksilver and Over Stratton. The dues are entered sometimes under the holding which paid them; but for Dulverton, South Petherton and Williton, under the manor that received them. No due is entered under both the paying and the receiving estate. The details are set out in the table on p. 212. It would seem that the usual contribution was at the rate of one ewe and its lamb per hide, but Oare, Allerford, Bossington, Brushford and Monksilver—Exmoor manors—paid 12 sheep per hide. There were complications, for Seaborough paid 4 sheep per hide and Bickenhall and Cricket St Thomas paid 1 each, all three payments in addition to renders of iron from each freeman. Over Stratton paid 8 sheep per hide in addition to a money render and Ashill paid only an amount in money.

There had belonged (*adjacuit*) to the manor of Old Cleeve (*103b*, 86b) the 'third penny' of the *burgherist* of the four hundreds of Carhampton, Williton, Cannington and North Petherton.[1] Maitland thought this might have been the equivalent of *burhgrith* (the fine for the breach of the peace of the *burh*) or of *burhgerihta* (*burh*-rights, borough dues). The word also occurs as one of the customs pertaining to Taunton, *burgerist* (*174*, 87b), and Ballard deduced that this was the responsibility of rural properties for repairing the walls of a borough.[2] The account of the manor of Taunton also sets out a variety of other customary dues rendered by a number of nearby holdings. The Exeter version reads as follows (*174–175b*):

These are the customary dues belonging to Taunton: burhright, theft, breach of the peace, house-breach, the hundred-pennies, Peter's pence, churchscot, holding the bishop's pleas three times a year without a summons; both from Tolland and from Oake and from Holford and Upper Cheddon and Cheddon Fitzpaine and from Maidenbrook and from Ford and Lydeard St Lawrence and Leigh and Hillfarrance and Hele and from East Nynehead, and all these are bound to go on military service with the bishop's men; and when the lords of the above lands die they must be buried in Taunton. Bagborough also owes this customary due to Taunton except military service and burial, and the above customary due is owed by Shopnoller and Stoke St Mary besides military service, and

[1] See p. 199 above.
[2] F. W. Maitland, *op. cit.* p. 88 n.; A. Ballard, *op. cit.* p. 100. See J. Tait, *op. cit.* p. 61 n.

Norton Fitzwarren and Bradford and Halse and Heathfield must attend the bishop's pleas three times a year and pay Peter's pence in Taunton and the hundred-penny, and all who live in the aforesaid lands, if they are obliged to make oath or give judgment, shall do so in Taunton...The customary dues and service of these lands [the sub-tenancies] always lay in Taunton.

The places are set out on Fig. 48.

Iron

Iron is mentioned as a render or customary due at six places. At Bickenhall, Cricket St Thomas and Seaborough (now in Dorset), a render of sheep to a royal manor was supplemented by a bloom of iron from each freeman (see above). The entries for the other three places were as follows:

Alford (*277b*, 92 b): *de villanis viii blomas ferri.*[1]
Lexworthy (*282*, 91 b): *Ibi ii molini reddunt ii plumbas ferri.*
Lexworthy (*432*, 94): *Ibi molinus reddit ii plumbas ferri.*
Lexworthy (*432 b*, 94): *Ibi molinus reddit ii plumbas ferri.*
Whitestaunton (*265b*, 91 b): *l acrae pasturae reddunt iiii blomas ferri.*

There is a reference to 8 smiths (*fabri*) at Glastonbury (*172*, 90), who may have been iron-workers. We are given no indication of the source of the iron.

Castles

Two castles are mentioned as having been built by Norman lords. On the hill overlooking the village of *Biscopestone*, a name subsequently superseded by that of the castle, the count of Mortain had raised a fortification called Montacute. The two versions of the entry read as follows:

E.D. (*280*): *In ista supradicta mansione est castrum comitis quod vocatur Montagut.*
D.B. (93): *ibi est castellum ejus quod vocatur Montagud.*

The other was at Dunster, and was the castle of William of Mohun, the sheriff of Somerset. Here are the references:

E.D. (*359*): *Ibi habet W. suum castrum.*
D.B. (95 b): *ibi est castellum eius.*

Of the boroughs, Bath alone had waste houses, but this fact does not seem to imply (as sometimes it does) the building of a castle.

[1] The Exeter text (*277b*) says that only i villein made the render: *i de supradictis villanis reddit viii blumas ferri.*

Customary Dues in Somerset
(*Exchequer version*)

Manor to which custom was due	Manor from which custom was due	Due
Curry Rivel	North Bradon (*268*, 92)	*unam ovem cum agno*
	North Bradon (*269 b*, 92)	*unam ovem cum agno*
	North Bradon (*269 b*, 92)	*ii oves cum agnis*
	Ashill (*268 b*, 92)	*xxx denarios*
	Donyatt (*270*, 92)	*v oves cum agnis*
	Hatch Beauchamp (*271*, 92)	*una ovis* [sic] *cum agno*
	Bickenhall (*270 b*, 92)*	*v oves cum totidem agnis et quisque liber homo unam blomam ferri*
Crewkerne	Seaborough (*154*, 87 b)†	*xii oves cum agnis et una bloma* [sic] *ferri de unoquoque libero homine*
Carhampton	Oare (*344*, 96 b)	*xii oves*
	Allerford (*463*, 97)	*xii oves*
	Bossington and Allerford (*510*)‡	*xxiiii oves (aut v solidi* in E.D.)
South Petherton (*88 b*, 86)	Cricket St Thomas§	*vi oves cum agnis totidem et quisque liber homo i blomam ferri*
	Over Stratton	*Reddit modo lx solidos in firma regis (et xxiiii oves* in E.D.)
Williton (*89*, 86 b)	Monksilver‖	*xviii oves*
Dulverton (*103 b*, 86 b)	Brushford	*xxiiii oves*

* Also mentioned under *Terrae Occupatae* (*514 b*).
† Now in Dorset—see p. 140 above. Also mentioned under *Terrae Occupatae*.
‡ Mentioned only under *Terrae Occupatae*. The 24 sheep included the 12 separately mentioned for Allerford.
§ Also mentioned in the Exeter text for Cricket St Thomas—see p. 412.
‖ This one had been added since 1066.

Other references

Only two mints are recorded. The one at Bath yielded 100*s*. (*114b*, 87) and that at Taunton, mentioned only in the Exchequer text, yielded 50*s*. (87b). All other boroughs were, apparently, without mints. This is all the more curious because there are coins from pre-Conquest mints at Axbridge, Bruton, Ilchester, Milborne Port, Langport and, possibly, Frome, as well as from Cadbury, Crewkerne, South Petherton and Watchet.[1]

References to topographical features are rare in the Domesday Book, but *Doneham* (Dunwear House), held by Walter of Douai, is described as lying 'between two waters'—*Haec est de illa terra quam rex dedit ei inter duas aquas (350,* 95). This may be a reference to the land between the Parrett and Brue rivers where Walter of Douai held a large number of manors. There was a park (*parcus*) at Donyatt (*270,* 92) and a garden (*ortus*) at Langport (*270,* 92).

REGIONAL SUMMARY

Somerset lies athwart a border region where the less resistant formations of the lowlands lie alongside the old core of Britain, so that it encompasses within its bounds bleak, damp uplands, where rocks of great age weather down to thin, sour soils, as well as fertile lands of gentler aspect and milder climate. Thus, despite large areas of alluvium which mask considerable tracts in central and north Somerset, there is no lack of variety in the physical resources of the county. A division into nine regions will serve to give a broad view of the county in the eleventh century (Fig. 50).

(1) *The Exmoor Region and the Quantocks*

This region corresponds broadly to outcrops of grits and slates, and comprises the main mass of Exmoor and the Brendon Hills together with the outlying Quantock Hills. They are mostly over 800 ft above sea-level, and considerable tracts are over 1,000 ft. A layer of peat is found over wide areas on the higher slopes, and there is a considerable extent of true moorland. In such an environment it is not surprising that settlements sought the more sheltered parts such as the valleys of its northern coastal margin and the interior valleys of the Exe and its tributaries. Some of

[1] R. H. M. Dolley (ed.), *op. cit.* pp. 131, 135, 146, and 147.

these valleys were very small. Cultivable land was scarce even in the valleys, so that over much of the area there was less than one team and under 4 people per square mile. Along the northern margin the density rose to about double these. As a result of the relative scarcity of arable land, mills were not numerous, and tended to be small. More than two thirds of them paid less than 5s. and those at Clatworthy and Winsford only 6d. each. Nearly four fifths of the villages in this region were credited with some woodland. Pasture was abundant, and very large quantities are recorded. Although the quantities appear under the names of the valley villages, the pasture itself was presumably situated on the higher ground. There were only small quantities of meadow along the narrow valleys, e.g. the half-acres at Chipstable and Hone. Sheep were the most numerous of the stock recorded for the region, but the average density was only about one sixth of that of the Bath area.

The outlying Quantock Hills are lower than the Exmoor country, but they reproduce on a smaller scale some of the same features. Their soils, for example, have weathered out from grits and slates of the same types as those of the western hills. But owing to the narrowness of the range, few villages had their territories entirely on the hills, and settlements tended to congregate along the spring-lines on either flank. The densities per square mile (1·7 for teams and 6·2 for population) were somewhat higher than those for the interior part of Exmoor and less than those for the lower northern part of Exmoor. Although the arrangement of parish boundaries makes any figures somewhat artificial, the general position is fairly clear. As in Exmoor, there were a number of villages without mills. The hill slopes seem to have been fairly well wooded, and to have provided abundant pasture for neighbouring vills; but there was little meadowland.

(2) *The Taunton Region*

In this tract of low-lying country, mostly under 400 ft above sea-level, rocks chiefly of Permian and Triassic age have weathered out to a considerable range of soils. Much of the vales of Taunton Deane and Wellington lies under light, free-working stoneless loams, interspersed with patches of sand and clay, while in the faulted corridor between the Exmoor mass and the Quantocks, it is not unusual to find within a single parish soils developed on marls, breccias, sandstones, pebble beds, and valley gravels. There is also a narrow belt of clay country along the coast in the north of the area.

This was one of the most prosperous parts of Domesday Somerset. Villages were spread thickly and evenly over the countryside. The lack of villages in the neighbourhood of Taunton is only apparent, for that enormous manor included at least fifteen vills which are not named in the

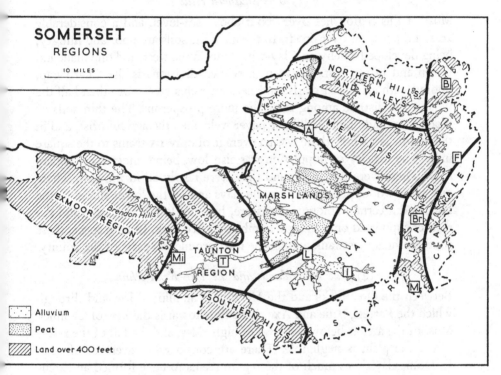

Fig. 50. Somerset: Regional subdivisions.

Domesday boroughs are indicated by initials: A, Axbridge; B, Bath; Br, Bruton; F, Frome; I, Ilchester; L, Langport; M, Milborne Port; Mi, Milverton; T, Taunton.

Domesday Book, and which do not, therefore, appear on Fig. 36. Over the region as a whole there was an average of 3 teams and over 10 men at work. Mills, although numerous, were generally small in size, frequently rendering less than 5s. Nearly three quarters of the vills had some wood on their territory, but amounts were normally small, and recorded predominantly in acres. Pasture was entered for many villages but probably much of it was actually situated on the western hills, on the Quantocks or on the slopes of Blackdown. Most of the holdings included some meadow, but acreages were not large. This region carried fewer livestock than might have been expected in view of its general prosperity; sheep, for example, were only moderately numerous.

(3) *The Southern Hills*

Most of this region lies over 400 ft above sea-level, and a considerable area in the west at over 600 ft. In the main the soils are poor and hungry, being developed chiefly on Clay-with-flints in the west, and on Chalk and Greensand in the east. Between these two divisions is the Chard gap, floored with clays and marls, and here were clustered more than half the vills of the district, among them the more prosperous. The thin soils of Blackdown and Staple Hill have never welcomed the agriculturist, and in the region as a whole there was an average of only 1·5 teams to the square mile. The density of population was also low, being under 5 per square mile. There were few mills, and they were nearly all in the east. There was some wood, presumably on the clays and on the lower slopes of the hills, and a certain amount of pasture, but meadow was scarce, and the district supported only meagre numbers of livestock. In all respects these southern hills must rank among the least productive parts of the county.

(4) *The Marshlands and the Yeo–Kenn Plain*

Between the Quantocks and the Mendips is an alluvial lowland through which the Parrett, Brue and Axe rivers flow towards the Bristol Channel. Most of the area is below the level of high tides, and the fall of the rivers across the plain is negligible. There are considerable stretches of peat. In Domesday times much of the region seems to have formed an extensive marsh broken into two separate tracts by the ridge of the Polden Hills consisting partly of Lias Clay and partly of Triassic marls. The tracts of marsh were diversified by 'islands' such as the large island of Wedmore and the small ones of Meare and Brent Knoll in the north, and the islands of Chedzoy and Westonzoyland in the south. Along the coast a belt of silt rises to about 20 ft above the general level of the marsh. Villages were located in these higher areas; and between, on the stretches of peat, there were tracts devoid of settlement. The density of teams and of population varied. In the area of the Polden Hills and along the coast there were under 2·5 teams and just under 8 people per square mile. The corresponding figures for the marshy areas to the north and south of the Polden Hills were below 1 for teams and about 4 for population. Scarcely any villages had mills, but this may have been the result as much of the low gradient of the streams as of the lack of grain. There was practically no woodland, apart from a strip of underwood along the Polden Hills, but a number of villages were credited with pasture. Many villages had fair quantities of meadow in the Polden district, in the silt belt, and around the margins of the upland. Some villages in the area and on the nearby upland are said to have moors (*morae*) but the references are disappointingly meagre; the lay-out of the parish boundaries would seem to indicate, however, that the marshes were not entirely unused. A number of places had fisheries, but, again, the record probably does far less than justice to the reality. Sheep were relatively more numerous than cattle, but, even so, their number was not great.

A similar area of marsh probably extended, in the north-east, around the lower valleys of the Yeo and Kenn rivers. Here, plough-team and population densities were about 3 and about 7 respectively. Although there were fewer people per square mile than on the silt belt, there were more teams. This may have been due in part to the fact that the territory of most parishes extended from the fen on to the surrounding limestone uplands. There were only 3 mills in this area, all at Banwell on its southern margin. Wood (mostly underwood), pasture and meadow appear in the entries for many villages. The close proximity of dry uplands and a lowland with a high water-table enabled the demesne farms to carry considerable numbers of sheep, cattle and swine.

(5) *The Lias Plain*

This region is mainly an undulating plain at about 100 ft above sea-level, but varying locally between 200 ft along its southern margin to less than 50 ft in the broad valley of the Yeo river. The typical soils of this region are heavy clays and marls derived from the Lower Lias formation, and are of considerable fertility, though difficult to work. Lighter soils occur only occasionally, mainly on local limestone outcrops, on patches of valley gravel, or on tongues of alluvium along the rivers.

Villages were frequent, and, apart from a tendency to avoid the Yeo marshes between Ilchester and Langport, they were spaced fairly evenly over the countryside. With some notable exceptions, such as the royal manor of Somerton, settlements were small, but two, Ilchester and Langport, had acquired the status of small boroughs. The density of teams ranged from 2·5 to 3·5 per square mile; that for population from over 7 to about 12. The lowest densities were in the areas nearest the central marshlands, and part of the reason may lie in the extensive tracts of alluvium in the upper valleys of the Cary, the Isle, the Parrett and their tributaries. Mills were fairly numerous, and were to be found mostly in areas away from the marshland, where stream gradients were steeper. The heavy clays of this area must once have carried a more or less continuous cover of wood, and much still remained. The high water-table and broad river valleys were suitable for meadows, and nearly every vill laid claim to some. Less than half the villages, on the other hand, possessed pasture. Sheep seem to have been fairly numerous.

(6) *The Northern Hills and Valleys*

This is one of the least homogeneous of the regions of Somerset. It is a deeply dissected area in which the rounded hill-tops are usually over 400 ft above sea-level, and the narrow valley bottoms usually under 200 ft. The soils of this region have weathered out from a wide variety of geological formations. Those developed on the Triassic marls of the south and west, and on the clay outcrops along the northern border of the county, are the most fertile; those on the limestones and Coal Measures tend to be poorer. But the formation of soils has been greatly complicated by downwash on the steep valley slopes, and it is impossible to generalise.

Vills were numerous and tended to be large. On the average 3–4 teams were working on each square mile of country, but doubtless most of these were to be found on the warm, red marls, for considerable tracts of shallow-soiled, limestone heathland were not cultivated until the end of the eighteenth century,[1] or later. The density of population ranged from about 7 to 10. Mills were numerous and clusters of 3 and over not uncommon. At Keynsham, there were as many as 6 rendering 60s. The progress of cultivation had by no means obliterated the woodland, large tracts of which still remained, and there were extensive pastures, many of them in the dry, limestone country. Another feature of the countryside was the abundance of meadow along the valleys of the Chew and other rivers. Finally, the district seems to have supported substantial flocks of sheep.

[1] See J. Billingsley, *A General View of the Agriculture of Somerset* (London, 1797), p. 76: 'A large portion will not admit of cultivation, the limestone rock being within two or three inches of the surface.'

(7) *The Mendips*

This upland lies for the most part over 600 ft above sea-level; much of it forms a plateau at between 800 and 900 ft. It consists mainly of lime-stone with a number of red sandstone inliers that rise to about 1,000 ft above sea-level. Between the main upland and the coast lies the outlier of Bleadon Hill, which rises to over 400 ft. Domesday Book records no villages on the main part of the upland, with its thin soils and lack of water, but shows them clustered on the warm, fertile soils derived from a belt of conglomerate at the foot of the southern slope. Here was spring water, good arable land and such appurtenances as wood, meadow and pasture within easy access, and this line of settlements included some of the largest and wealthiest in Somerset. Such were Wells, with 55 teams and 130 recorded people, and Cheddar, with 22 teams and 57 people. At the foot of the northern slopes of the Mendips there was a second line of pros-perous settlements, of which Chewton Mendip with 30½ teams and 85 men was the wealthiest. But the wide expanse of uninhabited country on the Mendips reduced the densities to between 1·5 and 2·0 teams and to be-tween 4·5 and under 7 people per square mile. Most villages had a mill or mills. They also had considerable tracts of wood and pasture recorded for them. The former were probably on the slopes and combe sides, and the latter on the plateau surface. Wells, for example, had pasture 3 leagues by 1 league; Chewton Mendip and Rodney Stoke each had 2 leagues by 1; Cheddar had 1 league by 1 league; and Litton 1,000 acres. In addition, nearly all the villages had some meadow, especially those to the north, in the valleys of the Bristol Avon tributaries. Some villages seem to have had substantial flocks of sheep.

(8) *The Oolitic Scarplands*

Much of eastern Somerset consists of the oolitic scarplands, which run from Bath into Dorset. In this region they are characterised by a rapid succession of limestones, clays and sands which in the north have been eroded to form a highly accidented topography. Here the Bristol Avon has cut a deep, but narrow, gorge through a limestone upland which rises in places to over 600 ft above sea-level. To the south the summits fall away somewhat, but the landscape is still one of flattened ridges and deep valleys. Throughout this region, the limestones forming the uplands weather to fertile and easily worked loams, but the valley soils are so extremely varied as to defy generalisation. All grades from clays to sands occur, more often than not within the bounds of a single parish, but it was precisely to this range of soils that the villages of the region owed their prosperity. Settlements were everywhere numerous, and, on the whole, large and wealthy. Four of them, Bath, Bruton, Frome and Milborne Port, were boroughs. The density of teams ranged between 3 and 4 per square mile, and that of recorded population between 10 and 12. The large number of mills in this region also bore witness to the extent of the arable land. And the other appurtenances recorded in the Domesday Book were hardly less abundant. Wood was especially plentiful, except in the Bath area, but here there was a considerable amount of underwood. Probably the limestone uplands in this part originally carried only a light woodland cover, but this was already a region of ancient settlement, and much wood must have been cleared. Most villages, especially those in the south, had some meadow along the various streams that crossed the belt. Many villages also had some pasture. Most villages seem to have had fairly substantial flocks of sheep.

(9) *The Clay Vale*

Eastern Somerset includes part of the Oxford Clay vale, the greater part of which lies in Wiltshire. Settlements shunned the sour and heavy soils of the clay, and sought sites either on the Cornbrash (e.g. Henstridge, Horsington, Cheriton and Wincanton) or on the Corallian ridge (e.g. Stoke Trister and Cucklington). Here were drier sites, supplies of spring water, and warm, well-drained arable soils. But the Cornbrash and the Corallian are narrow outcrops, and the greater part of the territory of each of these villages must have been situated down in the vale; hence the relatively low densities of plough-teams and population (falling to 1·3 and 5·1 per square mile respectively), and only a few mills were required to supply the needs of the area. The clays were still well wooded, and the high water-table of the vale was favourable for meadows; but amounts of pasture were small, and some villages were without any. There seem to have been moderate numbers of sheep.

BIBLIOGRAPHICAL NOTE

(1) It is interesting to note that John Collinson included a partially extended Latin transcription of the text of the Exchequer folios for Somerset in his *History and Antiquities of the County of Somerset* (Bath, 1791).

Although he did not reproduce any part of the text, William Phelps incorporated in his *History of Somerset* (London, 1836) a brief account of the material contained in the Domesday Book, as well as totals of some of the more important items.

(2) *The Victoria History of the County of Somerset*, vol. I (London, 1906) includes an introduction to the Somerset text (pp. 382–432) and a map of Domesday vills, both by J. H. Round, together with a translation by E. H. Bates (pp. 433–526). This is followed by the same author's translation of the Geld Accounts so far as they relate to Somerset (pp. 527–37).

(3) The first, and only, detailed investigation of the Somerset folios is R. W. Eyton's *Domesday Studies: an Analysis and Digest of the Somerset Survey*, 2 vols., London, 1880). Eyton's conclusions cannot now all be accepted, but his monograph is still of great interest as an early attempt to tabulate the information of Domesday Book, hundred by hundred.

(4) The following is a selection of the papers in the *Proceedings of the Somerset Archaeological and Natural History Society* (Taunton) that deal with various Domesday problems. They are set out in chronological order.

J. A. Bennett, 'Vestiges of the Norman Conquest of Somerset', XXV (1879), pp. 21–8.
E. Hobhouse, 'Domesday estates in Somerset', XXXV (1890), pp. 22–3.
E. Hobhouse, 'Remarks on Domesday map', XXXV (1890), pp. ix–x. Includes coloured map showing chief estates.
E. H. Bates, 'The five-hide unit in the Somerset Domesday', XLV (1899), pp. 51–107.
A. F. Major, 'The geography of the Lower Parrett in early times and the position of Cruca', LXVI (1921), pp. 56–65.
F. W. Morgan, 'The Domesday geography of Somerset', LXXXIV (1938), pp. 139–55.
R. Welldon Finn, 'The making of the Somerset Domesdays', XCIX and C (1954–5), pp. 21–37.
S. C. Morland, 'Some Domesday manors', XCIX and C (1954–5), pp. 38–48.
S. C. Morland, 'Further notes on Somerset Domesday', CVIII (1964), pp. 94–8.
S. Everett, 'The Domesday geography of three Exmoor parishes', CXII (1968), pp. 54–60.
S. C. Morland, 'Hidation on the Glastonbury estates: a study in tax evasion', CXIV (1970), pp. 74–97.

(5) Some other works of interest in the Domesday study of the county are:

T. W. WHALE, 'Principles of the Somerset Domesday', *Trans. Bath Field Club*, x (Bath, 1902), pp. 38–86.

T. W. WHALE, *Analysis of Somerset Domesday, Terrae Occupatae and index* (Bath, 1902). Includes reprint of the paper above.

F. W. MORGAN, 'Domesday woodland in south-west England', *Antiquity*, x (Gloucester, 1936), pp. 306–24. This contains a map of the Domesday woodland of Somerset.

R. LENNARD, 'A neglected Domesday satellite', *Eng. Hist. Rev.* LVIII (London, 1943), pp. 32–41.

R. LENNARD, 'Domesday plough-teams: the south-western evidence', *Eng. Hist. Rev.* LX (London, 1945), pp. 217–33.

H. P. R. FINBERG, 'The Domesday plough-team', *Eng. Hist. Rev.* LXVI (London, 1951), pp. 67–71.

R. LENNARD, 'The composition of the Domesday caruca', *Eng. Hist. Rev.* LXXXI (1966), pp. 770–5.

(6) For the Exeter Domesday and Geld Accounts, see p. 393 below.

CHAPTER IV

DEVONSHIRE

BY R. WELLDON FINN, M.A.

Devonshire is one of those south-western counties for which there are two Domesday descriptions—the Exeter and the Exchequer. The former version, however, is not quite complete. It lacks the 38 entries for the fiefs of Robert Bastard (113–113b), Richard fitzThorold (113b), Alfred the Breton (115b–116) and Hervey de Helléan (117), and also the last two entries for the fief of Robert of Albemarle (113) and the last six for that of Ruald *Adobatus* (115). The missing 46 entries refer to 45 places and one anonymous holding. The Exeter text, on the other hand, includes the entry for Shyttenbrook (*Sotrebroc, 459*) omitted from the Exchequer text. There are also ten other places named in the Exeter version but not in that of the Exchequer, although the latter tells us something about them.[1] The Exeter entries are more detailed than those of the Exchequer text. They usually distinguish between the hidage of the demesne and that of the peasantry, whereas the Exchequer text does this only in entries for Crediton (*117*, 101b) and Littleham near Exmouth (*184*, 104). They also enumerate demesne livestock, and also provide much other information omitted from Exchequer entries, e.g. the swineherds in 23 out of the 67 entries in which they are recorded, the pasture at Georgeham (*408*, 115b), the 12 cottars at Exminster (*83*, 100), and the values of a number of holdings. Moreover, the two versions often disagree in detail: the Exeter version (*179b*) attributes 10 plough-lands to Leigh (Romansleigh) whereas the Exchequer (103b) figure is only one; the Exeter value for Rockbeare is 12*d*. (*216*) whereas the corresponding Exchequer figure is 12*s*. (104b). Furthermore, the Exeter entries often give separate figures for demesne and villein teams which are combined into one total, not always accurately, in the Exchequer versions. The discrepancies that concern totals and general geographical information are set out in Appendix II. To what extent such differences are due to errors in one or the other text, we cannot say.

---

[1] Boode (*126b*, 102b); Battishill, Combebow, Ebsworthy, Fernworthy, Kersford, Way (*289*, *495*; 105b); Burntown, Warne, Wringworthy (*318*, *495*; 108b).

Except for the 46 missing entries, the Exeter text has been followed in the present analysis.

In addition to these two texts, there are auxiliary sources of information. Bound up with the Exeter text in the *Liber Exoniensis* is a series of entries, headed *Terrae Occupatae*, for Somerset, Devonshire and Cornwall, recording such matters as additions to and abstractions from manors, unsanctioned occupation of territory, and failure to pay customary dues. It is obviously constructed from Inquest material and from that which produced the Exeter Domesday, which at times it slightly supplements. It tells us, for example, of a waste plough-land in the hundred of South Molton (*500*). What is more, it refers to four settlements not otherwise mentioned in either text—Exeworthy (*Esseorda, 501b*), Chaffcombe (*Chefecoma, 499*), Widefield (*Witefelda, 499*) and Whitley (*Witelia, 503*).[1]

Also bound up with the Exeter Domesday are two lists of hundreds (*63*). The lists are not identical. The first names 32 hundreds and includes Molland and *Hertesberia*, not mentioned elsewhere in the *Liber Exoniensis*.[2] The second list names only 31 hundreds. The Walkhampton and Ermington of the first list are Roborough and *Aleriga* in the second; Walkhampton is a Domesday settlement in Roborough, but *Alerige* is not found elsewhere other than as a hundred-name. Furthermore, what the first list calls Molton, the second specifies as South Molton. Neither list, nor the Geld Accounts, mentions a hundred of North Molton, but this, strangely enough, is one of the few hundreds mentioned in the Domesday text itself. Yet another subsidiary document is the so-called Geld Rolls or Geld Accounts (*65–71*), which is the record of a geld levied at the rate of 6s. per hide at a time near that of the Inquest. This is of considerable use in identifying Domesday place-names and in allotting manors to eleventh-century hundreds, but it has obvious defects. The total assessment for its thirty-one hundreds comes to 1,026¼ hides, but, even allowing for obvious duplications, the Domesday total amounts to about 100 hides more.[3] O. J. Reichel's explanation was that some royal land formed what

[1] It may be worth noting that Exeworthy and Whitley were in the fief of Hervey de Helléan, who had died by the end of 1086. Chaffcombe was a constituent of the manor of Crediton, and the details for it may well be included with those of the manor itself.

[2] For the Devonshire hundreds, see O. S. Anderson, *The English Hundred-Names: the south-western counties* (Lund and Leipzig, 1939), pp. 73–104.

[3] The number of hundreds in the Geld Accounts happens to be the same as that in the second list on fo. *63*, but the hundred-names are not identical.

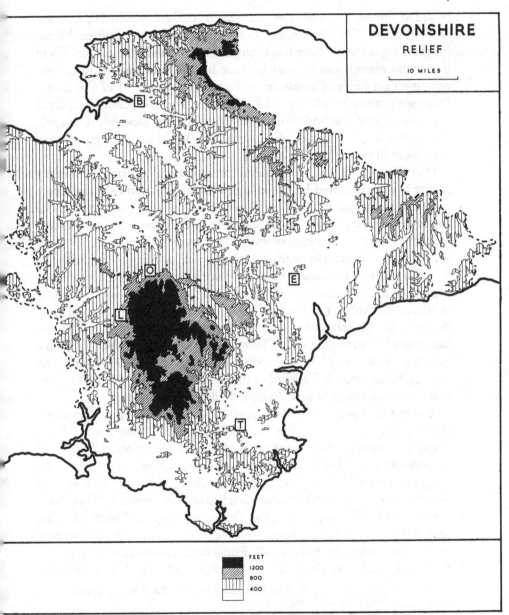

Fig. 51. Devonshire: Relief.

Domesday boroughs are indicated by initials: B, Barnstaple;
E, Exeter; L, Lydford; O, Okehampton; T, Totnes.

he called 'inland hundreds', as opposed to 'outland hundreds', and was ignored in the Geld Accounts.[1] But detailed collation raises problems, for the king is said to have had exempt demesne land in hundreds where he had no exempt manor, and in some hundreds he had more exempt demesne than hides of assessment.[2] Finally, near the end of the Exeter Domesday, among Summaries of various fiefs in the south-west, there is a Summary of the land of Glastonbury Abbey in Devonshire (*527b*); this consisted of the manor of Uplyme (*161*, 103b) and its figures agree with those of the Domesday account.

One difficulty in the geographical interpretation of many Domesday entries is that presented by those large composite manors that include a number of unnamed places. The Devonshire material, however, does not appear to present as many of these problems as do the other four south-western counties. There are, however, a number of large manors. Crediton (*117*, 101b), for example, with a recorded population of 407 and with 185 teams, must have included a number of settlements whose names do not appear in the Domesday text. Or again there were three manors belonging to Tavistock Abbey (*177–9*, 103b) the entries for which provide clear evidence of sub-infeudation—Tavistock itself, Hatherleigh near Oke-hampton and Burrington near Chulmleigh. It is also probable that such a manor as South Tawton (*93*, 100b), with 92 recorded inhabitants and with 44 teams, included a number of distinct settlements. The text pro-vides many instances of post-Conquest additions to manors.

The present county of Devon, in terms of which this study is written, is not exactly the same as the Domesday county. Churchstanton (*382*, 115b) was transferred to Somerset in 1896, and Thorncombe (*313*, 108) to Dorset in 1844. Maker was in Cornwall and surveyed in its folios (*256b*, 122), but part of it had belonged to the distant Devonshire manor of Walkhampton (*87*, 100b) from which it had been taken away.[3] There have been gains as well as losses. The two Domesday manors of Borough (*244*, 124) and Tackbear (*244b*, 125), surveyed in the Cornish folios, are now within the Devonshire parish of Bridgerule. Devonshire also gained Stockland (*44b*, 78) from Dorset in 1844, and Chardstock (77) and

[1] O. J. Reichel, 'The Devonshire Domesday and the Geld Roll', *Trans. Devon Assoc.* XXVII (Okehampton, 1895), pp. 165–98.
[2] In the hundreds of Axmouth (*68b*), Ermington (*70*), Lifton (*65*) and Silverton (*67*).
[3] It seems to have been restored to Devonshire in post-Domesday times, and was not finally transferred to Cornwall until 1844.

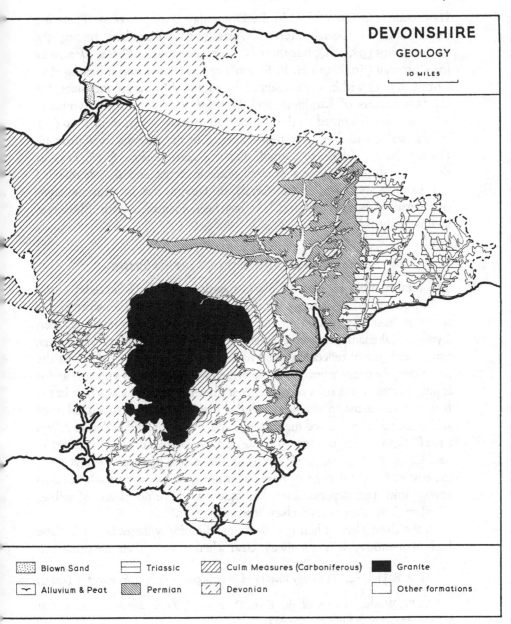

Fig. 52. Devonshire: Geology.
Based on Geological Survey Quarter-Inch Sheets 18 and 22.

Hawkchurch (not described in the Domesday Book) in 1869. There are three points of special interest. Werrington is described among the Devon folios (*98*, 101), but there is a reference to it in *Terrae Occupatae* for Cornwall (*508*), and H. P. R. Finberg has shown that it belonged to Cornwall until within a few years of the Inquest.[1] In the second place, the Cornish manors of Landinner and Trebeigh are said in the Devonshire folios to have belonged to the Devonshire manor of Lifton in 1066 (*93*, *495b*; 100b), and they have been included in the analysis for Cornwall. Finally, T. W. Whale suggested that the *Herstanahaia* (*398*, 113) of the Devonshire folios was entered there in error and that it referred not to Bluehayes (in Broad Clyst) in Devon but to Hurstonhayes (in Broadwindsor) in Dorset.[2] It has been identified with Bluehayes in this analysis.

SETTLEMENTS AND THEIR DISTRIBUTION

The total number of separate places mentioned in the Domesday Book and its associated documents for the area now within Devonshire seems to be at least 983, including the five boroughs of Barnstaple, Exeter, Lydford, Okehampton and Totnes. This figure, for a variety of reasons, cannot accurately reflect the total number of settlements in 1086. In the first place, there are a few entries in which no place-names appear,[3] and it is possible that these refer to holdings at places not named elsewhere in the text. With these may perhaps be included many of those unnamed holdings said to be added to named manors, e.g. the virgate of land added to Afton (*342b*, *503b*; 114b) or the two *terrae* (E.D. *maneria*) added to Burrington near Chulmleigh (*179*, *497b*; 103b). In the second place, as we have seen, the entries for some large manors cover not only the caput itself but also several unnamed dependencies, and again we have no means of telling whether these were named elsewhere in the text.

In the third place, when two or more adjoining villages bear the same basic name today, it is not always clear whether more than one existed in

[1] H. P. R. Finberg, 'The early history of Werrington', *Eng. Hist. Rev.* LIX (1944), pp. 237–51.

[2] T. W. Whale, 'History of the Exon Domesday', *Trans. Devon. Assoc.* XXXVII (Princetown, 1905), pp. 246, 250, 253.

[3] The relevant entries are: St Peter of Plympton, 2 hides (*86*, 100b); bishop of Coutances, 1 hide of land (*132*, 103); William *hostiarius*, one virgate of land (*475*, 117b); Alward Merta, half a virgate of land (*483*, 118).

the eleventh century.[1] There is no indication in the Domesday text that, say, the North Hole and South Hole (in Hartland) of today existed as separate villages; the Domesday information about them is entered under only one name *Hola* (*407*, 115), though they may well have been separate in the eleventh century, as they certainly were by 1300. When the holdings under such names were in different hundreds (as deduced from the Geld Accounts), we have counted them as separate villages. Thus there are two entries for *Tavi*, and we have counted Petertavy in Roborough hundred (116)[2] and Marytavy in Lifton hundred (*317b*, 108b). This is not satisfactory, because, if two or more such places lay in the same hundred, they have been counted as one in the present analysis, e.g. East Densham (*216b*, 104b) and North Densham (*367*, 111), both in Witheridge hundred. Moreover, the criterion of separate hundreds may not be infallible.

The problem is acute in four groups of places which took their respective Domesday names from the streams along which they were aligned. The table on p. 230 shows how these four groups of entries have been allocated on the basis of their name-forms and their respective hundreds as indicated by the Geld Accounts. Taken as a whole, these complications hardly invalidate the general pattern of the distributions.

Very occasionally, the Domesday text does distinguish between the units of a pair bearing the same name. Thus North and South Molton appear as *Nort Moltona* (*94b*, 100b) and *Sut Moltona* (*83b*, *194b*; 100, 104), Weare Gifford and Little Weare as *Wera* (*412b*, 115) and *Litelwera* (*376b*, 116b), Great and Little Rackenford as *Racheneforda* (*311*, 108) and *Litel Racheneforda* (*347*, 111b), Eastleigh and Westleigh as *Leia* (*399b*, 110) and *Weslega* (*420*, 113). In constructing our maps we have also distinguished between Abbotskerswell (*Carsuella*, *184*, 104) held by the abbot of Horton and Kingskerwell (*Carsewilla*, *85*, 100b) held by the king, both in Kerswell hundred.[3] The Exchequer text mentions Dinnaton and *altera* Dinnaton, but these refer to two holdings at one place (*85b*, 100b).

Quite different are those groups of places bearing the same Domesday

---

[1] For the place-names in this section, see J. E. B. Gover, A. Mawer and F. M. Stenton, *The Place-Names of Devon* (2 vols., Cambridge, 1931–2).

[2] The Exeter version is among those missing—see p. 223 above.

[3] Kingsteignton (*Teintona*, *84*, 100), held by the king, and Bishopsteignton (*Taintona*, *117*, 101b), held by the bishop of Exeter, have also been separately plotted, but they would be in any case because they were in separate hundreds—Teignbridge and Exminster respectively.

Four Groups of Devonshire Settlements

Clyst

Broad Clyst	*Clistona 95,* 101	Cliston hundred
Clyst St Lawrence, Clyst Gerred and Ashclyst	*Clist 213b, 301b, 457b;* 105, 107, 116b	Cliston hundred
Clyst St George	*Clisewic 339b,* 114	East Budleigh hundred
Clyst St Mary	*Clista 132b,* 102	East Budleigh hundred
Clyst Hydon	*Clist 301b,* 107	? Hemyock hundred
Clyst William	*Clist 487,* 118b	Silverton hundred
Sowton or Clyst Fomison	*Clis 131b,* 103	Wonford hundred
West Clyst	*Clista 309,* 107	Wonford hundred

Nymet (*Yeo, Mole*)

George Nympton	*Limet, Nimeta 95, 377;* 101, 116b	South Molton hundred
Bishop's Nympton	*Nimetona 119b,* 102	Witheridge hundred
King's Nympton	*Nimetona 98,* 101	Witheridge hundred

Nymet (*Yeo, Taw*)

Nymet Rowland	*Limet 295,* 106b	North Tawton hundred
Broadnymet, Nichol's Nymet, Nymet Tracey, Burston, Hampson, Middle Yeo, Natson, Walson and Zeal Monachorum	*Limet(a), Nimet, 125, 182, 295b, 296, 296b, 389b bis, 390b, 483 bis;* 102b, 103b, 106b *ter,* 112b *ter,* 118 *bis*	North Tawton hundred

Otter

Upottery, Mohun's Ottery and Combe Raleigh	*Otri 342b, 348, 348b;* 112 *bis,* 114b	Axminster hundred
Otterton	*Otritona 194b,* 104	East Budleigh hundred
Dotton	*Otrit 308b,* 107b	East Budleigh hundred
Rapshays	*Oteri 340, 403b;* 110b, 114	East Budleigh hundred
Waringstone and Deerpark	*Oteri, Otria, 338b, 342, 400b, 405;* 110b *bis,* 114, 114b	Hemyock hundred
Ottery St Mary	*Otri, 195,* 104	Ottery St Mary hundred
Hembury	*Otri(a), 342, 405;* 110b, 114b	Tiverton hundred

name but not lying adjacent. Thus the references to *Aisa* or *Aissa* cover nine widely separated places.[1] Or, again, the sixteen entries relating to *Bochelan, Bochelant* or *Bochelanda* refer to a variety of places, occasionally adjacent (East and West Buckland) but more often widely separated. The same is also true, for example, for the 21 entries for Leigh (*Lega, Liega, Weslega*) and for the 19 entries for Combe or Coombe (*Coma, Comba, Cumbe et al.*).

The total of 983 includes some two dozen places about which very little information is given. The record may be incomplete or the details may have been included within those for other manors. Thus all we hear of Sidmouth is an incidental reference to 'the land of St Michael of Sidmouth' in an entry relating to Ottery St Mary (*195*, 104). We are told scarcely anything of the four settlements mentioned only in *Terrae Occupatae*—Chaffcombe (*499*), Exeworthy (*501b*), Widefield (*499*) and Whitley (*503*). All we hear of Irishcombe (*109b*, 101b) is that it lay in (*adjacet*) Lapford. And again, apart from its assessment, we have no statistical detail for Boode (*126b*, 102b), which had been taken away from Braunton. All we are told, for example, of Broadaford (*219*, 105b) and Radish (*343b*, 114b) is their assessment, their plough-lands and their values. There are also other settlements with no teams or no population or neither entered for them.[2]

Devonshire resembles Cornwall in that as many as 60 per cent of its 983 Domesday place-names are represented today not by parish names but merely by those of hamlets, of individual farms and houses, or of topographical features. Thus *Bochiywis* (*407*, 115) is now the hamlet of Buck's Cross in Woolfardisworthy in Hartland hundred, and there is a Buck's Mills nearby; *Esnideleia* (*302*, 107) is that of Snydles in Chittlehamholt. *Derta* (*415b*, 111) appears in the name of Dart Farm in Cadeleigh, and *Rodeleia* (*131*, 103) in that of Rowley Farm in Parracombe. *Estandona* (*337b*, 114) survives only in the name Eastanton Down in Lynton, and *Reddix* (*343b*, 114b) in that of Radish Plantation in Southleigh. The parish of Braunton, for example, has seven Domesday names

[1] Ashwater (*121b*, 102) in Black Torrington hundred, Ash in Bradworthy (*335b*, 114), also in Black Torrington hundred, but some distance away, Ash in Petrockstow (*182*, 103b) in Shebbear hundred, Ash in South Tawton (*93b*, 100b) in South Tawton hundred, Ashreigney (*109b*, 101b) in North Tawton hundred, Ash Barton (*401b*, 110b) in Braunton hundred, Rose Ash (*310*, 108) in Witheridge hundred, Ash Thomas in Halberton (*394b*, 112b) in Halberton hundred, Ashford in Aveton Gifford (*182b*, 103b) in Ermington hundred.   [2] See pp. 245–6 below.

within its limits, apart from its own.[1] The parish of Plympton St Mary
has as many as thirteen additional Domesday names.[2] Some names have
entirely disappeared from the map. *Depdona* (*367b*, 111) is a lost 'Dewdon' in the parish of Widdecombe in the Moor, and the last record of it
is from 1737; *Wiborda* (*217*, 104b) was last recorded in 1630 as Wom-
bernford in Cotleigh; *Haletreu* (*421b*, 113) seems to have last been recorded about 1515 as *Halstowde* in Woodleigh; and *Deneord* (*84b*, 100b)
appeared last as long ago as 1330 in the form of *Deneworthi* in Membury.
It is possible to have divergent opinions about some identifications,
and thirteen names have been left unidentified in the present analysis.[3]
Whether they will yet be located or whether the places they represent
have completely disappeared, leaving no trace or record behind, we
cannot say.

On the other hand, about 40 or so parishes on the modern map are not
mentioned in the Domesday Book. Their names do not appear until the
twelfth century or later, and, presumably, if they existed in 1086, they are
accounted for under the statistics relating to other place-names. Thus, so
far as record goes, Chudleigh was first mentioned in 1150 and Thor-
verton in 1201. Some places not mentioned in the Domesday Book must
have existed, or at any rate borne names, in Domesday times, because they
are named in pre-Domesday documents and again in the twelfth or
thirteenth century. Such are Sandford, mentioned in 930; Halwell, mentioned in the tenth-century Burghal Hidage; and Dartmouth, mentioned
in 1049. Frequently, although a modern parish is not named in the
Domesday Book, it contains a name or names that are mentioned. Thus
Plymouth, not named before 1230, contains the Domesday names of
Lipson (*222*, 105b), Mutley (*328*, 109b), Sutton (*86b*, 100b) and Sutton

[1] Braunton (*83b*, *194b*; 100, 104), Ash Barton (*401b*, 110b), Beer Charter Barton
(*126b*, 102b), Boode (*126b*, 102b), Buckland Challons (*401b*, 110b), Lobb (*300b*, *414*;
107, 115), Saunton (*408*, 115b), Winsham (*128*, 102b).

[2] Plympton St Mary (*86*, 100b), Baccamoor (*331*, 110), Battisford (*417*, 111b),
Bickford (*488b*; 113b, 118), Challonsleigh (*Lega*, *417*, 111b), Elfordleigh (*Lega*,
333, 110), Hemerdon (115), Holland (*331b*, 110), Langage (*135b*, 103b), Loughtor
(*333*, 110), Torridge (*221*, *333*; 105, 110), Walford (*331b*, 110), Woodford (*333b bis*,
110 *bis*), Yealmpstone (115).

[3] *Alwinestona* (*213b*, *499b*, 105), *Assacota* (*416b bis*, 502, 111 bis), *Bernardesmora*
(*303b*, 107b), *Ciclet* (*391b*, 112b), *Eltemetona* (*311*, 108), *Ferding* (115b), *Liclemora* (*319*,
109), *Macheswelle* (115), *Madescama* (*404*, *501b*, 110b), *Magnetona* (*312b*, 108), *Metwi*
(113b), *Nochecota* (*459b*, 117), *Wedreriga* (*222b*, *505b*, 105). Three names, it will be
noticed, were on the missing pages of the Exeter text and so appear only in the Exchequer version.

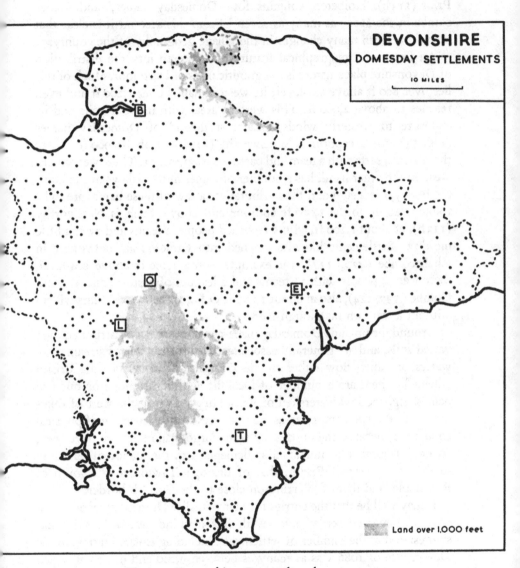

Fig. 53. Devonshire: Domesday place-names.

Domesday boroughs are indicated by initials: B, Barnstaple;
E, Exeter; L, Lydford; O, Okehampton; T, Totnes.

Prior (113b). Holbeton contains four Domesday names,[1] and Wide-
combe in the Moor as many as seven.[2] From this account it is clear that
there have been many changes in the village geography of the county.

An important geographical feature that entered into the distribution
of Devonshire place-names is the granitic mass of Dartmoor. Much of this
lies over 800 ft above sea-level; its western half is over 1,200 ft and even
reaches to above 2,000 ft. This western area, with its poor soils and its
exposure to westerly winds, is almost devoid of Domesday names
(Fig. 53). An exception is Willsworthy (115b), at about 900 ft, where
there were 4 serfs with a team and pasture 2 leagues by 1. The lower eastern
area, on the other hand, has a number of names in the valleys of the Dart,
the Bovey, the Teign and their tributaries; a few of these names are to be
found even up to the 1,000 ft contour and above. Such is Natsworthy
(113b), at about 1,200 ft, with 5 recorded people, 2 teams and even a little
meadow. Settlements likewise reached into Exmoor: at Radworthy in
Challacombe (415, 111), for example, some 1,200 ft above sea-level,
there were 4 people with 1¼ teams and 2 leagues of pasture; and at Lank
Combe (337, 114), also at about 1,200 ft O.D., there was a solitary villein
without a team on his plough-land.

Around the uplands Domesday place-names are widely distributed over
varied soils, and are generally associated with the numerous streams, the
waters of which flow either to the English Channel or to the Bristol
Channel. There are a number of localities with either none or but few
names, e.g. the low watershed of Broadbury Down to the west of Okehampton. But such empty areas are not necessarily a reflection of physical
conditions, because the entries for some large manors, as we have seen,
covered a number of unnamed settlements, e.g. the manor of Crediton
to the north-west of Exeter, that of Bishop's Tawton to the south of
Barnstaple, and that of Werrington close to the Cornish border.

It may well be that the entries for many smaller Domesday manors also
covered unnamed settlements. With this in mind, W. G. Hoskins has
suggested that the number of settlements, including isolated farms, in the
Devonshire of 1086 was as many as between 9,000 and 9,500, of which
a thousand or so were demesne farms and the remainder were occupied

---

[1] Battisborough (116), Flete Damarel (421, 113), Lambside (327b, 109b), Memb-
land ((327b, 109b).
[2] Blackslade (342, 114b), Dewdon (367b, 111), Dunstone (342, 114b), Natsworthy (113b), Scobitor (135b, 102), Sherberton (182b, 104), Spitchwick (97, 101).

by villeins.[1] This is on the assumption that we should 'allocate every villein to a separate farmstead'. At Honeychurch (292, 106), for example, there were five plough-lands with two teams on the demesne and one with the peasantry; and the recorded population comprised 4 villeins and 4 serfs. 'Throughout all the relevant documents of Honeychurch's history, down to the tithe award of 1839–41, there are five farms in the parish', scattered over its surface. From this it is but a step to envisage a demesne farm (on which the serfs worked) and four separate villein farmsteads. Very often, on the other hand, the number of villeins on a holding is the same as the number of plough-lands. 'In other words it looks as though when Domesday says of a Devonshire manor that 'x ploughs can till it' it is really saying 'there are x farms on this manor in addition to the demesne farm'. But the arithmetic is not always easy. There are sometimes more villeins than plough-lands and sometimes fewer; moreover, either figure often differs from the number of isolated farmsteads that appear on nineteenth-century surveys. It is clear that such interpretations involve many conjectures and assumptions and that, as Hoskins says, 'there are many exceptions for which an explanation is not easy'. Whatever be the situation in this or that locality, we are safe in envisaging Devonshire, with its broken landscape of hill and valley, to have been a land of hamlets and dispersed settlements in the eleventh century, as it very largely is today.

THE DISTRIBUTION OF PROSPERITY AND POPULATION

Some idea of the information in the Domesday folios for Devonshire, and of the form in which it is presented, may be obtained from the account of Lewtrenchard in the valley of the Lew to the west of Dartmoor, some ten miles south-west of Okehampton. It was held entirely by one owner and so is described in a single entry, both versions of which are set out on p. 236. This account does not include all the items of information that appear elsewhere in the folios for the county; but it is a fairly representative and straightforward entry, and it sets out the recurring standard items that are entered under most place-names. These are five in number: (1) assessment, (2) plough-lands, (3) plough-teams, (4) population, and (5) values. The bearing of these five items upon regional variations in the prosperity of the county must now be considered.

[1] W. G. Hoskins, *Provincial England* (London, 1963), pp. 20 ff., 33, 34, 43.

Lewtrenchard

Exeter Domesday (289–289 b)

Baldwin has one manor which is called Lewtrenchard which Brictric held on the day that King Edward was alive and dead, and it rendered geld for half a hide. Seven plough-teams can plough this. And now Roger de Meules holds it of Baldwin. There, Roger has 1 plough-team in demesne, and the villeins [have] 6 plough-teams. There, Roger has 12 villeins and 8 bordars and 6 serfs and 18 animals and 50 sheep and 30 acres of wood and 20 acres of meadow and 60 acres of pasture. This is worth £4, and when Baldwin received it, it was worth 60s.

Exchequer Domesday (106)

Roger de Meules holds Lewtrenchard of Baldwin. Brictric held it in the time of King Edward, and it paid geld for half a hide. There is land for 7 plough-teams. In demesne is one plough-team, and 6 serfs, and [there are] 12 villeins and 8 bordars with 6 plough-teams. There [are] 20 acres of meadow and 60 acres of pasture and 30 acres of wood. Formerly [it was worth] £3. Now it is worth £4.

(1) *Assessment*

The Devon assessment is stated in terms of hides, virgates and ferlings (or ferdings or fertings). Fractions of all three are encountered. An exceptional entry for Sherberton (*Aiserstone, 182 b*, 104) mentions geld acres: 1½ ferlings and 3 acres of land. The usual Exchequer formula is *geldabat pro n hidis*, but there are occasional variations, e.g. the entry for Exminster (*83*, 100) uses the phrase *defendit se pro una hida.*[1] The city of Exeter paid geld only when London, York and Winchester did, and then only half a silver mark (*88*, 100). Nothing is said about the geld of the boroughs of Barnstaple and Lydford (*87 b*, 100), but Totnes paid geld of 40d. when Exeter did (*334*, 108b); Okehampton was assessed in the usual way (*288*, 105b). Four virgates made a hide and 4 ferlings made a virgate. Various entries indicate these equations. Thus at West Putford (*412 b*), ½ hide and ½ virgate are equated with 2 virgates and 2 ferlings; and at Riddlecombe (*389*) 1 hide is equated with 2½ virgates and 6 ferlings.

Some people have thought that the hides of the south-west were small, and that they comprised not the usual 120 acres but only 40 or 48.[2] The

---

[1] The corresponding Exeter formula runs: *reddidit gildum pro n hidis.* An Exeter entry for Blakewell (*299 b*) interlines *defendebat se* over *reddidit.*

[2] See p. 80 above. Woodland (*Liteltrorilanda, 212*, 104b) paid geld for the third part of a ferling, which may indicate that the number of acres in a ferling was divisible by three.

solitary reference to acres at Sherberton, which we have already noted, throws no light on this, unless we can suppose that 3 acres was at any rate less than half a ferling. This would imply a reckoning different from that in Cornwall, where one ferling comprised 3 geld-acres.[1] On the other hand, for Devonshire as for Cornwall, the total assessment (in relation to, say, teams and population) is very low, much lower than that of the counties to the east—Dorset, Somerset and Wiltshire. To what extent this 'under-rating' reflects the poverty of the south-west or its previous history as 'West Wales', or both, we cannot tell.[2]

The Exchequer text sometimes states also the assessment of the demesne portion of an estate; the Exeter text not only very frequently does this, but in addition it states explicitly the assessment of the portion in the occupation of the peasantry. It is interesting to compare the sum of the demesne and other hidage with the total given for a holding as a whole. In most entries there is agreement, but in nearly fifty entries there are discrepancies between the stated total and the actual total. Some of these discrepancies may be due to simple clerical errors; others may be the result of unknown changes in the composition of manors. Thus both the Exchequer and Exeter assessments for Middlecott in Broadwood Kelly (*292b*, 106) amount to 1 virgate, but the Exeter details are 1 ferling in demesne and 2 with the peasantry. Again, both versions agree in stating the assessment of Lowley (*468b*, 117) to be 1 virgate, but the Exeter details are 1 ferling in demesne, 2 ferlings with the peasantry, and 1 ferling of underwood. This inclusion of underwood is very unusual.

Five manors are said never to have paid geld: the entries below are those of the Exeter text. The first three were, and had been, royal manors; the fourth had belonged to King Edward, but in 1086 was in the hands of Walter of Douai.

Silverton (*83b*, 100): *nescitur quot hidae sint ibi quia nunquam reddidit gildum.*

Axminster (*84b*, 100): *nescitur quot hidae sint ibi quia nunquam reddidit gildum.*

Axmouth (*85*, 100b): *nescitur quot hidae sint ibi quia nunquam reddidit gildum.*

Bampton (*345b*, 111b): *in ista terra quot hidae iacent nescimus quia numquam* [sic] *reddidit gildum.*

Buckfast (*183*, 104): *ista nunquam reddidit gildum.*

[1] See p. 306 below.
[2] See (1) F. W. Maitland, *Domesday Book and Beyond* (Cambridge, 1897), pp. 425, 463 and 467; (2) J. H. Round, *Feudal England* (London, 1895), pp. 63, 93.

None of the entries for these royal manors mentions the *firma unius noctis*, unlike similar entries for Wiltshire, Dorset and Somerset.[1] Nowhere for Devon are we told of *carucatae nunquam geldantes*.[2] We do, however, in three entries, hear of *carucatae*; but whether or not these refer to un-assessed land, we cannot say.[3] One of the entries—that for Heavitree (*343b*, 114b)—makes no reference to assessment.

There are only slight traces of the 5-hide unit, and these are to be found in the south and south-east, which may well have been the initial area of Saxon penetration and settlement. Only eleven settlements seem to have had a decimal assessment, and nine of these were in royal or ecclesiastical hands. Seven were rated at 5 hides, and one each at 10, 15, 20 and 25 hides. It is possible that some settlements were grouped together for the purpose of decimal assessment, but most Devon hidages were so small that the possibility of error in visualising such combinations is immense. It may, however, be of interest to note that thirteen out of the thirty-one hundreds of the Geld Accounts were said to be assessed in multiples of 5 hides, and it has been thought possible that some neigh-bouring pairs of hundreds were grouped together to produce a similar result: e.g. Colyton (25¾ hides) with Axmouth (9¼); Chillington (46) with Wonford (54); Shebbear (48) with North Tawton (42) or South Molton (22).[4]

It is clear that many assessments were artificial and bore no constant relation to the agricultural resources of a settlement. The variation among a representative selection of 1-hide holdings speaks for itself:

	Plough-lands	Plough-teams	Popula-tion	*Valuit*	*Valet*
Ashwater (*121b*, 102)	20	18	58	£7 10s.	£7 10s.
Bickleigh near Silver-ton (*214*, 105b)	8	8	30	40s.	£3
Brightston (*119*, 102)	5	4	9	£1	£1
Cheriton in Pay-hembury (*214b*, 105b)	2	1	4	15s.	15s.
Thrushelton (*316*, 108b)	14	14	51	£10	£10

[1] The *firma unius noctis* is, however, mentioned in an entry for three other royal manors—see p. 245 below. For Wiltshire, Dorset and Somerset, see pp. 30, 97, 179.
[2] See pp. 15, 83 and 153 above.   [3] See p. 240 below.
[4] Sir F. Pollock, 'A brief survey of Domesday', *Eng. Hist. Rev.* XI (1896), pp. 209–30.

An extreme example of low rating was the royal manor of North Tawton (*83*, 100). It answered for only half a virgate, yet there were 62 recorded people with 30 teams at work, and the annual value of the holding amounted to £15.

The total assessment amounted to 1,158 hides, 2 virgates, 2⅝ ferlings and 3 acres. This is for the area included in the modern county, and so is not strictly comparable with Maitland's total for the Domesday county.[1] The nature of some entries and the possibility of unrecognised duplicate entries make exact calculation very difficult. All that any estimate can do is to indicate the order of magnitude involved. We must also remember the five manors which had never paid geld.

(2) *Plough-lands*

Plough-lands are systematically entered for the Devon holdings, and the Exchequer formula normally runs: 'There is land for *n* plough-teams' (*Terra est n carucis*). The Exeter text usually says: *hanc terram possunt arare n carucae*. A reckoning in half-teams is not unusual; but there are only very occasional references to oxen, such as that for Shyttenbrook (*459*). There are also eight entries that make no reference to plough-lands although they state the number of teams at work. e.g. those for Buckfast (*183*, 104) and Kingsford (*303*, 107); these eight entries account for 10¾ teams. Not included in this total are the teams of those other holdings whose plough-lands seem to be recorded under those of parent manors, e.g. Luscombe (*368b*, 111). There are, moreover, six other entries which mention neither plough-lands nor teams; this is so, for example, in the account of Peek in Luffincott (*412*, 115). Many plough-land figures look suspiciously artificial, e.g. North Molton (*94b*, 100b), with 100 for 47 teams.

There are some curiosities. The Exchequer entry for Silverton runs: *Terra est xli carucis cum omnibus ibi pertinentibus* (100).[2] Four entries mention *carucatae* (*terrae*), and these are set out on p. 240. Whether they referred to unassessed land, we cannot say.[3] There is also an unusual entry

---

[1] F. W. Maitland, *op. cit.* p. 400, reckoned 1,119 hides.

[2] The Exeter version reads: *hanc possunt arare xli carrucae cum omnibus qui ibi pertinent* (*83b*). The figure of 41 here is a correction over an erasure. There was a total of 36 teams at work.

[3] See pp. 16, 82 and 153.

in the Exeter account of Bampton, but the Exchequer entry seems normal enough:

> E.D. (*345b*): *in ista terra quot hidae iacent nescitur quia numquam* [sic] *reddidit gildum; hanc terram possunt arare xxv carrucae. Inde habet W*[*alscinus*] *terram iiii carrucis in dominio, et ibi habet ii carrucas et villani habent terram xviii carrucis.* No villeins' teams are mentioned.
>
> D.B. (111b): *Haec terra nunquam geldavit. Terra est xxv carucis. In dominio sunt ii carucae et ii servi et xxxi villani et xx bordarii cum xviii carucis.*

The relation between plough-lands and teams varies a great deal in individual entries. They are equal in about 23 per cent of the entries which record both. Occasionally, we are explicitly told that this is so, as in the entry for Virworthy (*399*, 110): *Terra est i carucae quae ibi est*; in the Exchequer entry for Appledore in Clannaborough (106b): *Terra est i carucae et dimidia et tantum ibi est;*[1] or, again, in the Exchequer entry for Kenn (106b): *Terra est xxv carucis et totidem ibi sunt.* A deficiency of teams is much more common, and is encountered in 71 per cent of the entries which record both plough-lands and teams. Some of the deficiencies are very striking. Thus Hartland (*93b*, 100b), with 110, had only 45; Tawstock (*94b*, 101), with 80, had only 31 teams; and Werrington (*98*,

Devonshire: *Carucatae terrae*
(Exeter version)

Braunton (*83b*, 100): *defendit se pro i hida; hanc possunt arare xl carrucae. Inde habet rex i carrucatam terrae et i carrucam et villani habent xxx carrucas.* Carucates not mentioned in D.B.

Heavitree (*343b*, 114b): *In ea sunt ii carrucatae terrae et ii carrucae; i dominica est R*[*adulfi*] *et alia villanorum.* No mention of assessment either here or in D.B.

North Tawton (*83*, 100): *reddidit Gildum pro dimidia virgata T.R.E.; hanc possunt arare xxx carrucae. Ibi habet R*[*ex*] *iii carrucatas terrae et iii carrucas in dominio et villani habent xxvii carrucas.* Carucates not mentioned in D.B.

Chittlehampton (*484*, 118): *reddidit Gildum pro una virgata; hanc possunt arare xxx carrucae. De ea habet Godwinus v carrucatas in dominio et iii carrucas et villani habent x carrucas.* No specific mention of *terrae*. Carucates not mentioned in D.B. which reads: *In dominio sunt v carucae.*

---

[1] The Exeter entry (*295b*) mentions land for 1½ teams but records 1 team and 5 oxen at work.

101), with 86, had only 33 teams. The entry for the land of fifteen thegns added to the manor of Bovey Tracey (*135b*, 102) refers explicitly to a deficiency: *Terra est viii carucis et ibi sunt vi carucae.* The values of some deficient holdings had fallen, but on others they had remained constant, and on yet others they had even increased. No general correlation between plough-team deficiency and decreases in value is therefore possible, as the following table shows:

		Plough-lands	Teams	*Valuit*	*Valet*
Decrease	Arlington (*37t*, 115b)	15	7	£8	£3
in value	Roborough (*124b*, 102)	14	9	£4	£3
Same	Clayhidon (*304b*, 107b)	12	8	£4	£4
value	Upcott (*308*, 107b)	5	2	10s.	10s.
Increase	Stoke Rivers (*415*, 111)	12	5	20s.	40s.
in value	Martinhoe (*125*, 102b)	8	4	5s.	30s.

These figures are merely indications of unexplained changes in individual holdings.

An excess of teams is a good deal less frequently encountered, being found in only 6 per cent of the entries which record both plough-lands and teams. Nearly one half of the holdings with excess teams were in the fertile and relatively prosperous south and south-east. The values on some of these holdings had risen, e.g. from £10 to £18 at Otterton (*194b*, 104), where the excess was very striking—25 plough-lands with 46 teams. But very frequently this was not so, and other holdings with excess teams were valued the same at both dates, e.g. 60s. at Salcombe Regis (*118b*, 102), which had 6 plough-lands with 8 teams. The values of two holdings with excess teams had even dropped: from £4 to £3 at Thurlestone (*321b*, 109) and from £6 to 105s. at Edginswell (*461b*, 113b). Our attention is sometimes specifically drawn to excess teams. An entry for Butterleigh (*483b*, 118b), for example, reads: *Terra est ii carucis. Ibi sunt tamen iii carucae.* This phrase, or something like it, occurs altogether in 15 entries, but no entry hints at the reason for overstocking.[1]

[1] The entries are: Aller in South Molton (*377b*, 116b), Besley (*461b*, 116), Bicton (*472*, 117b), Buckland-Tout-Saints (*396*, 113), Burrington in Weston Peverel (*328b*, 109b), Butterleigh (*483b*, 118b), Cheldon (*459b*, 116), East Bradley (*472b*, 117b), Fenacre (*391b*, 112b), Grimston (116), Nymet in North Tawton hd. (*390b*, 112b), Shillingford (*468b*, 117b), Sutton in Halberton (*461*, 116), Torridge (*333*, 110), Whitnole (*471b*, 117).

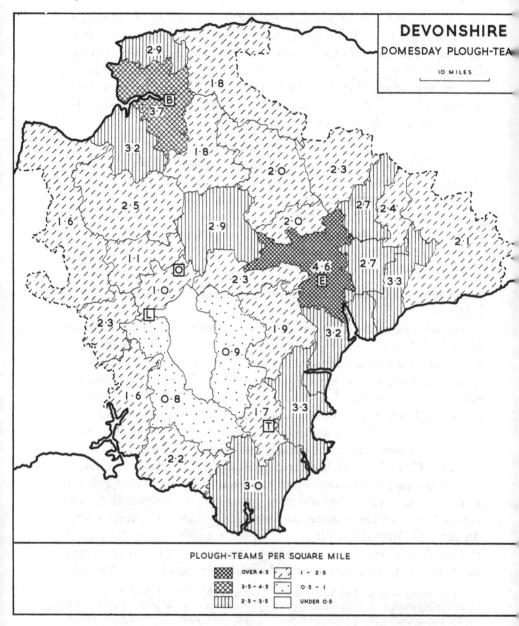

Fig. 54. Devonshire: Domesday plough-teams in 1086 (by densities).

Domesday boroughs are indicated by initials: B, Barnstaple; E, Exeter; L, Lydford;
O, Okehampton; T, Totnes. For the blank area of Lydford parish on Dartmoor,
see p. 255.

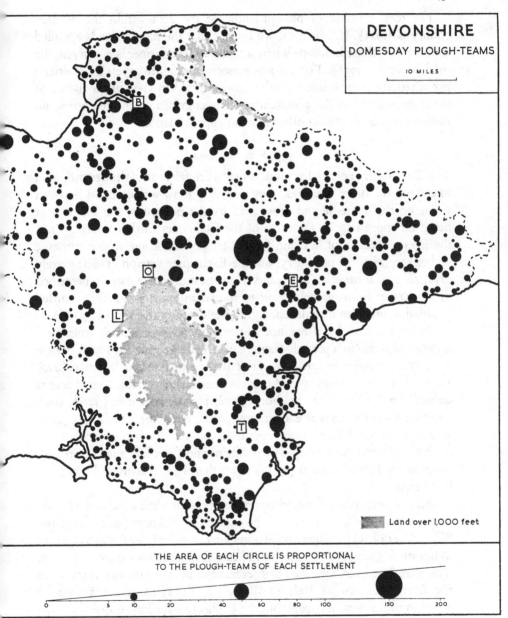

DEVONSHIRE
DOMESDAY PLOUGH-TEAMS
10 MILES

Land over 1,000 feet

THE AREA OF EACH CIRCLE IS PROPORTIONAL
TO THE PLOUGH-TEAMS OF EACH SETTLEMENT

0   5   10   20   40   60   80   100   150   200

Fig. 55. Devonshire: Domesday plough-teams in 1086 (by settlements).

Domesday boroughs are indicated by initials: B, Barnstaple;
E, Exeter; L, Lydford; O, Okehampton; T, Totnes.

The total number of plough-lands for the area within the modern county is 7,921¼. If, however, we assume that the plough-lands equalled the teams on those holdings where a figure for the former is not given, the total becomes 7,931⅞. For various reasons, the figure given by Maitland is not strictly comparable with the present total.[1] In view of the nature of some entries and of the possibility of unrecognised duplicate entries, no estimate can be anything other than an approximation.

(3) Plough-teams

The Devonshire entries for plough-teams are fairly straightforward, and, like those for other counties, they normally draw a distinction between teams on the demesne and those held by the peasantry, the Exeter text being far more regular than that of the Exchequer in this respect. Thus the Exeter account of Willand (*379b*) tells of one team on the demesne and of one with the peasantry, but the Exchequer version (116b) merely says: *Ibi sunt ii carucae.* The teams of the peasantry are usually described in the Exeter folios as belonging to the *villani* even when the recorded population includes no villeins, e.g. in the entry for Loventor (*320b*).

Half-teams and oxen are not uncommon. We hear sometimes of a number of oxen 'in a plough-team', e.g. at Killington (*127*, 102b), where there were *vi boves in carrucam*, or at Radworthy in Challacombe (*415*, 111), where there were two. An occasional variant in the Exeter text is *animalia in carrucam*, e.g. at Ilton (*322b*, 109) and Lupton (*320b*, 109). Another Exeter variant is encountered in the entry for Chichacott (*288*) where we read: *Villani arant cum ii bobus.* The Exchequer account of Potheridge (106) incorrectly says '6 villeins and 5 bordars with 2 bordars', whereas the Exeter version (*293b*) states that the 6 villeins and 5 bordars had 2 teams.

Occasionally, when the Exchequer text mentions half a team, the Exeter version speaks of 4 oxen, e.g. in the entries for Bickham (*420b*, 113) and Sigford (*414b*, 115). The two versions for Hareston (*221b*, 104b) and for Whitestone (*300b*, 107) likewise indicate that 4 oxen + 4 oxen = 1 team. There are, however, a number of discrepancies between the two texts. Thus the Exeter account of Hele in Ilfracombe (*128*) mentions 2 oxen in addition to the demesne team, but the Exchequer version records only the

[1] F. W. Maitland, *op. cit.* pp. 401, 410. Maitland's total of 7,972 was for the Domesday county, but, as in the present estimate, allowance was made for the missing plough-land figures.

team (102b). There are also other holdings for which the Exeter text mentions pairs of oxen that do not appear in the Exchequer text, e.g. Germansweek (*289*, 106), Goosewell (*417*, 111b) and Tetcott (*319*, 108b). Or, again, the single ox at Thorne (*122*, 103) and at Fernhill (*332*, 110) are likewise omitted from the Exchequer version. The discrepancy appears to be the other way round in the account of Blachford; here, the Exeter total of 2 teams 6 oxen (*327*) appears in the Exchequer entry as 3 teams (109b). Or, again, the Exeter account of Twinyeo mentions 3 oxen (*488b*), whereas the Exchequer entry records half a team (118b). Such anomalies led Reginald Lennard to believe that the accepted equation of 8 oxen to 1 team (as postulated by Round and Maitland)[1] was not always true in the south-west. He argued that the Domesday team here sometimes comprised 6 or even 4 oxen.[2] H. P. R. Finberg, in reply, argued in favour of a uniform team of 8 oxen and believed that any anomalies were due to the 'contempt for small fractions' shown by the Exchequer text.[3]

About 90 entries do not mention teams although they refer to ploughlands and sometimes to people as well as to resources such as wood and meadow. Such, for example, is the entry for Instow (*390b*, 112b) with eleven recorded inhabitants and with 2 plough-lands amongst other things. Another place without teams was Rewe (*131b*, 103) with 10 recorded inhabitants and 5 plough-lands. There are curiosities. An entry for Wallover (*129*, 102b) refers to 3 plough-lands and to 1 team and 4 recorded inhabitants, but assigns no teams to the demesne. An entry for Combe in South Pool (*324b*, 109), on the other hand, refers only to demesne teams. Occasional settlements seem to have possessed a number of teams beyond the capacity of their inhabitants. Thus 3 teams would seem to have been too many for the 3 recorded inhabitants of Owlacombe (*407b*, 115b) to manage without help.

The total number of plough-teams amounted to 5,758¼, but this refers to the area included in the modern county, and, in any case, a definitive total is hardly possible.[4] This count has been made on the assumption of an 8-ox team.

[1] J. H. Round, *op. cit.* p. 35; F. W. Maitland, *op. cit.* p. 417.
[2] R. Lennard, 'Domesday plough-teams: the south-western evidence', *Eng. Hist. Rev.* LX (1945), pp. 217–33. See also R. Lennard, 'The composition of the Domesday caruca', *Eng. Hist. Rev.* LXXXI (1966), pp. 770–5.
[3] H. P. R. Finberg, 'The Domesday plough-team', *Eng. Hist. Rev.* LXVI (1951), p. 68.
[4] F. W. Maitland's total for the Domesday county was 5,542 teams (*op. cit.* p. 401).

(4) *Population*

The bulk of the population was comprised in the three main categories
of villeins, bordars and serfs. In addition to these were the burgesses and
a miscellaneous group that included an unusually large number of swineherds together with coscets, salt-workers, cottars, coliberts and others.
The details of the groups are summarised on p. 247. There are two other
estimates, by Sir Henry Ellis[1] and R. Burnard,[2] but the present estimate is
not strictly comparable with these because it is in terms of the modern
county. In any case, an estimate of Domesday population can rarely be
definitive, and all that can be claimed for the present figures is that they
indicate the order of magnitude involved. These figures are those of
recorded population, and must be multiplied by some factor, say 4 or 5,
in order to obtain the actual population; but this does not affect the relative density as between one area and another.[3] That is all that a map such
as Fig. 56 can roughly indicate.

It is impossible to say how complete were these Domesday figures.
We should have expected to hear more, for example, of priests. Then there
are some 38 entries which do not mention inhabitants although they
sometimes refer to plough-lands, and even teams, and to such other
resources as wood or meadow. Thus Broadaford (*219*, 105 b) answered
for 1 virgate, had 2 plough-lands and was worth 5*s*., but we hear nothing
of population or of teams. Hook (*468*, 117) also answered for 1 virgate
and had 2 plough-lands, and was worth 4*s*.; it also had meadow and
pasture, but again no population or teams are recorded. Then there are
those other place-names for which hardly any information is given. We
cannot be certain about the significance of these omissions, but they (or
some of them) may imply unrecorded inhabitants.

Unnamed thegns are mentioned quite frequently as holders of estates in
1066, and they are often entered in small groups, e.g. the 2 at Poughill
(*215*, 104 b), the 4 at George Nympton (*377*, 116 b) and the 9 at Worling-

---

[1] Sir Henry Ellis, *A General Introduction to Domesday Book*, II (London, 1833),
pp. 435–6. His total for the county (excluding tenants, sub-tenants and the *ancilla*)
came to 16,954.

[2] R. Burnard, 'The ancient population of the Forest of Dartmoor', *Trans. Devon
Ass.* XXXIX (Axminster, 1907), p. 202. It is difficult to see why his estimate differed
so greatly from that of Ellis (and incidentally from the present count). His figure for
serfs, for example, amounted to 5,177, that is, 1,883 more than Ellis's figure for serfs.

[3] But see p. 368 below for the complication of serfs.

ton near Witheridge (*379*, 116b). Altogether, some 136 unnamed thegns are enumerated for 1066. *Terrae Occupatae* adds another 5 or so.[1] Unnamed thegns are not mentioned for 1086, except for the 15 at Bovey Tracey (*135*, 102). Both these and the named thegns, being landholders, have been excluded from our totals. So have the various *milites* and the 3 *francigenae* at Great Torrington (*376b*, 116b). We have also excluded those priests who were landholders.[2]

Recorded Population of Devonshire in 1086

A. Rural Population

Villeins	8,519
Bordars	4,876
Serfs	3,323
Miscellaneous	590
Total	17,308

There was also one bondwoman (*ancilla*).

Details of Miscellaneous Rural Population

Swineherds	. . 370	*Buri*	. . .	4
Coscets	. . . 70	Iron-workers (*ferrarii*)		4
Salt-workers	. . 61	Men (*homines*)	. .	3
Cottars	. . . 36	Fishermen	. .	2
Coliberts	. . . 32	Smiths (*fabri*)	. .	2
Bee-keepers	. . 5	Priest	. . .	1
		Total	. . .	590

B. Urban Population

BARNSTAPLE	67 burgesses, 2 *domus*, 38 *domus vastae* (see p. 282 below).
EXETER	399 *domus*, 51 *domus vastatae* (see p. 280 below).
LYDFORD	69 burgesses, 40 *domus vastae* (see p. 283 below).
OKEHAMPTON	4 burgesses, 31 villeins, 11 bordars, 18 serfs, 6 swineherds (see p. 284 below).
TOTNES	111 burgesses (see p. 284 below).

Note: The villeins, bordars, serfs and swineherds are included in both tables.

[1] Sometimes a man in the Domesday text appears in that of *Terrae Occupatae* as an anonymous thegn and vice versa. In arriving at our total we have included all those specifically described as thegns whether or not they are named elsewhere.

[2] See p. 278 below.

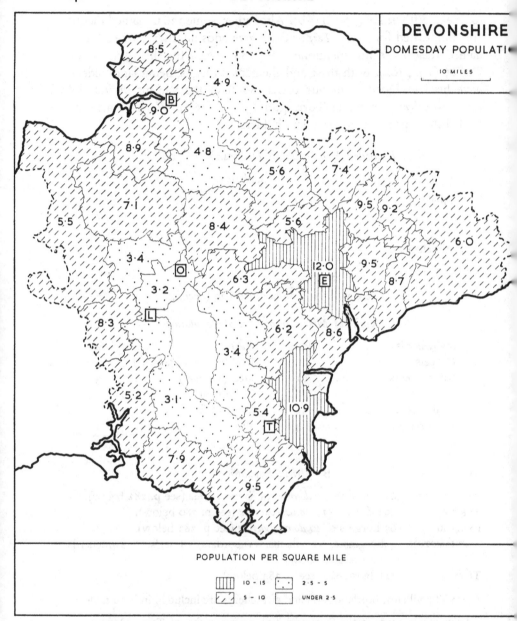

Fig. 56. Devonshire: Domesday population in 1086 (by densities).

Domesday boroughs are indicated by initials: B, Barnstaple; E, Exeter; L, Lydford; O, Okehampton; T, Totnes. For the blank area of Lydford parish on Dartmoor, see p. 255.

Fig. 57. Devonshire: Domesday population in 1086 (by settlements).

Domesday boroughs are indicated by initials: B, Barnstaple;
E, Exeter; L, Lydford; O, Okehampton; T, Totnes.

Villeins constituted the most important element numerically in the population, and amounted to 49 per cent of the total. A common Exeter formula states the hidage and plough-lands occupied by the villeins before going on to enumerate the categories of people on a holding; and the word *villani* is usually used in this connection even when no villeins appear in the subsequent enumeration. The entry for Kenn (*297*, 106b) speaks of *xlii inter villanos et bordarios*, and we have assigned 21 to each category. The Exeter text (*98*) assigns 116 villeins to the large manor of Werrington, but in the Exchequer version (101) the figure is only 16; we have adopted the former figure in this analysis. A half-villein is entered for each of two manors at Hembury (*Otria, 342*, 114b; *Otri, 405*, 110b); we have assumed that only a single villein was involved. At the unidentified *Madescama* (*404*, 110b) there was 1 plough-land, worth 30*d.*, which had been added to Buckerell (*Orescoma*); the only recorded inhabitant was 1 villein who, the Exeter text most unusually tells us, held all the land in demesne: *i villanus qui tenet totam terram in dominio*. The Exeter version of the account of Shilstone ends with an unusual statement about a virgate that lay in the manor: *nemo tenet eam modo et manent ii villani in ea et valet per annum iii solidos* (*458*, 116b). Villeins, either alone or on holdings with bordars and serfs, were able to rent manors, and sums *ad firmam* are recorded in Exeter entries for Bluehayes (*398*, 113) and Lympstone (*460*, 113). Next in numerical order came bordars, who amounted to 28 per cent; they call for no special comment, but we should note that the Exeter text says that the 3 bordars at Washford Pyne had one ferling (*367*, 111). Serfs amounted to about 19 per cent of the recorded population, and their incidence varied considerably. They amounted to some 28 per cent of the recorded population of the manors of Buckfast Abbey but to only about 12 per cent on the remaining ecclesiastical lands, and to about 13 per cent on the Ancient Demesne of the Crown. The Exeter text mentions a serf on 1 ferling of land at Buckland (*129b*, 102b) near North Molton: *in alio ferlino manet quidam servus*.[1] The only recorded population at Boasley (*288b*, 105b) were 7 serfs with one team, and at Buckfast (*183*, 104) there were 1 smith and 10 serfs with 2 teams.

The miscellaneous category was a very varied one. Its outstanding feature was the large number of swineherds (*porcarii*); nothing approaching the figure of 370 appears for any other county. Interestingly enough,

---

[1] The Exchequer text mentions a villein and 2 serfs. The Exeter text mentions only the serf who lived on another ferling.

the Exchequer text omits them for 23 of the 67 entries in which they appear in the Exeter text. One entry is unusual in that the swineherd and the villein on a ferling of land added to Tetcott appear as two villeins in the Exchequer version (*319*, 108 b). Most paid a fixed render in swine, e.g. at Fremington (*123b*) 13 swineherds rendered 21 swine, and at Coldridge (*125*) 3 swineherds rendered 50 swine. We hear of money rents in only two entries: at Kenton (*94b*) 4 swineherds paid 20*s*., and at Ottery St Mary (*195*) 5 swineherds paid 31*s*. 3*d*. There were 70 coscets[1] mentioned, for only 10 settlements, all of which were on Juhel's fief in the west, with the exception of Slapton (*120b*, 102), for which as many as 21 are entered. There were 36 cottars, of whom as many as 12 were on the manor of Exminster (*83*, 100); both these and the 5 at Brixham (*321*, 109) were omitted from the Exchequer text. The 32 coliberts were confined to two royal manors; there were 25 at Werrington (*98*, 101) and 7 at Broad Clyst (*95*, 101). The 61 salt-workers (*salinarii*) were at six coastal manors.[2] There were 4 *ferrarii* at North Molton (*94b*, 100 b) and a *faber* was to be found both at Buckfast (*183*, 104) and at Cruwys Morchard (*404*, 110 b). The 5 bee-keepers (*mellitarii*) were at Southbrook (*347*, 111 b), and they rendered 7 sesters of honey. The 4 *buri* were at Burrington near Chulmleigh (*179*, 103 b), and in the Exchequer text they are grouped with *porcarii*, and were entered separately from the villeins and bordars who had teams.[3] The 3 men (*homines*) were at Peadhill (*134b*, 103); they are the only recorded people on the manor, although the Exeter text refers to villeins with half a team. The 2 fishermen (*piscatores*) were at Dartington (*368b*, 111) on the River Dart, and they rendered 80 salmon. The solitary priest, that is, apart from those who were landholders, was at Instow (*390b*, 112 b). Finally, there was an *ancilla* (bondwoman) on an anonymous virgate (*475*, 117 b).

(5) *Values*

The value of an estate is generally given for two dates—for 1086 and for some earlier year. Some entries give only one value, and a very small number give no value at all. The following examples illustrate some of the variations to be found in the Exchequer text:

---

[1] For the Somerset coscets, see p. 166 above.
[2] See pp. 271–3 below.
[3] F. W. Maitland showed that *buri* and *coliberti* were, as he said, 'all one' (*op. cit.* pp. 36–8).

Stockleigh English (*215*, 104b): *Olim xl solidos. Modo valet xxx solidos.*
Haxton (*117b*, 101b): *Valuit et valet xxv solidos.*
Meddon (*482*, 118): *Valet xx solidos.*
Ashprington (*111*, 101b): *Reddit iiii libras.*
Boehill (*395*, 112): Value omitted (10s. in Exeter text).

Occasionally when the Exchequer text uses *valet* the Exeter text uses *reddit*, as in the entries for Leigh in Modbury (*326*, 109b) and Leonard (*395b*, 112).[1] The use of the words *valuit* and *olim* in the Exchequer folios leaves the exact date of the earlier valuation in doubt; but a few entries are more explicit and refer to the time when an existing holder received an estate (*quando recepit*), e.g. those for Ashburton (*119*, 102), Bishop's Tawton (*118*, 101b) and Paignton (*119*, 102). For Hartland (*93b*, 100b) and Woodbury (*96b*, 100b), the earlier date is given as *ante Baldwinum*, i.e. before Baldwin the sheriff began to farm the manors. The Exchequer entry for Werrington (101) gives the earlier date as *olim*, but the Exeter version (*98*) explicitly says *T.R.E.*, which is the only reference to a 1066 value in the Devon folios. The entry for Okehampton (*288*, 105b) is unusual and may result from an error in the transcription of the Exeter text: *Totum valet x libras. Cum recepit viii libras.*[2] In contrast to the Exchequer version, the Exeter text almost invariably gives the earlier date as *quando recepit*. Here are the two versions of the valuation of Germansweek:

E.D. (*335*): *valet per annum l solidos et quando R[adulfus] eam recepit valebat xl solidos.*
D.B. (113b): *Olim xl solidos. Modo valet l solidos.*

Very often when the Exchequer version states a value for only one date, the Exeter text gives two values, sometimes the same, sometimes different, as the following examples show:

Otri (Waringstone and Deerpark)
E.D. (*403*): *Valet xii solidos et quando W[illelmus] recepit valebat tantundem.*
D.B. (110b): *Valet xii solidos.*

---

[1] *Reddit* is interlined above *valet* in the Exchequer account of Lympstone (113) and the Exeter text specifically says that the render was *ad firmam* (*460*). Conversely, *reddit* is interlined over *valet* in the Exeter account of Stoke St Nectan (*456*) whereas the Exchequer text merely has *valet* (117).

[2] The Exeter text reads: *valet cum suis appenditiis x libras et quando B[aldwinus] recepit valebat viii libras.*

Leigh in Silverton

    E.D. (*214*): *Valet per annum x solidos et quando recepit valebat v solidos.*

    D.B. (104b): *Valet x solidos.*

Kingsteignton

    E.D. (*84*): *Haec reddit xiiii libras ad pensum et x solidos numero et quando B[alduinus] recepit reddebat x libras ad pensum.*

    D.B. (100): *Reddit xiiii libras ad pensum et x solidos ad numerum.*

These and other discrepancies are noted in Appendix II.

The amounts are usually stated in round numbers of pounds and shillings, e.g. £8 or 20s. Some amounts are very small, e.g. Buscombe (*127*, 102b) was valued at only 2s., and Chittleburn (*330*, 109b) at only 1s. Occasionally we hear of 20 pence to the ounce, e.g. at Wonford in Heavitree (*95b*, 100b). Occasional precise figures, however, suggest either a sum of rents or careful appraisal, e.g. the 7s. at King's Nympton (*98*, 101). In some entries, the silver mark entered into the valuation, e.g. that for Axmouth (*85*, 100). Two manors at Ottery St Mary and Rawridge, held by the canons of Rouen, together paid annually £70 of 'Rouen money': *reddunt hae ii mansiones lxx libras Rothomagensium denariorum per annum* (*195b*, 104).[1]

The Exeter text frequently gives more detail about the values of individual constituents of a manor than does the Exchequer version, e.g. in the account of the manor of Hatherleigh near Okehampton (*178*, 103b). Frequently the two texts are at variance, e.g. the 30d. for 20d. at Twinyeo (*488b*, 118b), or the 15s. for 20s. at Woolleigh (*294b*, 106b). In the entry for Villavin (*388*, 112b) we seem to see the result of careless copying, for the 10s. of the Exeter text becomes 100s. in the Exchequer version.

The method of valuing the royal manors (*83–111*, 100–101b) deserves special mention for three reasons. In the first place, we are given only one value for the majority. In the second place, the values of most are entered as renders (*reddere*, not *valere*, is the verb used). In the third place, the amounts had often been tested by assay, and are consequently qualified by such phrases as *ad pensum*, *ad pondus et arsuram*, *ad pensam et arsuram*; an occasional Exeter variant is *ad pondus et ad combustionem*. Sometimes the amounts are said to be by tale (*ad numerum*); this is especially true of the manors which had been held by Queen Mathilda, and the Exeter entries for these tell us that they were held *ad firmam*.

[1] 'Rouen pence were, at a later date, worth half the value of English pence' (*V.C.H. Devonshire*, I, p. 435 n.).

We obtain a glimpse of the archaic system of the night's *firma* in the Exeter text of the account of the three royal manors of Walkhampton, Sutton in Plymouth and King's Tamerton: *haec iii mansiones reddebant firmam hunius* [sic] *noctis cum suis appendiciis (86b)*. The account goes on to say that Maker in Cornwall used to contribute £5. 18s. 6d. to Walkhampton, from which it had been detached; the Exchequer version adds the information that the sum had been paid *in firma regis* (100b). In contrast to those royal demesne manors which elsewhere furnished, or contributed to, a night's *firma*, these three have assessments, although very small ones—a virgate or half a virgate.[1]

Values increased or remained the same; under a fifth had decreased. Occasionally the variations are striking, e.g. the rise from 20s. to £6 at Whipton (*404b*, 110b) and from £13 to £50 at Paignton (*119*, 102), or the fall from 40s. to 10s. at Tamarland (*411b*, 114b) and from £12 to £6 at Modbury (*217b*, 105b). Some decreases may have been due to a variety of causes, crop failure, animal murrain or bad management. There were yet other causes, such as the Irish raids along the south coast.[2] Some changes, however, may be more apparent than real, and may be explained by the fact that the composition of some manors had been altered between 1066 and 1086. Thus the value of Gappah (*341b*, 114b) had increased from 5s. to 30s., but *Terra Occupatae* (*502b*) shows that this was due to the addition of the land of four thegns.

Generally speaking, the greater the number of teams and men on a holding, the higher its value, but it is impossible to discern any consistent relationship, as the following figures for five holdings, each rendering £2 in 1086, show:

	Teams	Popula-tion	Other resources
Cheriton in Brendon (*337*, 114)	2	8	Pasture, wood
Dunterton (*289b*, 106)	5½	18	Meadow, pasture, wood
Feniton (*214b*, 105)	½	14	Meadow, pasture, wood
Ide (*117b*, 101b)	6	24	Underwood
Woolfardisworthy near Hartland (*481b*, 118)	4	8	Meadow, pasture

[1] In Wiltshire, Dorset and Somerset, the royal manors responsible for a *firma unius noctis* were not assessed.

[2] See p. 274 below.

It is true that the variations in the arable, as between one estate and another, did not necessarily reflect variations in total resources, but, even taking the other resources into account, the figures are not easy to explain. There are, moreover, many perplexities. Why should a manor at Cruwys Morchard (*132 b*, 103), with two teams and nine people and with meadow, pasture, wood and livestock, be worth only 12*s*. 6*d*.? Or why should a Tiverton manor (*460 b*, 113), with no teams on its 2 plough-lands and with 5 inhabitants and 2 acres of meadow, be worth as much as 30*s*.?

Conclusion

For the purpose of calculating densities, Devonshire has been divided into thirty-one units, based as far as possible upon the main physical and geological features. The boundaries of the density units have been drawn to coincide with those of civil parishes, and this gives the units an artificial character because many parishes extend across more than one kind of soil and country. This is true, for example, on the south-west flanks of Dartmoor where the parishes stretch from the valleys into the upland. The modern parish of Lydford, in the central part of Dartmoor, contains no Domesday place-name apart from that of the borough of Lydford in the extreme north-west; and it has been excluded from the calculation of densities. Another difficulty arises from the fact that the totals for some manors may cover a number of anonymous settlements that may not be contained within the same unit. The division into thirty-one units is therefore not a perfect one, but it may provide a basis for distinguishing broad variations over the face of the countryside.

Of the five standard formulae, those relating to plough-teams and to population are most likely to reflect something of the distribution of wealth and prosperity throughout the county. Taken together, certain common features stand out on Figs. 54 and 56. A prominent feature on both maps is the sparsely occupied mass of Dartmoor. To what extent the empty central portion was unused in the eleventh century, we cannot say; but surrounding it were areas with not much more than about 3 people and 1 team or under per square mile. The most densely occupied areas, on the other hand, seem to have been three in number: (1) the lower Taw and Torridge basins with fairly fertile soils, especially in the north; (2) the lower and middle parts of the Exe basin with their fertile red soils derived from the Permian New Red Sandstone; and (3) the coastal area of South Devon between the Teign and the Avon, with variable but generally fertile soils. In all three areas the density of population was between 9 and 12, and that of teams between 3 and 5 per square mile. Elsewhere, the general density of population is for the most part about 5–9 and that of teams about 1·5–2·5 per square mile even along the Exmoor border.

Figs. 55 and 57 are supplementary to the density maps, but it is necessary to make reservations about them. As we have seen on p. 228, some Domesday names covered two or more settlements, and several of the symbols should thus appear as two or more smaller symbols. This limitation, although important locally, does not affect the main impression conveyed by the maps. Generally speaking, they confirm and amplify the information on the density maps.

WOODLAND

Types of entries

The Devonshire folios give information both for wood itself and for underwood. The amount of each is normally expressed in one of two ways—either in linear dimensions or, more frequently, in acres. There are also a number of variant entries.

Woodland is called *silva* in the Exchequer folios and *nemus* in the Exeter folios, although the Exeter version uses *silva* in the entry for Anstey (*300*, 107). The linear dimensions are stated in terms of leagues and furlongs and, very rarely, of perches.[1] The exact wording varies slightly, but the following examples indicate the kind of entry that is encountered:

Drewsteignton
    E.D. (*304*): *i leuga nemoris in longitudine et iii quadragenariae in latitudine.*
    D.B. (107b): *Silva i leuua longa et iii quarentenis lata.*
Offwell
    E.D. (*314*): *v quadragenariae in longitudine et xx perticas in latitudine nemoris.*
    D.B. (108b): *Silva v quarentenis longa et xx perticis lata.*

Where the quantities are in furlongs, the number is normally less than the twelve usually supposed to make a Domesday league, but Halberton (*110b*, 101b) had wood 16 furlongs by 13; and amounts of 20 furlongs, as well as 12, are also recorded. The largest amount entered under a single name

---

[1] The three entries that mention perches are for Moor (*317*, 108b), Offwell (*314*, 108b) and Pyworthy (*318b*, 108b).

is the 5 leagues by a half for Crediton (*117*, 101 b). The exact significance of these linear dimensions is far from clear, and we cannot hope to convert them into modern acreages.[1] All we can do is to plot them diagrammatically as on Fig. 58, and we have assumed that a league comprised 12 furlongs. The measurements for very large manors, such as Crediton, may well cover a number of widely separated unnamed places, and this seems to imply, though not necessarily, some process of addition whereby the dimensions of separate tracts have been consolidated into one sum. There is one composite entry that involves wood—for Collacombe, Ottery in Lamerton, and Willestrew (*419*, 113); but this is peculiar and the symbol for 'other mention' has been marked for each place.[2]

Four entries state the amount of wood to be 'in length and breadth' or 'between length and breadth'. The Exchequer text omits any reference to wood in two of these entries, and for Chardstock, described in the Dorset folios, there is no Exeter version:

Abbotskerswell
E.D. (*184*): *xii quadragenariae nemoris inter longitudinem et latitudinem.*
D.B. (104): *Silvae xii quarentenae inter longitudinem et latitudinem.*
Chardstock (77): *Silva duae leugae inter longitudinem et latitudinem* (*inter* is interlined above).
Exminster (*83*, 100): *i leuga nemoris in longitudine et latitudine.*
King's Nympton (*98*, 101): *i leuga nemoris in longitudine et latitudine.*

Eyton thought that such variant formulae indicated 'areal leagues' or 'areal furlongs';[3] but one is tempted to assume that they imply the same facts as the more usual formula, and that they occur occasionally when length and breadth of woodland are the same.[4] We have followed this assumption in drawing Fig. 58.

Twenty or so entries give only one dimension, e.g. the 1 furlong of wood at Hollacombe in Kentisbury (*345*, 111 b), and the half league of wood at Belstone (*290*, 106). That these entries also imply that the length and breadth were the same may be indicated by the alternative versions for Clawton:

E.D. (*318 b*): *una leuga nemoris in longitudine et latitudine.*
D.B. (108 b): *una leuua silvae.*

[1] See p. 372 below.　　　　　　　　[2] See p. 258 below
[3] R. W. Eyton, *A key to Domesday . . . the Dorset survey* (London, 1878), pp. 31–5.
[4] For the evidence that points to this, see pp. 100 and 174–5 above.

On Fig. 58, however, such entries have been plotted as single lines, except that for Clawton itself, for which intersecting lines have been drawn.

The amount of wood is more frequently stated in acres. The figures range from half an acre at, for example, Spriddlescombe (*220b*, 105b) to 500 acres at Winkleigh (*109*, 101b), but the great majority do not rise above 30 acres. The figures are often in round numbers that suggest estimates, e.g. the 300 acres each at Great Torrington (*376b*, 116b) and Sidbury (*118b*, 102), but occasional entries seem to be precise, e.g. Slapton (*120b*, 102) had 31 acres of wood. No attempt has been made to convert these figures into modern acres, and they have been plotted merely as conventional units on Fig. 58. The wood on one holding in a settlement might be measured in acres and that on another holding in terms of linear dimensions. At South Brent, for example, there were 5 acres on one holding, and wood 1 league by 1 furlong on another (*183b bis*, 104 *bis*).

The acre seems to have been used as a linear measure in the combined entry for Collacombe, Ottery in Lamerton, and Willestrew (*419*, 113): *xx acrae silvae in longitudine et ii in latitudine*; this has been indicated under the category of 'other mention' at each place.[1] There are also a few other unusual entries. At Rawridge (*195*, 104) and at Smallridge (*343*, 114b) we hear of half a hide of wood; and at Doddiscombsleigh (*468b*, 117) of 1 virgate of wood. There were 2 acres of alder-wood (*alnetum*) at Ashleigh (*317*, 108b) and 4 acres at Loxhore (*298b*, 106b). There was a park for beasts (*parcus bestiarum*) at Winkleigh (*109*, 101b) as well as 500 acres of wood. Finally, an entry for Tiverton is a curiosity. The two versions read as follows:

E.D. (*98*): *i leuga in longitudine et vi quadragenariae et dimidia in latitudine et iiii quadragenariae nemoris.*
D.B. (100b): *iiii quarentenae silvae et una leuua in longitudine et vi quarentenae et dimidia in latitudine.*

It seems as if there were two stretches of wood, and on Fig. 58 we have followed the Exeter version and plotted a single line in addition to intersecting lines.

[1] It may be a scribal error; the Exeter version runs: *iiii quadragenariae et xx agri nemoris in longitudine et ii in latitudine.* For Dunsford (*347*, 111b) and Holsworthy (*93b*, 101), the acres of the Exchequer version appear as furlongs in the Exeter text, thus making one-dimensional entries. See p. 261 below for the acre of underwood at Meavy.

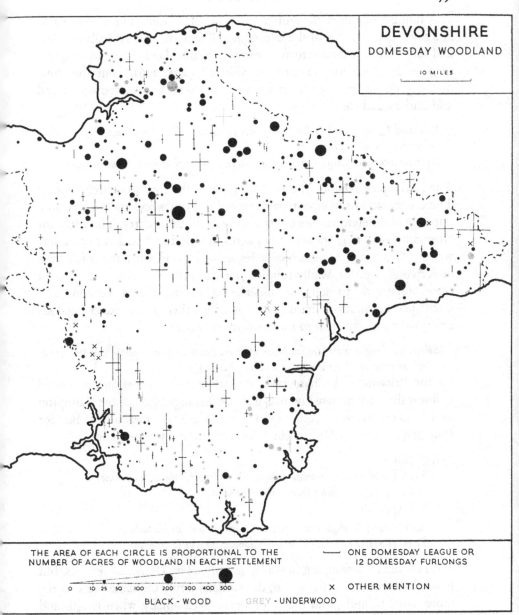

Fig. 58. Devonshire: Domesday woodland in 1086.

When the wood of a settlement is entered partly in acres and partly in some other way, both are shown. When wood and underwood are entered for the same settlement, both are shown.

A number of entries mention underwood: *nemusculus* in the Exeter text and *silva minuta* in that of the Exchequer. The amounts are usually in acres, and range from half an acre at, for example, Buckland-Tout-Saints (*396*, 113) to as much as 300 acres at Shirwell (*298*, 106b), but the great majority do not rise above 20 acres. Two Exeter entries mention wood and underwood together:

> Bridford (*319b*, 109): *i leuga nemoris et nemusculi in longitudine et i quadra-genaria in latitudine.*
> Burrington near Chulmleigh (*179*, 103b): *lx agri nemoris et nemusculi.*

The Exchequer version, in each case, records the total as entirely wood.[1] There were other settlements with record of both wood and underwood, but only in different entries or separately in the same entry. The entry for Northam (*194*, 104) speaks of '24 acres of wood and 30 acres of under-wood'. One version occasionally records as wood what the other calls underwood, e.g. for Wolborough (*313*, 108) the Exchequer 30 acres of wood appear as 30 acres of underwood in the Exeter version. The Exchequer account of Chardstock (for which there is no Exeter version) refers to wood and then to underwood elsewhere:

> *Silva duae leugae inter longitudinem et latitudinem et in alia parte iii quarentenas silvae minutae longitudine et ii latitudine* (77).

As for Bridford and Chardstock, underwood was sometimes measured in linear dimensions, usually in leagues or furlongs; though for Galmpton in Churston Ferrers (*462*, 113b) perches are also mentioned. But for Ipplepen and North Molton the two versions are not the same:

Ipplepen
> E.D. (*462*): *dimidia leuga nemusculi in longitudine et latitudine.*
> D.B. (113b): *dimidia leuga silvae minutae.*

North Molton[2]
> E.D. (*94b*): *i leuga nemusculi in longitudine et in latitudine.*
> D.B. (100b): *una leuga silvae in longitudine et latitudine.*

It is not safe to base an argument upon only a few Domesday entries, but it would seem that, as for wood, the variants imply the same facts as the more usual formula, and that they occur occasionally when length and

---

[1] On Fig. 58, 'other mention' of wood and of underwood is plotted for Bridford, and 30 acres of each for Burrington.

[2] Note also that one version speaks of underwood and the other of wood.

breadth are the same; the measurements for both Ipplepen and North Molton have been plotted as intersecting straight lines. Five other entries give only single dimensions and, following the practice adopted for some other counties, we have plotted them as single lines. For three of the entries both versions are in agreement—Dowland (*390*, 112b), Haccombe near Newton Abbot (*311 b*, 108), and Meeth (*294*, 106b). The other two entries have different versions each, and that of the Exeter Domesday has been plotted.

> Down St Mary
> E.D. (*182*): *viii quadragenariae nemusculi.*
> D.B. (103 b): *vii quarentenae silvae minutae.*
>
> Meavy
> E.D. (*329*): *dimidia leuga nemusculi.*
> D.B. (109 b): *dimidia acra silvae minutae.*[1]

Finally, there is an unusual entry for Lowley (*468 b*, 117), where there was 1 ferling of underwood.

The Forest of Dartmoor does not appear in the Domesday record, but there is a reference to Æthelraed the forester in the Geld Accounts (*69*); he must have been the Æthelraed who held Shapley in Chagford in 1066 (*297b*, 106b).[2]

Distribution of woodland

As Fig. 58 shows, wood was widely distributed over the varied lower-lying countryside. Fairly large amounts, measured in linear dimensions, were entered for many of the settlements around Dartmoor, but we cannot assess the significance of this, nor can we say whether this wood stretched up the valleys and on to the slopes of the upland. A fair number of settlements in the north-west were without wood, but generally speaking the county can be described as well wooded in 1086. The Culm Measures belt, with its outcrops of shale and clay, had substantial amounts of wood entered for many of its settlements.

---

[1] See p. 258 above for the acre apparently used as a linear measurement at Collacombe, Ottery in Lamerton, and Willestrew.

[2] He is also mentioned as a landholder in 1066 at Neadon (*312*, 108) and Clifford (*470*, 117) and in 1086 at Bickford (*488b*, 118) but is likewise not called a forester. All three places lie to the east and south of Dartmoor.

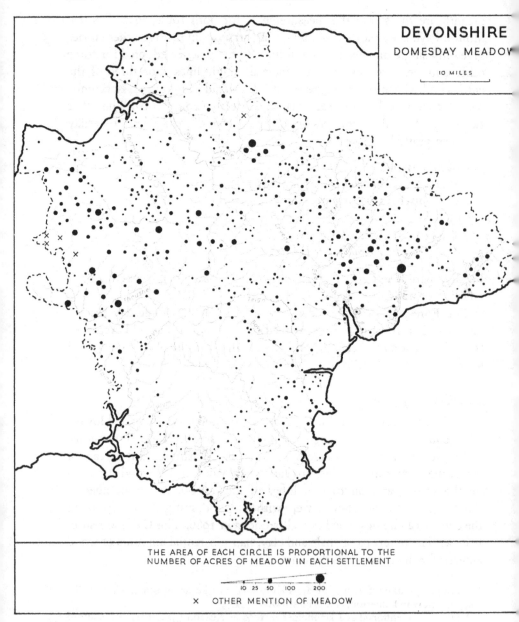

DEVONSHIRE
DOMESDAY MEADOW

10 MILES

THE AREA OF EACH CIRCLE IS PROPORTIONAL TO THE
NUMBER OF ACRES OF MEADOW IN EACH SETTLEMENT

10 25 50 100 200

× OTHER MENTION OF MEADOW

Fig. 59. Devonshire: Domesday meadow in 1086.

Areas of blown sand, alluvium and lowland peat are shown in grey;
rivers passing through these areas are not marked.

Types of entries

MEADOW

The entries for meadow in the Devonshire folios are comparatively straightforward. For holding after holding, the same phrase is repeated—'*n* acres of meadow' (*n acrae prati*). The amount of meadow entered under each place-name ranges from only half an acre at, for example, Whiddon (*377*, 116b) to 200 acres at Ottery St Mary (*195*, 104), which comprised an entire 25-hide hundred. Only about half a dozen entries record 100 acres or more, and these may well have covered more than one settlement, e.g. Bishop's Nympton (*119b*, 102) with 150 acres, Werrington (*98*, 101), and Hatherleigh near Okehampton (*178*, 103b), each with 100 acres.[1] By far the great majority of places had under 30 acres apiece. Round figures such as 10 or 20 acres suggest that they may be estimates, but there are also detailed figures that suggest precision, e.g. the 47 acres at Poltimore (*469b*, 117b), the 23½ acres at Crooke (*475*, 117b) and the many entries for 1, 2 or 3 acres. As for other counties, no attempt has been made to translate these figures into modern acreages, and they have been plotted as conventional units on Fig. 59. The 10 acres in the composite entry for Collacombe, Ottery in Lamerton, and Willestrew (*419*, 113) have been divided equally between the three places.

There are six unusual entries which measure in terms of linear dimensions. For one of these, the Exchequer version gives only one dimension whereas the Exeter account refers to length and breadth. For another, both versions give only one dimension. These differences recall the variations encountered in the measurement of wood.[2] The Exeter version of the third entry speaks of 'meadows', but we cannot assess the significance of these, especially as the manor involved was a small one with only two teams. Here are the entries:

Pyworthy
    E.D. (*318b*): *i leuga prati in longitudine et latitudine.*
    D.B. (108b): *Ibi i leuga prati.*

North Molton
    E.D. (*94b*): *ii leugae prati.*
    D.B. (100b): *Ibi ii leugae prati.*

---

[1] The 100 acres of meadow entered for Dartington (*368b*, 111) may possibly be a mistake for 100 acres of pasture. It is not a large manor, and there is little meadow entered for neighbouring settlements.   [2] See p. 257 above.

Willand

E.D. (*379 b*): *iii quadragenariae pratorum in longitudine et i quadragenaria et dimidia in latitudine.*

D.B. (116b): *Pratum iii quarentenis longum et i quarentena et dimidia latum.*

The two versions for the other entries that use linear dimensions agree and are straightforward—for Clawton (*318 b*, 108 b), Bridgerule (*411*, 114b), and Bradford in Pyworthy (*319*, 109).

Distribution of meadowland

Although amounts were small, meadow was widely distributed throughout lowland Devonshire (Fig. 59). Amounts often over 20 acres were to be found along the Axe, the Otter, the Clyst, the Culm, the Exe and their tributaries. Similar amounts were credited to many settlements along the upper streams of the Torridge system. The villages along the Thrushel, the Carey and other streams that flow into the Tamar also had substantial amounts by Devonshire standards. On the other hand, the lower Torridge and the various streams associated with the Taw, in the north, were bordered by settlements with much smaller amounts. Small amounts were also associated with the southern streams—the Bovey, the Teign, the Dart, the Avon, the Erme and the Plym.

PASTURE

Types of entries

Pasture is recorded for about three quarters of the settlements of Devonshire. It is entered in the Exeter text as *pascua,* and in the Exchequer version as *pastura.* Its amount is usually recorded in one of two ways—either in linear dimensions or, more frequently, in acres. The linear dimensions are given in terms of leagues and furlongs and, once, of perches. The exact wording varies slightly, but the following entry for Honiton shows the kind of phrasing that was used:

E.D. (*216b*): *i leuga in longitudine pascuae et v quadragenariae in latitudine.*
D.B. (104b): *Ibi pastura i leuua longa et v quarentenis lata.*

The exact significance of these linear dimensions is far from clear, and we cannot hope to convert them into modern acreages.[1] All we can do is to plot them diagrammatically as on Fig. 60, and we have assumed that a league comprised twelve furlongs. The dimensions (4 by 4 leagues) in

[1] See p. 372 below.

an entry for a large manor such as South Tawton (*93*, 100b) may possibly cover a number of separated places, and may imply, although not necessarily, some process of addition whereby the dimensions of separate tracts have been consolidated into one sum.

Twenty entries state the amount of pasture to be 'in length and breadth' or 'between length and breadth'. Judging from occasional variations between the Exeter and the Exchequer texts, the formulae seem to be interchangeable and they may both imply the same facts as the more usual statement. At any rate, that is what the following examples seem to indicate:

Anstey
E.D. (*300*): *i leuga pascuae in longitudine et in latitudine.*
D.B. (107): *una leuua pascuae in longitudine et latitudine.*

Littleham near Exmouth
E.D. (*184*): *vi quadragenariae pascuae inter latitudinem et longitudinem.*
D.B. (104): *Pastura vi quarentenae in longitudine et latitudine.*

There is thus no need to invoke the 'areal leagues and furlongs' of Eyton.[1] All these entries have been plotted in the normal manner as intersecting lines. The Exeter text of the entry for King's Nympton is unusual in that it speaks of the dimensions as being 'every way':

E.D. (*98*): *i leuga pascuae ab omni parte.*
D.B. (101): *una leuua pasturae in longitudine et latitudine.*

The Exchequer versions of eleven out of the twenty entries, give only one dimension, e.g.

Clawton
E.D. (*318b*): *i leuga pascuae in longitudine et latitudine.*
D.B. (108b): *una leuua pasturae.*

Southbrook
E.D. (*347*): *x quadragenariae pascuae inter latitudinem et longitudinem.*[2]
D.B. (111b): *x quarentenae pasturae.*

These have all been plotted as intersecting lines. As many as 46 other entries give only one dimension for pasture in both the Exchequer and the Exeter versions, e.g. the 2 leagues at Brendon (*337*, 114) and the 7 furlongs at Hillersdon (*378b*, 116b). The entry for Staverton (*120*, 101b) gives

[1] See p. 257 above.
[2] This entry is unusual in that it puts breadth before length.

the unusual quantity of 30 furlongs. It is possible that all these one-dimension entries imply the same facts as the usual formula when the length and breadth are equal. On Fig. 60, however, they have been plotted as single lines.

Measurements in acres are much more frequent. Amounts are generally below 50 acres, and they range from a single acre at Bradaford (*317b*, 108b) to 500 acres at Upottery (*342b*, 114b) and Werrington (*98*, 101). The figures are often in round numbers that suggest estimates, and they are frequently in multiples of 5 or 10. On the other hand, some seem to be precise, e.g. the 42 acres at Lifton (*93*, 100b) and the 53 acres on a Poltimore manor (*469b*, 117b). Like the acreages for meadow, they have all been treated as conventional units. Occasionally both types of measurement appear in entries for one place-name, e.g. West Putford with 10 acres of pasture (*399*, 110) and with pastures half a league by 2 furlongs and 1 furlong by 1 furlong (*336*, *412b*; 114, 115).

There are a number of unusual entries. Four use, or seem to use, the acre as a linear measure. Here are the Exeter entries, with which those of the Exchequer text agree:

Abbotskerswell (*184*, 104): *v quadragenariae pascuae in longitudine et xxx agri in latitudine.*
Kilmington (*97*, 101): *xii quadragenariae et xii agri pascuae.*
Murley (*393b*, 112): *i pertiqua et xxx agri pascuae.*[1]
Throwleigh (*458b*, 113b): *dimidia leuga pascuae in longitudine et iiii agri in latitudine.*

Five entries measure pasture in terms of hides or virgates. There was a virgate of pasture each at Lipson (*222*, 105b) and at Washbourne in Halwell (*396*, 113); there were 3 virgates at Cadbury (*415b*, 111); there was a hide at Combe Raleigh (*Otri*, *348b*, 112), and as many as 8 hides in the manor of Ottery St Mary (*195*, 104). A sixth entry may perhaps throw light on these. Mullacott (*469*, 117) was a half-hide manor which is described in the usual way; we are then told of a ferling of this half-hide upon which was a bordar, the remainder of the ferling lying uncultivated as pasture: *De supradicta dimidia hida tenuit quidam tegnus i ferlinum...Ibi habet G[odebold] i bordarium et alia terra iacet vastata ad pasturam.* The Exchequer version tells of the ferling and the bordar but omits any reference to pasture.

---

[1] The only mention of perches in connection with the Devonshire pasture.

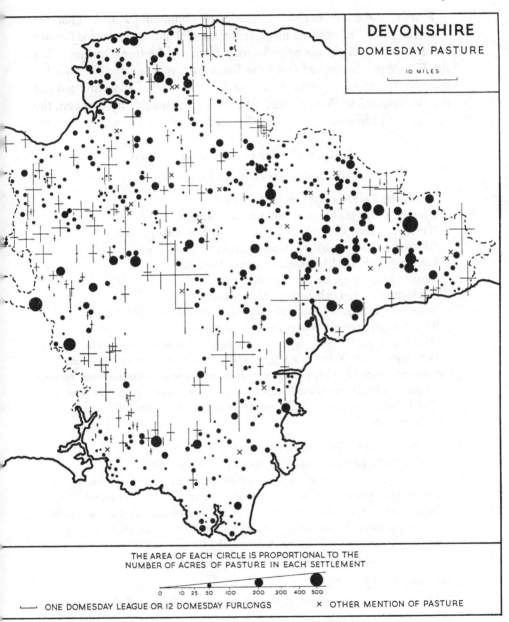

THE AREA OF EACH CIRCLE IS PROPORTIONAL TO THE
NUMBER OF ACRES OF PASTURE IN EACH SETTLEMENT

0   10 25 50   100   200 300 400 500

ONE DOMESDAY LEAGUE OR 12 DOMESDAY FURLONGS   × OTHER MENTION OF PASTURE

Fig. 60. Devonshire: Domesday pasture in 1086.

When the pasture of a settlement is entered partly in acres
and partly in some other way, both are shown.

Finally, there are twelve references to common pasture. That for
Tiverton and Worlington is measured in acres; that mentioned for Benton
and for Haxton lay in the neighbouring manor of Bratton Fleming. That
for Woodbury is omitted from the Exchequer version (100b). That for
Newton Tracey belonged to the township in the time of King Edward and
for five years of William's reign, when it was 'withheld' by Colsuen, the
man of the bishop of Coutances. The relevant entries are set out below.

Common Pasture in Devonshire in 1086

The entries are those of the Exeter text.

Axminster (*405b*, 111): *communis pascua.**
Benton (*117b*, 101b): *communa pascua Bratonae.*
Burrington near Chulmleigh (*179*, 103b): *communis pascua.*
Gillscott (*390b*, 112b): *communis pascua.*
Haxton (*117b*, 101b): *communa pascua Bratonae.*
Meeth (*294*, 106b): *pascua communis.*
Ruckham (*471b*, 117b): *communis pascua.*
Stockleigh in Meeth (*294*, 106): *pascuae communis ibi habet satis.*†
Tiverton (*98*, 100b): *xl agri pascuae communis.*
Woodbury (*96b*): *valet per annum xx solidos cum communi pascuo.*
Worlington near Witheridge (*379*, 116b): *xxx agri communis pascuae.*
Newton Tracey (*388b*, 112b): *haec mansione aufert Colsuenus homo episcopi*
  Constantiensis communam pascuam quae pertinebat villae tempore regis
  Edwardi et postquam W. rex habuit Anglicam habuit eam Goscelmus quietam
  per v annos.‡

* The wording is incomplete, but this is clearly meant.
† The Exchequer text reads: *pastura communis sufficiens.*
‡ The Exchequer text reads: *Colsuen homo episcopi Constantiensis aufert ab*
hoc manerio communem pasturam quae ibi adjacebat T.R.E. et etiam T.R.W.
quinque annis. V.C.H. Devonshire, vol. I, p. 495, translates *aufert* as 'withholds',
but it could mean 'took away', i.e. transported into another manor. *Quieta*
could mean 'undisturbed' or 'without paying dues'.

Distribution of pasture

The fact that pasture was recorded for three quarters of the settlements
of Devonshire means, as Fig. 60 shows, that it was widely distributed
throughout the county. The pasture of many of the settlements around
Dartmoor was measured in terms of linear dimensions. There are occa-

sional substantial amounts, e.g. the 4 leagues by 4 for South Tawton (*93*, 100b) and the 3 leagues by 1 for Bradworthy (*335b*, 114). To what extent the pasture of these surrounding settlements lay on Dartmoor itself we cannot say. Large amounts were not as frequent as might be expected, certainly not as frequent as around the Cornish moorlands. The pasture of many settlements near Exmoor was also measured in linear dimensions, as was that of the settlements in the north-west of the county. The settlements of the Exe–Culm–Otter basin had their pasture measured largely in acres, which may indicate smaller amounts. Finally, an appreciable number of places along the south coastal area, between the Dart and the Tamar, were without any record of pasture.

FISHERIES

Fisheries are specifically mentioned in connection with only fourteen places, but their presence at two other places is implied (Dartington and Hollowcombe). No renders are entered for some, but others rendered money (from 1*s.* to 25*s.*) or salmon, and at one place 3 salt-workers rendered, amongst other things, a load of fish. Only half a fishery is recorded for Weare Gifford, and there is no clue to the other half.[1] The fishery at Bideford is said to have been there in 1066, but, as it is valued in the present tense, we may assume it was there in 1086 also. The fisheries at Dartington and Woodleigh appear only in the Exeter versions of their respective entries; and that at Efford only in the Exchequer version. The references to all sixteen places are set out on p. 271.

As Fig. 61 shows, the places with fisheries were on the lower courses of rivers.

SALT-PANS

Salt-pans (*salinae*) are mentioned in connection with twenty-two places in Devonshire, and they are implied in connection with six other places for which salt-workers (*salinarii*) are mentioned, making a total of twenty-eight places associated with the making of salt. The number of pans on a holding is stated, and usually their annual render in money. No render is given for the salt-pan at, for example, Lobb (*414*), which is not mentioned in the Exchequer version (115). Most places had only one salt-pan, yielding between 10*d.* and 60*d.* each. At Northam (*194*, 104) there were 2 pans; at Bere Ferrers (*222b*, 105) there were 7; and at Bishopsteignton

---

[1] We know from non-Domesday sources that the other half was at Monkton: P. E. Dove (ed.), *Domesday Studies* (London, 1888), I, p. 20.

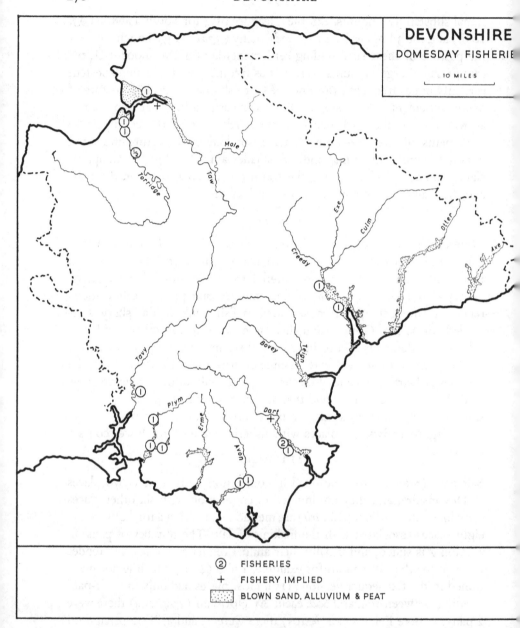

Fig. 61. Devonshire: Domesday fisheries in 1086.

The figure in each circle indicates the number of fisheries. Rivers flowing across blown sand, alluvium and lowland peat are not marked.

Domesday Fisheries in Devonshire

The entries are those of the Exeter text except that for Efford,
which appears only in the Exchequer text.

Ashprington (*111*, 101 b): *ii piscatoriae.*
Bickleigh near Tavistock (*417b*, 111 b): *i piscatia quae valet per annum v solidos.*
Bideford (*108b*, 101): *huic mansioni adiacebat quaedam piscatio ea die qua
rex E. fuit vivus et mortuus quae reddit per annum xxv solidos.*
Buckland Monachorum (*417b*, 111 b): *i piscatio quae valet per annum x
solidos.*
Cornworthy (*323b*, 109): *i piscatoria quae reddit per annum xxx salmones.*
Dartington (*368b*, 111): *ii piscatores qui reddunt lxxx salmones.*
Efford (113 b): *Ibi piscaria reddit xii denarios.*
Exminster (*83*, 100): *i piscatoria quae reddit per annum xx solidos.*
Haccombe in Exeter (*401*, 110): *i piscatoria.*
Heanton Punchardon (*298b*, 106 b): *i piscatura quae valet per annum ii solidos.*
Hollowcombe in Fremington (*408*, 115 b): *iii salinarii qui reddunt per annum
iiii solidos et ix denarios et v summas salis et i summam piscis.*
Loddiswell (*321b*, 109): *i pescaria qui* [sic] *reddit xxx salmones.*
Northam (*194*, 104): *i piscatio quae reddit per annum xxx denarios.*
Weare Gifford (*412b*, 115): *dimidia piscatura quae valet per annum xl denarios.*
Woodford in Plympton St Mary (*333b*, 110): *i piscatoria.*
Woodleigh (*421*, 113): *i piscatoria.*

(*117*, 101 b) there were as many as 24; each of these three groups of pans
yielded 10s. a year. The 4 pans at Hollowcombe in Ermington (*218*,
105 b) yielded 40d. and 2 loads of salt (*ii saginae salis/ii summae salis*). The
inland manor of Ottery St Mary (*195*, 104) had a salt-pan paying 30d.
a year 'in the land of St Michael of Sidmouth', and it seems as if the salt-
pan was physically on the coast at the estuary of the River Sid.[1]

A total of 61 salt-workers is recorded. No render appears for the 33
at Otterton (*194b*, 104); but the 8 at Kenton paid 20s. (*94b*, 100b); the
4 at Holcombe in Dawlish (*336b*, 114) paid 6s. 5d.; the 3 at Hollowcombe
in Fremington (*408*, 115 b) paid 4s. 9d. and 5 loads of salt and one load of
fish; the 11 at Seaton rendered only 11d. (*Fluta*, *184*, 104).[2] The 2 salt-
workers at Honiton (*216b*, 104b), some distance inland, paid 5s. a year

---

[1] For salt-pans at Sidmouth within the parish of Ottery St Mary in 1061, see
V.C.H. Devonshire, I, p. 435 n.
[2] These *salinarii* of the Exeter text appear as *salinae* in that of the Exchequer; they
have been counted as salt-workers in the present analysis.

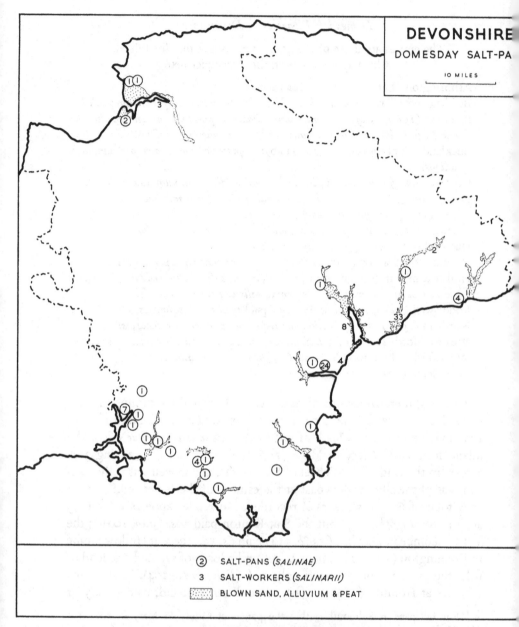

Fig. 62. Devonshire: Domesday salt-pans in 1086.
The figures indicate the number of salt-pans and salt-workers respectively.

by way of rent (*de firma*). These salt-workers may have had their pans at Beer on the coast; we are told that 4 pans had been taken away from the abbot of Horton's holding at Beer (*184b, 503b*; 104), and that in 1086 they were held by the count of Mortain's tenant, Drogo, who also held Honiton. It is interesting to note that the entry for Salcombe Regis (*118b*, 102) makes no reference to salt-pans, although the name means 'salt valley' and is mentioned in a pre-Domesday charter; the other Salcombe (west of Dartmouth) is not mentioned in the Domesday texts.

Fig. 62 shows that the salt-making areas were associated with a number of estuaries along the south coast and with the Taw–Torridge estuary in the north. One can only suppose that salt-pans entered for somewhat inland places lay physically along the coast, e.g. the salt-pan for Kenton on the Creedy. As we have seen, the salt-pan in the manor of Ottery St Mary was most probably at Sidmouth, and the pans of the two salt-workers entered for Honiton may well have been at Beer, also on the coast.

WASTE

There are occasional references to holdings that were waste in 1086, usually only in the Exeter text. Furzehill and Melbury were completely waste (*penitus vastata*); so, apparently, was Coombe in Templeton (*vacua*). We can also assume that Culbeer was at least partially waste because, although it rendered 3s. per annum, we are told that there was nothing (*nichilum*) there. There was an acre of meadow at Lower Credy but the rest of the land was so wasted that it rendered only 2s. One half of the virgate at Higher Molland, in North Molton, had its complement of resources but the other half was entirely waste (*tota vastata*). There was also a complement of resources on one ferling at Mullacott, but on the other ferling there was but 1 bordar, and the rest of the land was wasted and was used as pasture. *Terrae Occupatae* refers to an unnamed holding (in the hundred of South Molton) of 1 ferling with 1 plough-land which was entirely waste and which no one claimed. The entries are set out on p. 275. We might also remember those other holdings without teams or people, yet for which no waste was specifically entered (Fig. 82).

There were also a number of holdings which, we are told, had been waste when their tenants received them. Those at Buscombe and Coleridge in Egg Buckland were being tilled, but no men or teams are mentioned for Whitefield in Challacombe although it was worth 6s. in 1086. The

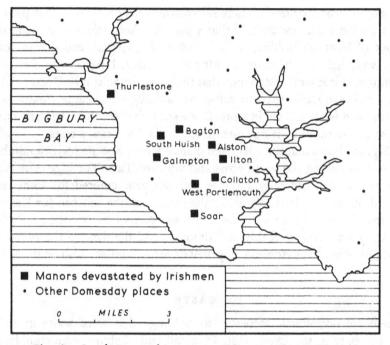

Fig. 63. South Devonshire: Manors devastated by Irishmen.

(*321 b–323*, 109)

	Valuit	*Valet*
Bagton	15s.	15s.
South Huish	25s.	25s.
Galmpton in South Huish	40s.	30s.*
Thurlestone	£4	£3
Alston	20s.	10s.
Soar	40s.	20s.
Collaton in Malborough	20s.	5s.
West Portlemouth	40s.	10s.
Ilton	20s.	5s.

* The Exchequer text reads 50s.

Exeter text (*323*) tells us, in a marginal note, of nine manors, in the extreme south of the county, which had been laid waste by Irish raiders: *hae ix predictae mansiones sunt vastatae per Irlandinos homines* (Fig. 63). Two manors had recovered to their former values, but the other seven had far from recovered, and three were worth only a quarter of their earlier values, as the table above shows.

Domesday Waste in Devonshire

A. *Waste in 1086*

Coombe in Templeton (*133b*): *est vacua* (1 virgate and 1 ferling, 1 plough-land). D.B. (103) does not mention *vacua*.

Culbeer (*314b*): *nichilum* (interlined) *et valet per annum iii solidos* (3 virgates, 2 plough-lands). D.B. (108) does not mention *nichilum*.

Furzehill (*366*): *est penitus vastata* (½ virgate; 3 plough-lands). D.B. (111) says *est vastata*.

Higher Molland in North Molton (*409*): *alia medietas virgatae est tota vastata* (the other half-virgate is entered normally). D.B. (115b) does not mention waste.

Lower Credy (*340b*): *Ibi habet Willelmus agrum prati et alia terra est ita vastata quod non valet nisi ii solidos et quando R. recepit tantundem* (1 ferling; 1 plough-land). D.B. (114b) does not mention waste.

Melbury (*415b*): *est penitus vastata* (1 virgate). D.B. (111) says *vastata est tota*.

Mullacott (*469*): *Ibi habet G. i bordarium et alia terra iacet vastata ad pasturam* (1 ferling). D.B. (117) does not mention waste.

South Molton hundred (*500*): *In hundreto Moltonae est 1 ferdinus terrae quae potest arare una carruca qui penitus vastatus iacet. Hunc nullus hominum* [sic] *clamat.*

B. *Formerly waste*

Badgworthy (*402b*): *valet per annum x solidos et quando recepit erat vastata* (1 virgate, 2 plough-lands, 2 teams, 1 villein, 1 serf, livestock, meadow, pasture). D.B. (110b) makes no mention of former waste.

Buscombe (*127*): *valet ii solidos et quando episcopus eam recepit erat vastata* (1 ferling, 2 plough-lands, 1 villein, ½ team, meadow, pasture). D.B. (102b) makes no mention of former waste.

Coleridge in Egg Buckland (*329*): *valet per annum xv denarios et quando Juhel recepit erat vastata* (½ virgate, 2 plough-lands, 2 bordars, 1 ploughing beast, wood). D.B. (109b) says *Olim vastata est*.

Lupridge (*397*): *valet per annum v solidos et quando W. recepit erat vastata* (1½ virgates, 2 plough-lands, 1 villein, 1 team, wood, meadow, pasture). D.B. (112b) makes no mention of former waste.

Whitefield in Challacombe (*127*): *valet vi solidos et quando episcopus recepit erat vastata* (½ virgate, 2 plough-lands). D.B. (102b) says *Vasta fuit*.

MILLS

Mills are mentioned in connection with only 80 of the 983 Domesday settlements in Devonshire, and also in an entry for one anonymous holding.[1] This is strikingly fewer than in Somerset and Dorset. The number of mills on a holding is stated, and normally their annual value, ranging from a mill at Hatherleigh near Okehampton that rendered 6d. a year[2] to one at Columbjohn (*469*, 117b) rendering 25s. Such sums as 20d. (i.e. the *ora*), 40d. and 50d. were frequent; so were sums that amounted to eighths of a pound. The one place with a render of an uneven amount was described in the Dorset folios—Stockland (*44b*, 78), where 3 mills rendered 37d. The only fraction is the half-mill entered for Buckerell (*Orescoma, 338b*, 114), and there is no record of the corresponding half. The mill at Lyn (*402*, 110b) is said to be a new one (*novum molinum*). Mills at two places are said to be for the use of the hall or of the demesne; the versions of each entry differ:

> Tavistock
> E.D. (*177*): *i molendinum qui servit abbatiae.*
> D.B. (103b): *Ibi molinum serviens curiae.*
>
> Stoke Fleming
> E.D. (*348b*): *i molendinum qui tantummodo servit domui suae.*
> D.B. (112): *Ibi molinum in dominio serviens.*

The majority of settlements with mills had only one apiece. Ten had 2 each, and five had as many as 3: Silverton (*83b*, 100), Otterton (*194b*, 104), Ottery St Mary (*195*, 104), Stockland (*44b*, 78), and Weycroft (*343, 343b*; 114b *bis*). The Domesday text does not tell us whether any or all of these manors comprised a number of separate settlements. There are a number of strange features about the incidence of mills. Why, for example, should only one mill yielding but 30d. be entered for the large manor of Crediton (*117*, 101b) with 185 plough-teams and a recorded population of as many as 407?

As Fig. 64 shows, most of the settlements with mills were in the south-east, along the streams of the Exe, the Culm, the Clyst, the Otter and the

---

[1] One mill, paying 20s., on a hide belonging to the bishop of Coutances (*132*, 103).
[2] This is assigned to Walter in the Exeter text (*178*), but to Geoffrey in the Exchequer version (103b). We have assumed that the same mill is meant, and Hatherleigh has been counted as a place with only 1 mill.

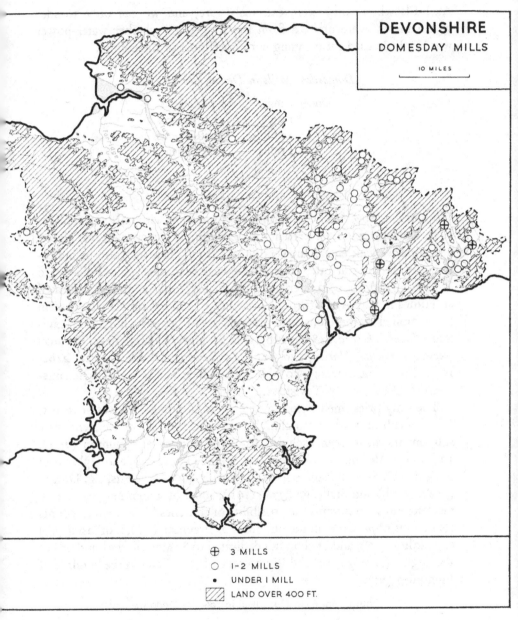

Fig. 64. Devonshire: Domesday mills in 1086.

There were also mills at *Alwinestona*, at *Bernardesmora*, and on an anonymous holding
—see p. 276. Areas of blown sand, alluvium and lowland peat are shown in grey; rivers
passing through these areas are not marked.

Axe. Elsewhere, they were extremely rare, and we can only wonder whether they were omitted from the Inquest or whether water-power was but little used for grinding corn over most of the county.

Domesday Mills in Devonshire in 1086

Under 1 mill	1 settlement
1 mill	64 settlements
2 mills	10 settlements
3 mills	5 settlements

CHURCHES

Churches are specifically mentioned in connection with only seven places in Devonshire, and their presence is clearly implied at another two places. The nine relevant entries are set out on p. 279. There may also be indications of churches at other places. We hear that Juhel gave Follaton to St Mary (*325b*, 109 b), a reference, presumably, to the church so dedicated at Totnes nearby. Or, again, the existence of some churches would seem to be implied by the Domesday place-names St James's Church in Heavitree (*Jacobeschercha, 487*, 118 b) and St Mary Church near Torquay (*ecclesia Sanctae Mariae, 120*, 101 b, or *Sanctae Mariae Cherche, 216b*, 105). Then, there were two abbey churches, named only as tenants-in-chief: Buckfast and Tavistock.

The only priest mentioned among the rural population was at Instow (*390b*, 112 b), for which no church is recorded. Other priests, who were sub-tenants, have been excluded from our count—at Braunton (*194b*, 104), South Molton (*194b*, 104), Swimbridge (*194b*, 104), and Uplowman (*394*, 112 b). Several manors had been held in 1066 by priests, e.g. Dotton (*308b*, 107 b) and Stallenge (*378*, 116 b), some of whom may have been resident and so may imply the possibility of churches.[1] The Geld Accounts also tell us of *presbiteri* in the hundreds of Braunton (*66*), Exminster (*69*) and Molton (*66*), and of churches dedicated to St Mary in the hundreds of Ermington (*Alleriga, 70*) and Plympton and to St Peter in the hundred of Plympton (*70*).

---

[1] Dotton appears as *Otrit*, held in 1066 by Dodo the priest.

Domesday Churches in Devonshire

The entries are those of the Exeter text.

A. *Churches mentioned*

Exeter (*120b*, 101b): *Episcopus habet i ecclesiam in Essecestra quae reddit per annum i marcam argenti.*

Exeter (*196*, *505b*; 104): *Abbas Batailliae habet in Essecestra i ecclesiam de Sancto Oilaf.*

Exeter (*222b*, 104b): *comes de Moritonio habet in Essecestra i ecclesiam.*

Exeter (*307b*, 107): A fourth church may be inferred from the fact that we hear of the canons of St Mary, though Exeter is not mentioned by name.

Axminster (*85*, 100): *Ecclesiae de Alsemenistra adjacet dimidia hida terrae illius mansionis.*

Colyton (*85*, 100b): *haec predicta mansio habet i ecclesiam ubi iacet dimidia virga.*

Cullompton(*195*,104): *Abbas de Praelio* (D.B. *Ecclesia de Labatailge*) *habet i hidam terrae et i ecclesiam in Colitona quam tenuit Torbertus ea die qua rex E. fuit vivus et mortuus.*

Kingskerswell (*85*, 100b): *in ecclesia istius villae est dimidia virga terrae.*

Pinhoe (*95b*, 101): *de hac mansione tenet abbas de Batallia ecclesiam et adjacet illi i virgata prescriptae terrae.*

Woodbury (*96b*, 100b): *Inde habet Abbas Sancti Michaelis de Monte ecclesiam et terram quam tenuit sacerdos ea die qua rex E. fuit vivus et mortuus; hoc* [sic] *est dimidia hida et i virgata et dimidium ferlinum.*

B. *Churches implied*

Plympton St Mary (*86*, 100b): *habent canonici Sancti Petri de Plintona ii hidas.*

Yealmpton (*86b*, 100b): *et sacerdotes* (D.B. *clerici*) *istius villae habent i hidam terrae.*

URBAN LIFE

Five places seem to have been regarded as boroughs in Domesday Devonshire. The county borough of Exeter is described at the beginning of the Exchequer account of the shire, and there are also several references to it later in the body of the text. Brief statements about Barnstaple and Lydford appear among the folios dealing with *Terra Regis*. A very similar statement about Totnes begins the account of Juhel's fief. Finally, the account of Baldwin the sheriff's large manor of Okehampton contains a reference to a small and probably newly created borough. The information relating to all five boroughs is very scanty and we can form no clear picture of their respective economic activities. No mints are recorded for any of them, although we know from other evidence that there were pre-Conquest mints at all except Okehampton.[1] On the other hand, Okehampton is the only borough for which the Devonshire text records a market and a castle. We can, however, infer the recent building of castles at Exeter, Barnstaple and Lydford from the references to waste houses at each of these places. Churches are mentioned only for Exeter. The evidence, slender as it is, is set out below.

Exeter

The Domesday information about Exeter is given partly in the opening paragraph of the Devonshire folios and partly in a number of separate entries scattered through the body of the text. The opening paragraph (*88*, 100) says that the king had 285 houses that paid customary dues and that another 48 had been laid waste (*vastatae*) since King William came to England. The devastation, presumably, had been to make room for the building of Exeter castle in 1068, but no castle is mentioned in the Domesday texts.[2] The separate entries enumerate the houses owned in Exeter by various tenants-in-chief, and they occur mostly either at the beginning or at the end of the description of a fief, 'as if', wrote Ballard, 'the commissioners had noted them down when entering the rural properties, added them up, and then entered them in one total'.[3] Almost all these houses of the separate entries were each liable for a due of 8*d.* or some multiple of 8*d.* None of these entries says to which rural manor the town house belonged, with the exception of that relating to Baldwin the sheriff, who had 19 houses, 12 of which were attached to Kenn.[4] Three other entries relating to rural manors also mention houses in Exeter; the manors were those of Tawstock, Bishopsteignton and Ilsington, but we cannot say whether or not those for Bishopsteignton were included in the total for the bishop of Exeter. The details are set out on p. 281. Most of the houses seem to have been liable for annual dues at the rate of 8*d.* per house, but the account of *Terrae Occupatae* notes the fact that a number of these dues had not been paid for some years (*505–6b*). The sum of the

[1] R. H. M. Dolley (ed.), *Anglo-Saxon Coins* (London, 1960), p. 144.
[2] For a discussion of the Conqueror at Exeter, see J. H. Round, *Feudal England*, pp. 431–55.
[3] A. Ballard, *The Domesday Boroughs* (Oxford, 1904), p. 27.
[4] A separate entry for the rural manor of Kenn records also 11 burgesses and it is possible that these were included in the 12 houses—see p. 281.

properties in these separate entries amounts to 114 houses[1] and 3 waste houses. The grand total of recorded houses for Exeter amounts to 399 together with 51 devastated houses.[2] The figure of 399 houses may imply a total population of about 2,000 people in 1086.

Properties in Exeter belonging to tenants-in-chief

The king (*94b*, 101)	5 (in entry for Tawstock)*
Bishop of Exeter (*120b*, 101 b)	48 and 2 waste (*vastatae per ignem*)†
Bishop of Exeter (*117*, 101 b)	9 (in entry for Bishopsteignton)‡
Bishop of Coutances (*136*, 102)	3 and 1 waste (*vastata*)
Bishop of Coutances (*136*, 102)	6
Abbot of Tavistock (*180b*, 103 b)	1
Abbot of Battle (*196*, 104)	8
Robert of Mortain (*222b*, 104b)	1
Baldwin the sheriff (*315*, 105 b)	7 and 12 attached to Kenn
Baldwin the sheriff (*297*, 106b)	11 burgesses (in entry for Kenn)§
Juhel of Totnes (*334b*, 108b)	1
William 'Capra' (*406*, 110)	2
Walter of Douai (*349b*, 112)	11
Richard fitz Thorold (113b)	1 (not in E.D.)
Ralph Paynel (*460*, 113b)	1 (in entry for Ilsington)
Ralph of La Pommeraye (*344*, 114b)	6
Ruald 'Adobatus' (115)	1 (not in E.D.)
Theobald fitzBerner (*410*, 115b)	1
Alfred the Breton (116)	1 (not in E.D.)
Osbern of Saussay (*462b*, 117)	1
Godebold (*473*, 117)	2

* These may be included with the 285 of the main entry, and so have been excluded from our count.

† The Exchequer version says 47 houses and 2 waste.

‡ These may be included in the other 48 houses of the bishop of Exeter, and have not been counted in our total.

§ These eleven burgesses (*burgenses*) may be included in the 12 houses said to be attached to Kenn in the earlier entry for Baldwin the sheriff, and have not been counted in our total.

[1] Or 139 if 5 houses at Tawstock, 9 houses at Bishopsteignton and 11 burgesses at Kenn are not regarded as duplicate entries.

[2] On the other hand, some people have assumed that 114 houses of the separate entries were included within the 285 of the main entry (*V.C.H. Devonshire*, I, p. 397).

Some indication of the importance of Exeter may be given by the fact that it was called a city (*civitas*) and that it paid geld only when London, York and Winchester did (*88*, 100). It was also the seat of a bishopric. Three churches are mentioned, held respectively by the bishop of Exeter (*120b*, 101b), the count of Mortain (*222b*, 104b) and the abbey of Battle (*196*, 104); the last is said to be that of St Olave. We may infer the existence of at least another from the fact that the canons of St Mary are mentioned (*307b*, 107). We hear of various dues and services rendered by the city, but nothing of the economic activity that sustained its importance. There is no mention, for example, of a mint or a market, although we know from other sources that there was a mint there[1] and there must have been a market. We are told, however, that the burgesses had land for 12 plough-teams outside the city (*extra civitatem*), and that this was subject to no customary payment save to the king himself; the bishopric of Exeter had 2½ acres of land (*120b*, 101b) which lay with the land of the burgesses (*jacent cum terra burgensium*). We are also told of a fruit orchard (*virgultum*) formerly in the demesne of King Edward, but belonging to the count of Mortain in 1086 (*222b*, 104b). But of the agricultural resources that went with the 12 plough-lands we hear nothing.

Barnstaple

The opening entry in the account of *Terra Regis* tells us that the king 'has' the borough of Barnstaple (*87b*, 100):

Rex habet burgum Barnstaple. Rex E. habuit in dominio. Ibi sunt intra burgum xl burgenses et ix sunt extra burgum...Ibi sunt xxiii domus vastatae postquam rex venit in Angliam.

A number of other entries also refer to properties in Barnstaple, and the evidence as a whole is summarised on p. 283. The presence of houses 'outside the borough' may imply either a rural community engaged in non-urban activities or burgesses attached to rural manors. On the assumption that the properties in the entries for Fremington and Shirwell were not counted among the figures of their respective tenants-in-chief, the total amounts to 67 burgesses, 2 houses and 38 wasted houses; this may imply a population of some 350 people in 1086. The mention of wasted houses suggests the presence of a castle, but none is mentioned. Nor are we given any hint of the economic life that sustained the community. We know, for

[1] See p. 280 above.

example, that there was a pre-Domesday mint at Barnstaple, but this does not appear among the Domesday entries; nor is there mention of a market, which must have played an important role in a place like Barnstaple set at the head of the estuary of the Taw. When there was an expedition by land or by sea, Barnstaple rendered the same service as Lydford or Totnes, and the service of all three together equalled that of Exeter (*87b, 88*; 100).

Properties in Barnstaple belonging to tenants-in-chief.

The king (*87b*, 100): 40 burgesses within; 9 burgesses without; 23 wasted houses.

Bishop of Coutances (*136*, 102): 10 burgesses; 7 wasted houses; ½ virgate of land; a mill rendering 20*s*.

Bishop of Coutances (*123b*, 102): 1 burgess (in the entry for Fremington).

Bishop of Coutances (*128b*, 102b): 1 garden (*ortus*) rendering 4*d*. (in the entry for Bray).

Baldwin the sheriff (*315*, 105b): 7 burgesses; 6 wasted houses.

Baldwin the sheriff (*298*, 106b): 2 houses (*domus*); rendered as 2 *mansurae* in Exeter text (in the entry for Shirwell).

Robert of Albemarle (113): 2 wasted houses.

Lydford

In the account of *Terra Regis* we hear that Lydford (*87b*, 100) belonged to the king:

Rex habet burgum Lideforde: Rex E. tenuit in dominio. Ibi sunt xxviii burgenses intra burgum et xli extra. Inter omnes reddunt regi lx solidos ad pensum et habent ii carucatas terrae (Exeter, terra ad ii carrucas) extra burgum (Exeter, foras civitatem). Ibi sunt xl domus vastae postquam rex venit in Angliam.

The total thus amounts to 69 burgesses and 40 wasted houses; this may imply a population of some 350 people in 1086; that is, about the same number as in Barnstaple. We know from other evidence that some, at any rate, of the burgesses outside the borough were at Fernworthy, 8 miles to the east, on the other side of Dartmoor.[1] Others may have cultivated land immediately around the borough. The reference to devastated houses suggests the presence of a castle, but none is mentioned. Nor are we given any hint of the economic life that sustained the community. We know, for example, that there was a pre-Domesday mint

[1] H. P. R. Finberg, *Tavistock Abbey* (Cambridge, 1951), p. 74.

at Lydford, but this does not appear among the Domesday entries; nor is there mention of a market. When there was an expedition by land or by sea, Lydford rendered the same service as Barnstaple or Totnes, and the service of all three together equalled that of Exeter (*87b, 88*; 100).[1]

Totnes

The Exchequer account of the fief of Juhel of Totnes begins with a description of the borough from which he took his name:

Judhel tenet de rege Totenais burgum quod rex Edwardus in dominio tenebat. Ibi sunt c burgenses v minus et xv extra burgum terram laborantes (334, 108b).

The total of 110 burgesses may imply a population of some 550. We are given no hint of the commercial activity that helped to sustain the community. We know, for example, that there was a pre-Domesday mint at Totnes, but this does not appear among the Domesday entries; nor is there mention of a market. When there was an expedition by land or by sea, Totnes rendered the same service as Barnstaple or Lydford, and the service of all three together equalled that of Exeter (*87b, 88*; 100).

The account of Walter of Claville's manor of Lupridge (*397,* 112b) mentions a burgess, and, since Walter held no property in Exeter, the reference may be to Totnes, some seven miles to the north-east of Lupridge. On the other hand, the reference follows a mention of one villein, and *burgensis* could be a slip for *bordarius*; but we have not so regarded it; which makes a total of 111 burgesses. There is one incidental reference to Totnes. King William is said to have released Juhel from the payment of 20*s.* which Totnes had formerly paid to the manor of Langford in Ugborough by way of *firma (97b,* 101).

Okehampton

Baldwin the sheriff's manor of Okehampton (*288,* 105 b) was a substantial one, with 24 teams at work and with a recorded population of 31 villeins, 11 bordars, 18 serfs and 6 swineherds. Following the main account of the manor, the Exeter version goes on to say:

in ista terra sedit castellum Ochenemitona. Ibi habet B. iiii burgenses et i mercatum qui reddit iiii solidos per annum.

[1] One of the manors of Ralph of La Pommeraye is called *Lidefort* in the Exeter text (*335*) but *Tideford* (114) in that of the Exchequer.

The Exchequer text gives the same information but somewhat differently arranged. It is impossible to say how the agricultural population was distributed between the borough and the rest of the manor. One receives the impression of an incipient borough developing in connection with the building of a castle and the establishment of a market.

LIVESTOCK

The entries of the Exeter text record information about livestock—not about total stock but only about that on the demesne. We cannot say how complete even these restricted figures were, and it may well be that some demesne animals were omitted. A number of entries for holdings with demesne land make no reference to livestock, e.g. those for Twitchen (*371*) and for Churchill (*366b*). Conversely, stock is occasionally entered for holdings without recorded demesne, e.g. for Chichacott (*288*) and for Rowley (*131*). A further complication affects the Devonshire statistics. Some parts of the Exeter folios are missing and other parts are damaged, so that in any case our information for demesne stock is incomplete.[1] The table on p. 286 shows the variety of stock mentioned, and also the number in each category.

Horses are mentioned in a variety of ways. Rounceys (*roncini* or *runcini*) were widely distributed, and were most probably pack-horses. A distinction is made between *equae indomitae* and *equae silvestres*, but we may wonder whether there was any difference between the two categories. *Equae indomitae* are recorded for only four entries;[2] and, out of a total of 155, as many as 104 were at Brendon (*337*) in the Exmoor region, and so perhaps may have been Exmoor ponies. *Equae silvestres* are recorded in only eleven entries;[3] and, out of a total of 162, as many as 72 were at Lynton (*402*), also in the Exmoor region. The mare was at Potheridge

---

[1] When the Record Commission edition of the *Liber Exoniensis* appeared in 1816, part of the entry for Warbrightsleigh (*488*) was illegible, and so we do not know the number of sheep in demesne. Today, the whole of the entry, like a few others, is wholly indecipherable.

[2] Bradworthy (*335b*), Brendon (*337*), Cornwood (*218b*), and Musbury (*313*).

[3] Appledore in Burlescombe (*395*), Ashmansworthy (*293*), Ayshford (*395*), Bratton Clovelly (*288b*), Colaton Raleigh (*96b*), Lynton (*402*), Northlew (*108*), Ottery in Axminster hd. (*348*, *348b*), Ottery St Mary (*195*), and Villavin (*388*).

($293b$),[1] and the two donkeys (*asini*) were at Dipford in Bampton (346).

Demesne Livestock in Devonshire in 1086

Sheep (*oves*)	49,884
Wethers (*berbices*)	155
Swine (*porci*)	3,682
She-goats (*caprae*)	7,263
Animals (*animalia*)	7,357
Cows (*vaccae*)	23
Rounceys (*roncini* or *runcini*)	159
Unbroken mares (*equae indomitae*)	155
Forest mares (*equae silvestres*)	162
Mare (*equa*)	1
Donkeys (*asini*)	2

Animals (*animalia*) presumably included all non-ploughing beasts. Cows (*vaccae*) appear infrequently and never on the same holding as *animalia*; the cows and animals at Hatherleigh near Okehampton (178) were in different sub-tenancies. There were as many as six cows at Fremington ($123b$), but out of the thirteen entries that mention a total of 23 cows, ten entries record only one apiece. Demesne cattle seem to have been most abundant in the north-west of the county.

She-goats (*caprae*) appear quite frequently in the Devonshire entries. The flocks were much smaller than those of sheep, and the number entered for the demesne of a manor was usually less than 50 and often less than 10. Only on three large widespread manors did the number reach 100 or slightly over: Crediton (117), Hartland ($93b$) and Oakford (404). The entry for Soar ($322b$), however, gives a combined figure for sheep and goats (*ccxl inter oves et capras*), and the table above assigns 120 to each group.

Swine were recorded usually in smaller numbers. There were as many as 60 at Otterton ($194b$) and 57 at Crediton (117), but both these were large and scattered manors. Swine were often recorded for holdings for which no wood was entered, e.g. for Ilfracombe (301) with 15, and for

---

[1] This appears as *i qua*. The *V.C.H.* translation renders it as 'one cottager' (*V.C.H. Devonshire*, I, p. 451), but, from its position in the entry, it is more likely to be 'one mare'. It may be of interest to note that the account of Impington in the *Inquisitio Eliensis* mentions *i quam* in MS. A but *i equam* in MSS. B and C (N. E. S. A. Hamilton (ed.), *Inquisitio Comitatus Cantabrigiensis...subjicitur Inquisitio Eliensis*, London, 1876, p. 113).

DEVONSHIRE
DOMESDAY SHEEP

10 MILES

THE AREA OF EACH CIRCLE IS PROPORTIONAL
TO THE NUMBER OF SHEEP IN EACH SETTLEMENT

0   10 25 50   100   200   400   600

Fig. 65. Devonshire: Domesday sheep on the demesne in 1086.

Wethers (*berbices*) have been included with sheep. Forty-six entries are missing from
the Exeter text (see p. 223), and so the map cannot give a complete picture. The
missing entries refer mainly to places in the south of the county between the Tamar
and the Dart.

Chivelstone (*325b*) with 13. An unusually large number of swineherds is recorded for Devonshire,[1] but, interestingly enough, some of the manors with swineherds had no swine, or only a few, entered for them. Thus at Kenton four swineherds paid 20*s.*, but we hear only of *animalia*, sheep and goats (*94b*). Or, again, at Bampton fifteen swineherds rendered 106½ swine; one wonders what happened to the other ½ swine. Only 6 swine are included in the list of livestock (*345b*). Clearly, the Exeter text tells us of only a fraction of the total swine in the county.

Sheep were, far and away, the most numerous of all the recorded livestock, and flocks of many hundreds could frequently be encountered. It is true that the large figures for some complex manors may be deceptive in the sense that they covered separate unnamed places, e.g. the 700 at Hartland (*93b*), the 560 at Berry Pomeroy (*342*), and the 500 each at Tawstock (*94b*) and Shebbear (*94*). The figures are frequently in round numbers, but the count sometimes seems to have been made with precision, e.g. there were 146 sheep at Bickleigh near Tavistock (*417b*) and 231 at Afton (*342b*). Wethers (*berbices*) are mentioned only twice, for Northwick (*481*), where there were 30, and for Ingsdon in Ilsington (*460*), where there were 125. Fig. 65 shows that demesne sheep, at any rate, were widely distributed throughout the county. They seem to have been particularly numerous in parts of the south and in the lands around the Taw and Torridge lowland in the north. Most of the settlements around Dartmoor, on the other hand, were not well endowed with demesne sheep.

MISCELLANEOUS INFORMATION

Markets

Markets are recorded for only two places, Okehampton and Otterton; that for the latter place was a Sunday market and it is not mentioned in the Exchequer text. Here are the entries of the Exeter text:

Okehampton (*288*, 105 b): *Ibi habet B[aldwinus vicecomes] iiii burgenses et i mercatum qui reddit iiii solidos per annum.*

Otterton (*194b*): *Abbas Sancti Michaelis de Monte habet i mansionem quae vocatur Otritona et ibi est mercatum in dominica die.*

Okehampton is thus the only one of the five boroughs with a market, but clearly there were other unmentioned markets.

---

[1] See p. 250 above.

Moors

There are two references to moors, for which the Exeter entries read as follows:

> Molland (*95*, 101): *In mansionem quae vocatur Mollanda pertinet tertius denarius hundretorum Nortmoltonae et Badentonae et Brantonae et tercium animal pascuae morarum.*[1] *Hanc consuetudinem non habuit rex postquam ipse habuit Angliam.*[2]
> Sherford in Brixton (*329b*, 109b): *i agra morae.*

Molland is near Exmoor and Sherford lies between Dartmoor and the south coast.

Other references

The only castle (*castellum*) mentioned is that of Baldwin the sheriff at Okehampton (*288*, 105b), but the presence of devastated houses at Barnstaple, Exeter and Lydford suggests that castles had also been raised at these boroughs.

There was a park for beasts (*parcus bestiarum*) in the royal manor of Winkleigh (*109*, 101b), and a portion of the manor was held by Norman the park-keeper (*custos parci*).

There was a render of honey at Southbrook (*347*, 111b), where 5 bee-keepers paid 7 sesters of honey.

There are two references to a garden (*ortus*). One is in the account of the manor of Ottery St Mary (*195*, 104); and the other, described under the manor of Bray (*128*, 102b), is said to render 4d. and to be in Barnstaple. There was also a fruit orchard (*virgultum*) at Exeter (*222b*, 104b).

There are two references to the render of the 'third penny' from certain hundreds. One was to the manor of Molland from the three hundreds of North Molton, Bampton and Braunton (*95*, 101), and the entry is set out under 'Moors' above. The other was to Moretonhampstead from the hundred of Teignbridge: *huic predictae mansione adjacet tertius denarius hundreti de Taignebrige* (*96*, 101).

[1] For the due of every third beast, see *V.C.H. Devonshire*, I, p. 409 n.
[2] The entry is virtually repeated in the account of *Terrae Occupatae* (*499b*).

Customary dues

A large number of manors were liable for customary dues comprising sums of money, but many of these had not been rendered for some years previously. The manors were either royal or had recently been royal. Parford (*456, 496*; 116b) and Taw Green (*475, 495b*; 117b) could each substitute an ox for its due of 30*d*.

REGIONAL SUMMARY

The outstanding feature of the geography of Devonshire is the granite massif of Dartmoor. From this upland radiate a variety of streams, and the lower country through which they flow may be divided into a number of regions based largely, but not entirely, on geological features. Altogether, including Dartmoor and its borders, seven main regions may be recognised, and these will serve to give a broad view of the county in the eleventh century (Fig. 66).

(1) *The Exmoor Border*

The dark red soil produced by the Devonian sandstone outcrop characterises the region, but there is no uniformity about its other features. The surface ranges from the hilly country north of the Taw estuary to the bare peaty slopes of Exmoor over 1,000 ft above sea-level, rising to 1,618 ft just within the Devon border, and with an annual rainfall that amounts to over 60 in. in places. The density per square mile of teams showed a corresponding variation from 1·8 to nearly 3; that of population likewise ranged from nearly 5 to over 8. Some small settlements lay above the 1,000 ft contour, e.g. Lank Combe and Radworthy. There are very few hints of the moorlands of Exmoor. One is the reference to dues of beasts rendered to the manor of Molland from the hundreds of North Molton, Bampton and Braunton on account of 'the pasture of the moors'. Another may be the reference to common pasture held by Benton and Haxton at Bratton Fleming among the foothills of Exmoor. Yet another possible indication may be the substantial amounts of pasture entered for the settlements bordering Exmoor. Taking the region as a whole, demesne sheep were more common in the west than among the settlements near Exmoor. Many settlements had a little wood, and a small number had a little meadow.

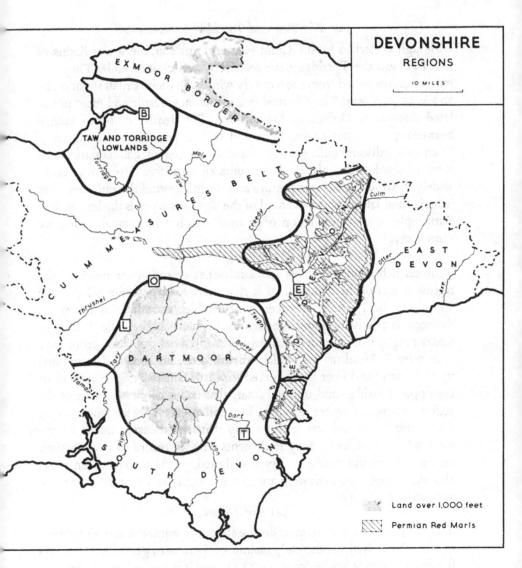

Fig. 66. Devonshire: Regional subdivisions.

Domesday boroughs are indicated by initials: B, Barnstaple;
E, Exeter; L, Lydford; O, Okehampton; T, Totnes.

(2) *The Taw and Torridge Lowlands*

Although underlain by the Culm Measures outcrop, the lower basins of
the Taw and the Torridge have relatively well-drained soils. They are,
moreover, protected from the westerly winds and their rainfall is not much
above 35 in. a year. They formed one of the most fertile and most populated districts in Domesday Devonshire. The density of teams ranged
between 3 and 4 per square mile, and that of population was about 9.
Even so, hardly any mills were recorded for the region, and its settlements
were credited with only small amounts of meadow. Wood was fairly
widely distributed; so were pasture and sheep. Around the estuaries were
fisheries and salt-pans. At the head of the Taw estuary was the borough of
Barnstaple, with a population of at least 350, but how much more, we
cannot say.

(3) *The Culm Measures Belt*

A broad belt of clays, shales and sandstones extends over much of the
northern part of the county, and is characterised by infertile cold, heavy
and wet soils. There is, however, considerable variation, and where the
drainage is relatively good, the fertility is much higher. The density of
teams ranged from about 1 to nearly 3 and that of population from over
3 to over 8. Meadow was widely distributed, and substantial amounts
up to 50 acres and over were entered for settlements along the streams of
the upper Torridge and upper Tamar, that is, in the western part of the
region, where soils were heaviest; but mills were very infrequent along
these streams. Wood was fairly widely distributed except in the north-
west, where there were many settlements without record of any. Pasture,
on the other hand, while widely distributed, seems to have been most
abundant in this north-western area; here, too, cattle were relatively more
important than sheep.

(4) *Red Devon*

This name is often used to describe the stretch of warm red soils developed
on the 'New Red Sandstone', mainly of Permian age. Almost all this
fertile countryside lies below 400 ft O.D., and it is drained by the Exe and
the Culm with their tributaries. The density of teams ranged from over
2 to nearly 5 per square mile, and that of population from over
7 to 12. Settlements with mills were frequent along the Exe, the Culm and
the Clyst. Salt-pans and an occasional fishery were also to be found along
the lower Exe. Meadow, pasture and demesne sheep were fairly evenly

distributed. Fewer settlements had wood entered for them but, even so, it was fairly well distributed. Set in the valley of the Exe was the city of Exeter, the seat of a bishopric, and with a castle and a population of perhaps 2,000; clearly it was one of the most important cities in Domesday England.

(5) *East Devon*

This is a varied countryside characterised by flat-topped hills that lie mostly between 400 and 800 ft above sea-level and that are developed mainly on Upper Greensand which is covered in places with coarse gravel. Separating the hills are deeply cut valleys in the underlying Keuper Marl. Other formations include Lias Clay and Chalk in the east and Triassic sandstones and pebble beds in the west. This mixed background of light infertile soils and heavier fertile clays not unnaturally produced somewhat average densities—between 2 and 3·3 teams per square mile and from 6 to about 9 people. A large number of settlements had moderate amounts of wood and nearly all had meadow, pasture and demesne sheep. Settlements with mills were frequent. There were also salt-pans along the estuaries of the Axe and the Otter.

(6) *South Devon*

Broadly speaking, this is an area of Devonian sandstones, with outcrops of limestone and with igneous rocks. The varying soils are generally more fertile than those of the Devonian rocks of the north of the county, and it is also climatically a more favoured area. Moreover, the region as a whole lies almost entirely below 600 ft O.D., although occasional localities reach almost to 700 ft O.D. The surface is broken by the valleys and estuaries of the Dart, the Avon, the Erme, the Plym, the Tavy and the Tamar. The district between the Plym and the Dart is traditionally known as the South Hams; the density of teams ranged between 2 and 3 per square mile, and that of population between about 8 and 9·5—all well above the average for the county. Elsewhere, in the middle valley of the Dart and in the valleys of the Tavy and the Tamar, the figures dropped to under 2 and about 5 respectively. Many settlements in the region as a whole were without wood. Most settlements, on the other hand, had pasture and meadow, but the amounts, particularly of meadow, were small; sheep were fairly common. Very few mills were recorded, as for Devonshire as a whole. Fisheries and salt-pans were to be found along the river valleys and estuaries. The largest settlement was the borough of Totnes, situated on the Dart, with a population of perhaps between 500 and 600.

(7) *Dartmoor and its borders*

Much of the granitic mass of Dartmoor lies above 1,200 ft O.D., and in places it reaches to over 2,000 ft. Its unfavourable aspect is emphasised by its rainfall, which is well over 60 in. a year. The modern parish of Lydford contains no Domesday place-name apart from that of the borough of Lydford in the extreme north-west. The parish includes the ancient area known as Dartmoor Forest, the boundaries of which were defined in a perambulation of 1240. There is no Domesday evidence relating to the condition and utilisation of this infertile area, but a few incidental references may provide hints. There is, for example, mention of an acre of moor (*mora*) at Sherford in Brixton to the south. Yet another possible indication of the moor may be the fairly substantial amounts of pasture entered for many bordering settlements.

As might be expected, the marginal areas were neither fertile nor populous; they had mostly under 1 team and about 3 people per square mile. Some small settlements lay about the 900 ft contour, e.g. Willsworthy on the west and Natsworthy on the eastern side. On the eastern side of Dartmoor, the well-stocked manors in the neighbourhood of Moretonhampstead brought the densities up to nearly 2 for teams and over 6 for population. There were substantial amounts of wood, which may have been located in the valleys that radiated from the massif, but there were only small amounts of meadow, few sheep and scarcely a mill.

BIBLIOGRAPHICAL NOTE

(1) The Exeter text, together with the Exchequer version, was edited by a committee of the Devonshire Association—*The Devonshire Domesday and Geld Inquest: Extensions, Translations and Indices*, 2 vols. (Plymouth, 1884–92). Unfortunately the order of the Exeter entries was altered to make it correspond with that of the Exchequer text.

The Victoria County History of Devonshire, vol. I (London, 1905), includes a translation of the Exeter text and an introduction by O. J. Reichel. Many of his deductions and theories, and some of his identifications, have now been superseded.

(2) The following is a selection of papers in the *Transactions of the Devonshire Association* that deal with various Domesday problems:

Sir F. Pollock, 'The Devonshire Domesday', xxv (Torquay, 1893), pp. 286–98.

R. N. WORTH, 'Identification of Domesday Manors of Devon', XXV (Torquay, 1893), pp. 309–42, and XXVII (Okehampton, 1895), pp. 374–403.

O. J. REICHEL, 'Some Suggestions to Aid in Identifying the Place-Names in Devonshire Domesday', XXVI (South Molton, 1894), pp. 133–69.

O. J. REICHEL, 'The Devonshire Domesday and the Geld Roll', XXVII (Okehampton, 1895), pp. 165–98.

T. W. WHALE, 'Analysis of the Devonshire Domesday', XXVIII (Ashburton, 1896), pp. 391–463.

O. J. REICHEL, 'Domesday Identifications', XXXIII (Exeter, 1901), pp. 603–49; XXXIV (Bideford, 1902), pp. 715–31; XXXVI (Teignmouth, 1904), pp. 347–79.

J. J. ALEXANDER, 'The Saxon Conquest and Settlement', LXIV (Totnes, 1932), pp. 75–112.

F. W. MORGAN, 'The Domesday Geography of Devon', LXXII (Exeter, 1940), pp. 305–31.

O. J. REICHEL, 'The Hundreds of Devon', in ten parts extra to LX–LXX (Plymouth, Exeter, Torquay, 1928–38); 'Index of Personal and Place Names in the "Hundreds of Devon"' (Torquay, 1942).

R. WELLDON FINN, 'The Making of the Devonshire Domesdays', LXXXIX (Exeter, 1957), pp. 93–113.

(3) Other works of interest in the Domesday study of the county are:

F. W. MORGAN, 'Domesday Woodland in south-west England', *Antiquity*, X (Gloucester, 1936), pp. 306–24. This contains a map of the Domesday woodland of Devonshire.

R. LENNARD, 'Domesday plough-teams: the south-western evidence', *Eng. Hist. Rev.* LX (London, 1945), pp. 217–33.

H. P. R. FINBERG, 'The Domesday plough-team', *Eng. Hist. Rev.* LXVI (London, 1951), pp. 67–71.

H. P. R. FINBERG, *Tavistock Abbey* (Cambridge, 1952).

W. G. HOSKINS and H. P. R. FINBERG, *Devonshire Studies* (London, 1952).

R. LENNARD, 'The composition of the Domesday caruca', *Eng. Hist. Rev.* LXXXI (London, 1966), pp. 770–5.

(4) For the Exeter Domesday and Geld Accounts, see p. 393 below.

(5) A valuable aid to the study of the county is *The Place-Names of Devon* (ed. J. E. B. Gover, A. Mawer and Sir F. M. Stenton; Cambridge, 2 vols., 1931–2).

CHAPTER V

CORNWALL

BY W. L. D. RAVENHILL, M.A., PH.D.

Cornwall is one of those south-western counties for which there are two Domesday descriptions, the Exeter and the Exchequer. The former text is much the more detailed. It usually distinguishes between the hidage of the demesne and that of the peasantry, whereas the Exchequer version does this only for the royal manors. It mentions the settlement of St Tudy (*204b*), unnoticed in the Exchequer folios. It enumerates the demesne livestock. It also provides much other information not included in the Exchequer version, e.g. in the account of the holdings that had been part of the royal manor of Winnianton (*99–100b, 224b–8*; 120) which had passed into the hands of the count of Mortain;[1] and of that of the manors which the count had taken from the church of St Petrock (*204b–5*; 121). Only occasionally does the Exchequer version provide additional information, e.g. that there was a plough-team at St Keverne (*205b*, 121). The two versions often disagree in detail. Thus the Exeter text (*249*) for Brannel mentions 6 teams with the villeins, whereas as many as 20 are recorded in the Exchequer version (121 b). There are also many minor differences between the Exeter and Exchequer figures for teams and oxen. Furthermore, the Exeter entries often give separate figures for demesne and villein teams which are combined into a total, not always accurately, in the Exchequer version. Other categories of information likewise show differences. At Milton (*232*, 123) 13 villeins appear as 14 in the Exchequer version; and for Liskeard (*228*, 121 b) 300 acres of wood appear as 400. Or, again, the wood at Thurlibeer (*233b*, 124) was described as underwood by the Exchequer clerks. The discrepancies that concern totals and general geographical information are set out in Appendix II. To what extent such differences are due to errors in one or the other text, we cannot say. The Exeter version has been used in the present analysis.

In addition to these two texts there are auxiliary sources of information. Bound up with the Exeter text in the *Liber Exoniensis* is a series of entries, headed *Terrae Occupatae*, for Somerset, Devonshire and Cornwall, re-

[1] For differences between the two Exeter accounts themselves, see p. 309

cording such matters as additions to and abstractions from manors, un-sanctioned occupation of territory, and failure to pay customary dues. It is obviously constructed from Inquest material, and from that which produced the Exeter Domesday. The Cornish section (*507–8*) does not add to the geographical information of the main texts.

Also bound up with the Exeter text are two lists of hundreds (*63b*). The first contains eight names including that of St Petrock interlined. The second list of seven names omits the name of St Petrock, and we may wonder to what extent this was a hundred-unit. It is said to have 30 hides that never paid geld in yet another subsidiary document included in the *Liber Exoniensis*, the so-called Geld Rolls or Geld Accounts (*73*).[1] This is the record of a geld levied at the rate of 6*s.* per hide at a time near that of the Inquest, and for all hundreds but one (Stratton) it is possible to reconcile its figures with those of the Domesday texts in a reasonable manner that helps to confirm the likelihood of some identifications. Finally, near the end of the Exeter Domesday, among Summaries of various fiefs in the south-west, there is a Summary of the St Petrock holdings which sets out totals of hides, teams, population and values (*528b*). These are given first for a group of seven demesne manors and then for a group of ten manors held of the Church by the count of Mortain. There is also reference to a further nine manors taken away from the Church by the count. The totals of each of the first two groups can be collated approximately with the details given in the body of the text.[2]

One of the great difficulties in the geographical interpretation of the Cornish material is that presented by those large composite manors that include a number of unnamed places. We can hardly suppose that Pawton (*199b*, 120b), the largest of the Cornish manors, and the only one mentioned in the hundred of that name, was the sole inhabited place in the hundred—what with its 44 hides, its 60 plough-lands, its 43 teams, its recorded population of 86, and its value of £24. There are other large manors, such as Connerton (*111b*, 120) with 31 teams and 70 people and Binnerton (*112b*, 120b) with 18 teams and 67 people. Some of these large manors are in the west of the county where we would certainly not expect to find large nucleated villages. Many may well have consisted of a centre

---

[1] Both the Exeter and the Exchequer texts say that it never paid geld except to the Church of St Petrock (*205*, 121).

[2] For a discussion, see R. Welldon Finn, *Domesday Studies: The Liber Exoniensis* (London, 1964), pp. 126–8.

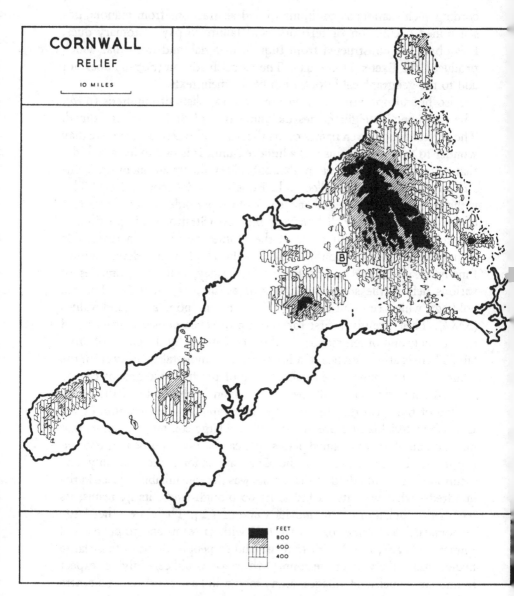

Fig. 67. Cornwall: Relief.

B indicates the borough of Bodmin.

with outlying hamlets and farms, the names of which were not recorded by the Domesday clerks.

The present-day county of Cornwall, in terms of which this study is written, is not exactly the same as the Domesday county. Some territory has been gained from Devon. In the south, along the Tamar estuary, part of Maker was in Cornwall and surveyed in its folios (*256b*, 122), but part had previously belonged to the distant Devonshire manor of Walkhampton (*87*, 100b) and was not transferred to Cornwall until 1844. Territory has also been lost. The two Domesday manors of Borough (*244*, 124) and Tackbear (*244b*, 125), surveyed in the Cornish folios, are now within the parish of Bridgerule in Devonshire. Werrington is of special interest. It is described among the Devon folios (*98*, 101) but there is a reference to it in *Terrae Occupatae* for Cornwall (*508*), and H. P. R. Finberg has shown that it was part of Cornwall until within a few years of the Inquest.[1]

SETTLEMENTS AND THEIR DISTRIBUTION

The total number of separate places mentioned in the Domesday Book for the area now forming Cornwall seems to be at least 330, including the borough of Bodmin.[2] This figure, for a variety of reasons, cannot accurately reflect the number of settlements in 1086. In the first place, there are a few entries in which no place-names appear,[3] and it is possible that some of these refer to holdings at places not named elsewhere in the text. In the second place, as we have seen, the entries for some large manors must have covered not only the caput itself but also several unnamed dependencies, and again we have no means of telling whether these are named elsewhere in the text.

In the third place, when two or more adjoining places bear the same basic name today, it is not always clear whether more than one existed in the eleventh century. There is no indication in the Domesday text that, say, the Higher and Lower Tregantle (in Antony) of today existed as

[1] H. P. R. Finberg, 'The early history of Werrington', *Eng. Hist. Rev.* LIX (1944), pp. 237–51.

[2] This total includes Constantine, not specifically mentioned as a place, but as the half-hide held by St Constantine (*207*, 121).

[3] The relevant entries are: Walter de Claville's 'virgate of land' (*112*, 120b) and the ablations from the Church of St German (*201*, *507*; 120b) and from the Church of St Piran (*206b*, 121).

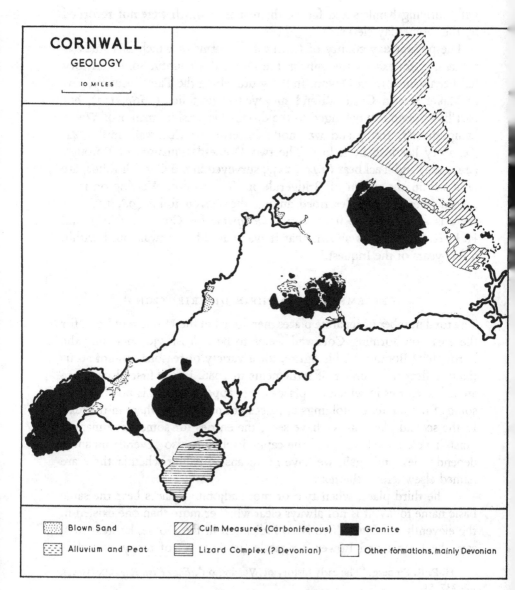

Fig. 68. Cornwall: Geology.

Based on Geological Survey Quarter-Inch Sheets 18, 21 & 25, and 22.

separate villages; the Domesday information about them is entered under only one name (*Argentel, 256b*, 122), though they may well have been separate in the eleventh century, as they were in later times. In the same way, for example, we have not distinguished between Great Draynes and West Draynes (*231, 232*; 124b *bis*) or between Higher and Lower Polscoe (*228, 230b*; 122, 125). The components of each of such groups almost invariably belong to the same hundred, so that any attempt to separate them on the basis of hundreds (as deduced from the Geld Accounts) does not greatly help.[1] For some counties the Domesday text occasionally differentiates the units of these groups by designating one unit by such a term as *alia* or *altera*, but none of the Cornish groups is distinguished in this way.

Quite different are those groups of places bearing the same Domesday name but not lying adjacent. Thus, of two entries bearing the name *Trethac*, one refers to Trethake in Lanteglos by Fowey (*235*, 123) and the other to Tretheake in Veryan (*250*, 123). Or, again, the various entries bearing the name *Trenant* refer to four separate places, to Trenance in Mullion (*99, 225b*; 120), to Trenance in St Keverne (*224*, 124b), to Trenant in Fowey (*251b, 252b*; 123, 124), and to Trenant in Duloe (*230*, 124). The absence of hundredal rubrication increases the difficulty of identifying each holding, but the Geld Accounts and later manorial history help to show which holding belongs to which village.[2]

One further difficulty in this land of Celtic place-names is that a modern name so often bears little or no resemblance to the Domesday form, e.g. Blisland (*Glustona, 101b*, 120), Crantock (*Langorroc, 206*, 121), Gothers in St Dennis (*Widewot, 254b*, 124b), Kea (*Landighe, 254b*, 124b), Perranuthnoe (*Odenol, 255*, 124b), and Trevalga (*Melledam, 112*, 120b).

The total of 330 includes over a dozen places about which very little information is given. All we are told, for example, of Treworder, Treal and Trevedor (*100–100b*; 120) is their assessment, their plough-lands and their values.[3] The same is true of a holding at Pendavy which had

---

[1] The description of St Germans is in two parts, and the Geld Accounts show that the bishop's part was in Rillaton hundred and the canons' part was in Stratton hundred. We have therefore plotted the former at St Germans itself and the latter at South Petherwin, not mentioned in the Domesday texts, but clearly episcopal property in 1086.

[2] The difficulty is further accentuated by the similarity of many names, e.g. *Trewent*, which is, Trewince in St Martin in Meneage (*224*, 124b), and *Trawint* or Trewint in Altarnun (*260b*, 125).

[3] These belonged to the Winnianton complex which had passed to the count of Mortain—see p. 309 below.

been taken from the manor of Blisland (*101b*, 120). The same is also true of three holdings that had been taken from the manor of Pendrim (*101b*, 120): *Pennadelwan* (Bonyalva), *Botchonoam* (Bucklawren) and *Botchatuuuo* (Bodigga), for which joint totals are given. Holdings at seven places had been taken away from the church of St Petrock and acquired by the count of Mortain (*204b–5*; 121), and for these only the hidages and former renders to the church are stated. St Michael's Mount is recorded only as a landholder. Here might also be mentioned those other villages with no teams or no population or neither entered for them.[1]

Cornwall is unusual in that as many as some four fifths of its Domesday place-names are represented today not by parish names but merely by those of hamlets or individual farms and houses. Thus *Wadefeste* (*233*, 123) is now the hamlet of Wadfast in Whitstone; *Karsalan* (*254*, 124b) is that of Carsella in St Dennis; and *Pennehele* (*102b*, 120) is Penheale Manor in Egloskerry. The parish of St Gennys, for example, has five Domesday names represented within its limits, apart from its own.[2] It is difficult to locate the exact modern equivalents of some Domesday names, but their approximate sites can sometimes be inferred from documentary evidence, and no appreciable error is introduced by plotting such approximations on small-scale distribution maps. Very little evidence is to be found on the ground of the important hundred-centre of Tybesta (*247*, 121b) with a recorded population of 61, but its site in the parish of Creed is known. *Heslant* (*262*, 121b) seems to have lain near what has become the borough of Liskeard. The site of *Chienmarc* (*263b*, 124b) was probably in what is now the parish of St Kew, called *Lannohoo* or *Lantloho* (*101*, *238*; 120, 123b) in the Domesday text. *Avalda* (*234*, 124) was very likely Havet in Liskeard. Physical changes in the landscape sometimes make exact identification difficult. But *Widewot* (*254*, 124b) can be equated with Gothers among the china-clay pits and dumps of the parish of St Dennis; and *Delio* (*261*, *263*; 125 *bis*) lay in the much altered landscape of the slate quarries around Delabole. The Celtic place-names of Cornwall present very considerable difficulties. It is unlikely that any two scholars will agree about the identification of some of them. Many of the identifications in the present analysis differ from those of the *Victoria County History*;

---

[1] See pp. 316–17 below.
[2] St Gennys (*101*, *238b*; 120, 123b), Crackington (*240*, 123b), Dizzard (*238b*, *243*; 123b, 124b), Pengold (*238b*, 125), Rosecare (*240*, 123b) and Trefreock (*240*, 123b).

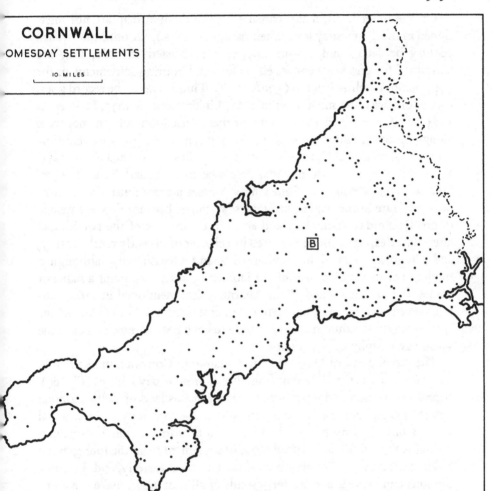

CORNWALL

OMESDAY SETTLEMENTS

10 MILES

Fig. 69. Cornwall: Domesday place-names.

B indicates the borough of Bodmin. The dot in the bay,
off the south-west coast, represents St Michael's Mount.

there are four names which remain unidentified.[1] Whether they will yet
be located, or whether the places they represent have completely disappeared, leaving no trace or record behind, we cannot say.

[1] The names that have remained unidentified in this analysis are as follows: *Languer*
(*257b*, 122), *Tregrebri* (*240b*, 123b), *Treurghen* (*264b*, 123b) and *Trenidered/Trewderet*
(*245b*, 124b).

On the other hand, many parishes on the modern map are not mentioned in the Domesday text. Their names do not appear until the twelfth century or later; and, presumably, if they existed in 1086, they are accounted for under the entries either for neighbouring settlements or for large manors such as Pawton (*199b*, 120b). Thus, so far as the record goes, Camelford was first mentioned in 1205, Linkinhorne in 1235, St Veep in 1284, and Luxulyan in 1329.[1] Many of these post-Domesday names were along the coast, and their appearance may reflect the growing maritime interests of medieval Cornwall—e.g. Fowey, first mentioned about 1255, Hayle in 1265, Boscastle in 1302, Penzance in 1332, and Bude in 1400. Some of the places not mentioned in the Domesday text must have existed, or at any rate borne names, in Domesday times, because they are named in pre-Domesday documents and again in documents of the twelfth and later centuries, e.g. Roseland, named in a charter of the early ninth century, and Tregoney, named in another of 1049.[2] Occasionally, although a modern parish is not named in the Domesday text, it contains a name or names that are mentioned. Thus Altarnun, first mentioned in 1160, contains seven such names,[3] and Menheniot, first mentioned in 1260, contains six.[4] From this account it is clear that there have been many changes in the village geography of the county.

The distribution of Domesday place-names in Cornwall, unlike that in many counties, is very different from that of present-day villages (Fig. 69). It reflects the scattered nature of settlement in this land of Celtic agrarian arrangements. Domesday Cornwall was a land not only of substantial villages but of many hamlets and scattered farmsteads. The outstanding feature is the almost complete absence of settlement from the four granitic areas of the county. The elevation of the two most easterly, and the infertile soils and exposure to westerly winds of all four areas, make them unattractive to settlement. Bodmin Moor, in the east, lies mostly over 800 ft above sea-level. Hensbarrow Downs, to the west, lie generally between

[1] The dates of the place-names in this section are from E. Ekwall, *The Concise Oxford Dictionary of English Place-Names* (Oxford, 4th ed. 1960).

[2] *V.C.H. Cornwall*, I, part 8 (London, 1924), p. 91, identifies *Tregoin* (*251*, 123) as Tregoney, but it is more likely to be Trewoon in St Mewan.

[3] Bowithick (*243b*, 124), Halvana (*Hirmeneu*, *264*, 124b), Landinner (*261b*, 121b), Penpont (*243b*, 124), Tredaule (*242b*, 123b), Trevague (*242b*, 123b) and Trewint (*260b*, 125).

[4] Cartuther (*228b*, 122b), Penhawger (*256*, 122), Tregrill (*263b*, 124), Tregarrick (*Hroscarec*, *245*, 124), Trehawke (*257b*, 122) and Trewolland (*234b*, 123).

400 and 700 ft; they are adjoined on the north by the upland of the St Breock Downs, not made of granite but above the 400 ft contour and also devoid of settlements. Further west, the granite masses of Carnmenellis and of Land's End peninsula are also above 400 ft. Although lower than the two eastern masses, their infertile soils and their exposure to wind resulted in but few settlements. Around these infertile areas, mostly of granite, settlements took advantage of favoured locations along the many small streams of the county. Generally speaking, settlements became less numerous towards the west. There was, however, a surprising number of settlements in the Lizard peninsula, although, as we have seen, the majority were without recorded plough-teams, and some were also without population (Fig. 82).

THE DISTRIBUTION OF PROSPERITY AND POPULATION

Some idea of the information in the Domesday folios for Cornwall, and of the form in which it is presented, may be obtained from the account of Trelaske, situated in the heart of the peninsula, some 4 miles to the south-west of Launceston and today represented by a house in the parish of Lewannick. It was held entirely by one owner, and so is described in a single entry, both versions of which are set out below.

Exeter Domesday (263b)

The count [of Mortain] has one manor which is called Trelaske, which Osfel held on the day that King Edward was alive and dead, and in which is half a hide, and it rendered geld for one virgate. Eight plough-teams are able to plough this. Osfel [holds] this of the count. There, Osfel [has] in demesne half a virgate and half a plough-team, and the villeins [have] the remaining land and two teams. There, Osfel has 3 villeins and 12 bordars and 2 serfs, and 5 animals and 4 pigs and 60 sheep and 10 acres of wood and 50 acres of pasture. And it is worth 20s. a year, and when the count received it, it was worth 60s.

Exchequer Domesday (124)

Offels holds Trelaske of the count [of Mortain]. He himself held it in the time of King Edward, and it paid geld for one virgate of land. There is, however, half a hide there. Land for 8 plough-teams. There are 2½ plough-teams and 2 serfs and 3 villeins and 12 bordars, and 10 acres of wood and 50 acres of pasture. Formerly [it was worth] 60s. It is now worth 20s.

This account does not include all the items of information that appear elsewhere in the folios for the county; but it is a fairly representative and straightforward entry, and it sets out the recurring standard items that are entered under most place-names. These are five in number: (1) assessment, (2) plough-lands, (3) plough-teams, (4) population, and (5) values. The bearing of these five items of information upon regional variations in the prosperity of the county must now be considered.

(1) *Assessment*

The Cornish assessment is stated in terms of hides, virgates, and ferlings (or fertings or ferdings) and of geld-acres. Four virgates made a hide and 4 ferlings made a virgate, but the Cornish geld-acres differed from those of the rest of England in that 3 geld-acres made a virgate.[1] Various entries indicate these equations. Thus Polsue (*248*) was rated at one hide, and we are told that 1 virgate was in demesne and 3 with the villeins. Tredinnick (*256b*) was rated at 1 virgate, and 2 ferlings were in demesne and 2 with the villeins. Trelan, first entered with the king's estates (*100*) as comprising 4 geld-acres, is entered again under the count of Mortain's fief (*227*) as 1 virgate 1 acre. Or, again, the entry for Clinnick (*231b*) says that it was assessed at 2 geld-acres, that half a ferling was in the demesne and the rest (*alia terra*) with the villeins, thereby confirming that 2 geld-acres were clearly larger than half a ferling. Fractions of all four units are frequent; thus Halliggye paid geld for $1\frac{1}{3}$ virgates (*100*, *226b*; 120) and Lesnewth for half a geld-acre (*242*, 124b). We cannot say how, if at all, the Cornish evidence fits into the theory of a small geld-hide in the south-west.[2]

The Exeter version usually distinguishes between the hidage of the demesne and that of the peasantry, whereas the Exchequer text does this only for the royal manors. In almost one half of the entries, the hidage of the *terra villanorum* is not specifically stated, but we are merely told that the villeins had the remaining land (*alia terra*). Where, however, separate figures are given, their sum is normally in agreement with the stated total, but there are occasional discrepancies, as the following examples indicate:

	Stated total	Sum
Gulval (*Landicla*, *200b*, 120b)	$1\frac{1}{2}$ hides	2 hides
Penheale (*102b*, 120)	$2\frac{1}{2}$ hides	2 hides
Trezance (*228b*, 122)	2 hides	1 hide

---

[1] There is, incidentally, frequent reference to one third of a virgate.
[2] See pp. 236–7.

We can only assume that these discrepancies are the result of clerical errors, that is, unless there had been unrecorded changes in the composition of the manors involved.

A distinction is almost invariably drawn between the assessment of a holding and the number of hides for which it actually paid geld.[1] The wording varies slightly but the following entries are characteristic:

Boconnoc
    E.D. (*229b*): *In ea est dimidia hida et reddidit gildum pro i virgata.*
    D.B. (124): *et geldabat pro una virgata. Ibi tamen est dimidia hida.*

Tregear
    E.D. (*199*): *In ea sunt xii hidae. Iste reddiderunt gildum pro ii hidis tempore regis Edwardi.*
    D.B. (120b): *T.R.E. geldabat pro ii hidis sed tamen sunt ibi xii hidae.*

Maitland suggested 'with some hesitation' that this double entry for geld indicated an assessment current in the Confessor's day, and an increase therein since the Conquest.[2] On all general grounds this is unlikely. While the Domesday Book contains many examples of reduction in assessment, it has no record of an increase. What is more, the meaning of the difference between the two figures becomes clear if it is compared with the amount of land in the Geld Accounts said to be 'in demesne'. What the Domesday entries are in fact telling us is that a manor is assessed at so many hides, but that when fiscal demesnes are exempt, it gelds for a smaller amount. Thus the count of Mortain's manors in the hundred of Connerton were Alverton and Tehidy (*255*, 121b). Each was rated at 3 hides and each paid geld on 2 hides, making a total difference of 2 hides. This is the exact amount of fiscal demesne the Geld Accounts record for the count's land in the hundred (*72*).[3] We should, however, be over-optimistic if we expected to find perfect correspondence between the figures derived from the Geld Accounts and those of the Domesday Book, but L. F. Salzman's totals calculated in terms of fiefs are impressive.[4] The two sets of figures for

[1] Occasional examples can also be found in Wiltshire and Somerset—see pp. 15 and 152. For examples from other counties, see R. Welldon Finn, *The Domesday Inquest and the Making of Domesday Book* (London, 1961), pp. 152–8.
[2] F. W. Maitland, *Domesday Book and Beyond* (Cambridge, 1897), pp. 410, 462.
[3] It should be noted that fiscal demesne and manorial demesne are not necessarily the same. In each of the two manors of Alverton and Tehidy, there was only half a hide of manorial demesne. Very occasionally fiscal and manorial demesnes were identical, e.g. at Penpoll (*257b*).
[4] L. F. Salzman in *V.C.H. Cornwall*, 1, part 8, pp. 48–9.

the reconstructed hundreds also show a remarkable correspondence between the 'non-gelding hides' of the Cornish folios and the 'hides in demesne' of the Geld Accounts.

Such collation suggests, moreover, that, since the Conquest, some manors had lost the privilege of retaining the geld for which the demesnes were liable. In the Winnianton geld account (*72*), St Constantine's entire half-hide was fiscal demesne. The Domesday entry (*207*, 121) says that *T.R.E.* it was 'free of all service' but that since the count received the land, 'it has rendered geld unjustly as if it were villeins' land'.

Some entries merely state the number of gelding hides, e.g. for most of the Tavistock manors (*180b*, 120–121 b). We can only suppose that these holdings paid their full geld, especially as the figure is exactly equal to the sum of the demesne and villein hides where given, e.g. Treliever (*199*) 'rendered geld for 1½ hides', and there was half a hide on the demesne and 1 hide with the villeins.

It would seem that some manors in ecclesiastical hands had never paid geld to the king, but were privileged to collect and retain it for their own purposes. Thus we are told that the canons of St Petrock 'never paid geld except to the Saint' (*205*, 121); and the entries for several of the Saint's manors make reference to the fact (*202*, 120b). Thus Bodmin had 'at no time paid geld'; Padstow had never paid geld except for the use of the church (*nisi ad opus ecclesiae*); and Rialton was 'free from all service'.[1] St Petrock's land in St Enoder (*203*, 121) had likewise never paid geld, but we are told this fact only in the Exchequer version. Lancarffe belonged to the count of Mortain, but it was of the honour (*de honore*) of St Petrock and had also never paid geld (*241b*, 123b). St Michael's Abbey had 2 hides in Truthwall in Ludgvan (*208b*, 120b), which, again, had never paid geld; one of these had been seized by the count of Mortain, and it is described with the count's fief but it is said to render geld to St Michael (*258b*). Land which the count had taken from St Neot had also never paid geld (*207*, *230b*; 121, 124). There were yet other churches that held their local estates free of geld: St Germans (*199b*, 120b), St Probus (*206*, 121), St Stephen by Launceston (*206*, 120b), and St Crantock, noted only in the Exchequer version (121). St Buryan (*207*, 121) was said to hold his land freely (*libere*), so was St Piran at *Lanpiran* or Perranzabuloe (*206b*, 121). One of St Piran's manors at the unidentified *Tregrebri* was in the count's hands, but it was of the honour of St Piran, and we are told that it never

[1] Not necessarily the same as free from geld but implying that fact.

paid geld (*240b*, 123b). As we have seen above, St Constantine also held half a hide 'free of all service'; the Geld Accounts show this half-hide to have been in Winnianton hundred (*72*). Collation with the Geld Accounts suggests that there may have been yet other ecclesiastical landowners with geld-free land not mentioned in the Domesday texts, e.g. the demesnes of St Ewe and St Goran in Tybesta hundred (*72*).

The description of the royal manor of Winnianton presents some interesting features. It contained 15 hides, 4 of which belonged to Winnianton itself and which were fully described (120).[1] The other 11 hides spread over 22 manors had passed to the count of Mortain and are mentioned in *Terrae Occupatae* (*508*). The Exeter text includes two very different accounts of these manors:

(1) Under *Terra Regis* (*99–100b*) is a summary of each manor—the name of its tenant, its assessment, its plough-lands and its values. The statement about assessment is of two kinds. For each of eleven manors we are told that it 'paid geld' for so much; each of the remaining entries tells us how much 'is there'. The Exchequer version, more accurately as we shall see, never speaks of geld being paid but says each time that so much 'is there'.

(2) Under the fief of the count of Mortain (*224b–7b*) there is a fuller description of 18 of these manors, which gives details about inhabitants, plough-lands, teams and resources. It also gives a double entry for the geld of 16 of these, the larger figure being always the same as that of the *Terra Regis* entry, irrespective of the form of the latter, e.g.

Lizard (*99b*): *reddidit gildum T.R.E. pro i hida.*
Lizard (*226*): *In qua est i hida terrae et reddidit gildum pro dimidia hida.*
Tredower (*100*): *Ibi est i hida terrae.*
Tredower (*227b*): *In qua est i hida terrae et reddidit gildum pro i ferling.*

For the two remaining manors, Crawle and Roscarnon, the 'geld paid' alone is entered, and these two entries also differ from the rest in that neither says that the manor was of the king's demesne manor of Winnianton. What is even more interesting is that the Exchequer account of the Mortain fief omits these expanded entries, with the exception of those for Crawle and Roscarnon. Without the Exeter text we should know nothing

---

[1] For an account of the general interest presented by the entry, see V. H. Galbraith, *The Making of Domesday Book* (Oxford, 1961), pp. 117–20.

of the resources and inhabitants of thirteen manors.[1] Here is a striking example of the perils of the Exchequer version.

As might be expected in a county that is Celtic rather than Anglo-Saxon, examples of the 5-hide unit are rare, certainly not more than might occur by chance in any group of entries. Some half-dozen or so places were assessed at 5 hides, e.g. Climsom (*102*, 120), and a few more at multiples of 5 hides or at 2½ hides, e.g. the 2½ at Penheale (*102b*, 120). Five-hide groups of villages might be constructed, but these would have little meaning as there are so many 1-hide vills that could be combined to form any group. L. F. Salzman drew attention to the frequent occurrence of a 3-hide unit in the assessment itself, e.g. the 3 at Towan (*100b*, 120), the 6 at Treglasta (*237*, 121b) and the 12 at Tregear (*199*, 120b).[2] On the other hand, it must be noted that only 27 places seem to have been assessed on a 3-hide basis: eleven places at 1½ hides, ten at 3 hides, two at 6 and four at 12 hides.

Whatever may have been the unit, if any at all, it is clear that the assessment was by 1066 largely artificial in character, and bore no constant relation to the agricultural resources of a vill. The variation among a representative selection of 2-hide holdings speaks for itself:

	Plough-lands	Plough-teams	Population	*Valuit*	*Valet*
Launcells (*244*, 124)	9	3½	16	20s.	40s.
Rosecraddock (*234*, 123)	15	4½	29	60s.	40s.
Treknow (*203b*, 121)	8	3½	17	25s.	15s.
Trevisquite (*260*, 122b)	12	6	25	30s.	25s.
Tywardreath (*247*, 122b)	12	7	33	£4	40s.

The total assessment, including non-gelding hides, amounted to 423 hides, 3 virgates, 1 ferling and 45 acres. The number of hides that actually paid geld is 130 hides, 2 virgates, 2 ferlings and 1 acre. These figures relate to the area included within the modern county and so are not strictly comparable with Maitland's totals for the Domesday county.[3] The nature of some entries and the possibility of unrecognised duplicates make exact

[1] We also know, from the Exeter version, the details of the plough-lands and values of another five manors.

[2] *V.C.H. Cornwall*, I, part 8, p. 53.

[3] F. W. Maitland's figures were 399 and 155 hides respectively (*op. cit.* pp. 400–1). The corresponding totals for the Geld Accounts are (1) 400 hides, 3 virgates, 1 ferling and (2) 124 hides, 3 ferlings (*V.C.H. Cornwall*, I, part 8, p. 48).

calculation very difficult. All that any estimate can do is to indicate the order of magnitude involved. Cornwall, like Devonshire, was very lightly assessed as compared with the counties to the east.[1]

(2) Plough-lands

Plough-lands are systematically entered for the Cornish holdings, and the Exchequer formula normally runs: 'There is land for *n* plough-teams' (*Terra est n carucis*). The Exeter text usually says: *hanc terram possunt arare n carrucae*. A reckoning in half-teams occurs in the entries for Westcott (*239b*, 123b) and West Curry (*264b*, 123b), but there is never a reckoning in terms of oxen. There are many round numbers, e.g. 10, 20, and 30 plough-lands, which may make us suspect that they are estimates rather than actual figures. Two entries for Whitstone (*255b*, 125) and St Neot (*207*, 121) make no reference to plough-lands, although they state the number of teams at work; these account for half a team and 1 ox. There are, moreover, a number of other entries which mention neither plough-lands nor teams, e.g. those relating to the holdings that had passed from the church of St Petrock to the count of Mortain (*204b–5*; 121).

There are some anomalies. The manor of Trembraze is one of those described both under *Terra Regis* and within the fief of the count of Mortain.[2] The first entry (*99b*, 120) refers to one plough-land, but this is not mentioned in the second entry (*226*).[3] The manors of Poundstock and St Gennys are also described in parallel entries. The first, under *Terra Regis*, says of both manors that '12 plough-teams can plough the whole land (*totam terram*)', but under the count of Mortain's fief the manors are described separately and are credited with 6 and 10 plough-lands respectively (*238*, *238b*; 123b *bis*). Another anomaly is presented by the entry for Ludgvan. At the foot of folio *260*, the Exeter scribe wrote '15 plough-teams can plough this'; but when he turned over the page, he repeated himself, giving the same details about hidage and then going on to say that '30 plough-teams can plough this'. The Exchequer scribe was unable to make up his mind which figure was correct, and so wrote 'land for 15 plough-teams or 30 plough-teams' (122b).[4]

---

[1] See p. 237 above.   [2] See p. 309 above.

[3] Both entries say the holding was waste—*vacua* (*99b*, 120), *vastata* (*226*); see p. 333 below.

[4] As there were only 12 teams at work, this has been taken to be 15 plough-lands for the purpose of our count.

The relation between plough-lands and teams varies a great deal in individual entries. They are equal in number in about 5½ per cent of the entries which record both. Occasionally the Exchequer text explicitly says that this is so, as in the entry for Cann Orchard (*244*, 124): *Terra ii carucis, quae ibi sunt*. A deficiency of teams can be encountered in as many as 94 per cent of the entries which record both plough-lands and teams. The values of some deficient holdings had fallen, but on others they had remained constant, and on others they had even increased. No general correlation between plough-team deficiency and decrease in value is therefore possible, as the following table shows:

		Plough-lands	Teams	*Valuit*	*Valet*
Decrease in value	Milton (*232*, 123)	20	8	£3	30s.
	Launcells (*244*, 124)	9	3½	£5	£2
Same value	Morton (*237b*, 123b)	3	2	10s.	10s.
	Woolston (*240b*, 123)	6	4½	20s.	20s.
Increase in value	Norton (*237*, 123b)	5	3½	20s.	25s.
	Week St Mary (*259*, 122b)	8	3	20s.	30s.

These figures are merely indications of unexplained changes in individual holdings.

An excess of teams is encountered in only one Cornish entry—that for Antony (*180b*, 121), where there were 6 plough-lands with 7 teams and where the value had remained unchanged.

The total number of plough-lands for the area within the modern county is 2,562. If, however, we assume that the plough-lands equalled the teams on those holdings where a figure for the former is not given, the total becomes 2,562⅜. The figures by Sanders,[1] Maitland[2] and Salzman[3] are not comparable with the present total because they refer to the Domesday county. In view of the nature of some entries, and of the possibility of unrecognised duplicate entries, no estimate can be anything other than an approximation.

---

[1] W. B. Sanders counted 2,585 plough-lands: *A literal translation of the part of Domesday Book relating to Cornwall* (Southampton, 1875), in prelims (unpaginated).

[2] F. W. Maitland counted 2,377 plough-lands (*op. cit.* p. 401).

[3] L. F. Salzman counted 2,572 plough-lands (*V.C.H. Cornwall*, I, part 8, p. 51).

(3) Plough-teams

The Cornish entries for plough-teams are fairly straightforward, and, like those for other counties, they normally draw a distinction between teams on the demesne and those held by the peasantry, the Exeter text being far more regular than that of the Exchequer in this respect. Thus the Exeter account of Tremoddrett (*250 b*) tells of 2 teams on the demesne and 3 teams with the peasantry, but the Exchequer version (123) merely says 'there are 5 teams there'. The teams of the peasantry are usually described in the Exeter folios as belonging to the *villani* even when the recorded population includes no villeins, e.g. in the entry for Pendrim (*101 b*, 120).

Half-teams and oxen are not uncommon. Occasionally when the Exchequer version mentions half a team, the Exeter version speaks of 4 oxen, e.g. in the entry for St Buryan (*207*, 121); the Exeter formula very often is *iiii boves in carrucam*, as in the entry for Trelowia (*234 b*, 123). The same equation is apparent in the entry for Treveniel, where the single team of the Exchequer entry (124) appears as 6 oxen + 2 oxen in the Exeter version (*244 b*). There are, however, a number of discrepancies between the Exeter and Exchequer texts, as the examples in the table below show. Such anomalies led R. Lennard to believe that the accepted equation of 8 oxen to 1 team (as postulated by Round and Maitland)[1] was not always true in the south-west.[2] He argued that the Domesday

Comparison of some Exeter and Exchequer plough-team figures

	Exeter	Exchequer
Ashton (*258 b*, 122)	3 oxen	½ team
Alvacott (*239 b*, 123 b)	1 team + 3 oxen	1½ teams
Trewince (*224*, 124 b)	½ team + 1 team + 2 oxen	2 teams
Trewolland (*234 b*, 123)	½ team + 2 oxen	1 team
Draynes (*231*, 124 b)	1 ox + 4 oxen	½ team
Trenant in Fowey (*252 b*, 124)	7 oxen	1 team
Tolcarne in North Hill (*181 b*, 121 b)	2 oxen	omitted
Westcott (*239 b*, 123 b)	2 oxen	omitted

[1] J. H. Round, *Feudal England*, p. 35; F. W. Maitland, *op. cit.* p. 147.
[2] R. Lennard, 'Domesday plough-teams: the south-western evidence', *Eng. Hist. Rev.* LX (1945), pp. 217–33.

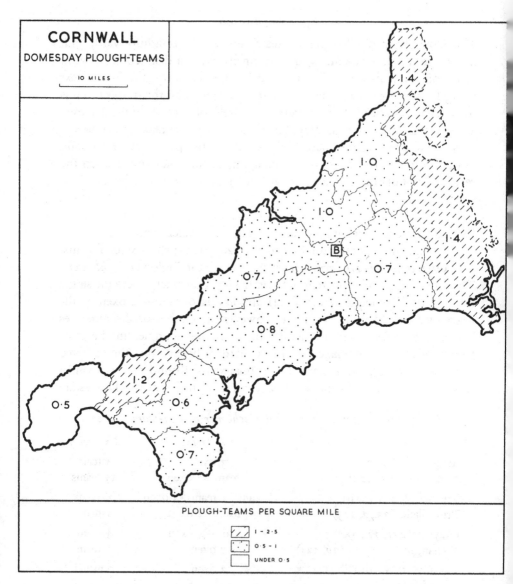

CORNWALL

DOMESDAY PLOUGH-TEAMS

10 MILES

PLOUGH-TEAMS PER SQUARE MILE

1 – 2·5

0·5 – 1

UNDER 0·5

Fig. 70. Cornwall: Domesday plough-teams in 1086 (by densities).
B indicates the borough of Bodmin.

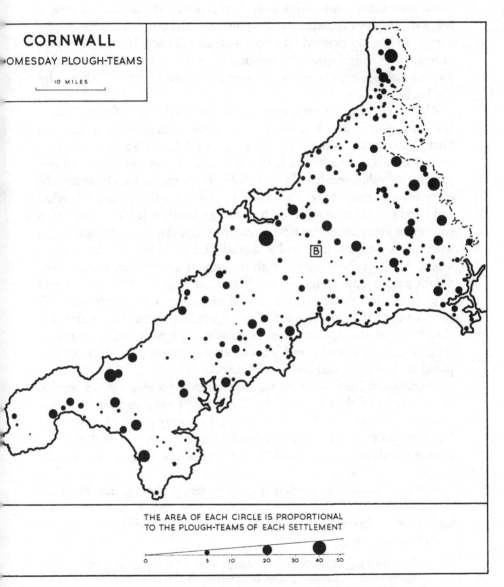

Fig. 71. Cornwall: Domesday plough-teams in 1086 (by settlements).

B indicates the borough of Bodmin.

teams were often composed of only 6 or 4 oxen, and that there were also teams of 7 or even 10 oxen. He also pointed to the fact that the Exchequer version frequently omitted odd oxen and pairs of oxen. H. P. R. Finberg, in reply, argued in favour of a uniform team of 8 oxen and believed that any anomalies were due to the 'contempt for small fractions' shown by the Exchequer version.[1]

About 60 entries do not mention teams although they refer to plough-lands and often to people as well as to resources such as wood or meadow. Such, for example, is the entry for Langunnett (229b, 122), where there were 8 bordars, 3 serfs, and land for 2 teams as well as woodland and pasture; its value, incidentally, had fallen from 10s. to 5s. Or, again, the entry for Clinnick (231b, 123) records, amongst other items, 2 bordars and 1 serf but again no teams on its 2 plough-lands. A few of these entries enumerate livestock amongst their resources, e.g. that for Tregeagle (253), where there were 15 sheep and 1 'animal' and also a villein, a bordar and a serf. At all of these places (with or without demesne livestock), how did the people support themselves? Were their settlements mainly pastoral in spite of their plough-lands? Or did oxen from other settlements till these lands? Or is the absence of any record of teams due to the fact, already noted above, that the clerks frequently ignored the presence of single oxen or pairs of oxen? This absence of teams (and sometimes of population) is a marked feature of Cornwall (Fig. 82).

On the other hand, some manors seem to have possessed teams beyond the capacities of their inhabitants. There was 1 team each at Buttsbear (244, 124) and at Pigsdon (397b, 125), yet the only population recorded for these places was 1 serf and 1 bordar respectively. Could these have obtained assistance from outside over and above that given by their respective families?

The total number of plough-teams amounts to 1,217$\frac{5}{8}$, but this refers to the area included within the present county, and, in any case, a definitive total is hardly possible.[2] This count has been made on the assumption of an 8-ox team.

[1] H. P. R. Finberg, 'The Domesday plough-team', *Eng. Hist. Rev.* LXVI (1951), pp. 67–71. See also R. Lennard, 'The composition of the Domesday caruca', *Eng. Hist. Rev.* LXXXI (1966), pp. 770–5.

[2] The estimate by W. B. Sanders (*op. cit.*, unpaginated) came to 1,218 teams; that of F. W. Maitland (*op. cit.*, p. 401) to 1,187 teams; and that of L. F. Salzman (*op. cit.* p. 51) to 1,220 teams. All these figures refer to the Domesday county and were compiled on the basis of an 8-ox team.

(4) *Population*

The bulk of the population was comprised in the three main categories of bordars, villeins and serfs. In addition to these were the burgesses of Bodmin and a small miscellaneous group comprising coliberts and *cervisarii*. The details of the groups are summarised on p. 320. There are three other estimates, by Sir Henry Ellis,[1] W. B. Sanders[2] and L. F. Salzman,[3] but the present estimate is not strictly comparable with these because it is in terms of the modern county. In any case, an estimate of Domesday population can rarely be definitive, and all that can be claimed for the present figures is that they indicate the order of magnitude involved. These figures are those of recorded population, and must be multiplied by some factor, say 4 or 5, in order to obtain the actual population; but this does not affect the relative density as between one area and another.[4] That is all that a map such as Fig. 72 can roughly indicate.

It is impossible to say how complete were these Domesday figures. We should have expected to hear more, for example, of priests. Then again there are 26 entries which do not mention inhabitants although they sometimes refer to plough-lands and even teams and to such other resources as wood or meadow. Thus the Exchequer entry for St Keverne (121) mentions a team as well as plough-lands and pasture, but does not refer to any people who might have worked the team. Or, again, the entry for St Stephens by Launceston (*206b*, 120b) is a fairly full one, and it mentions, among other items, 3 demesne teams and 6 teams with 'the villeins', but neither villeins nor any other people are enumerated. We cannot be certain about the significance of any omissions, but they (or some of them) may imply unrecorded inhabitants.

The only reference to unnamed thegns is to the seventeen who, in 1066, had held parts of the manor of Winnianton later abstracted by the count of Mortain (*99*, 120). Most, if not all, of these seem to be included among the named persons of 1086. These latter, as sub-tenants, have been excluded from our count. So have the priests who were landholders.[5]

---

[1] Sir Henry Ellis, *A General Introduction to Domesday Book*, II (London, 1833), p. 432. His total for the Domesday county (excluding tenants and sub-tenants) came to 5,334.

[2] W. B. Sanders, *op. cit.*, unpaginated.

[3] L. F. Salzman, *V.C.H. Cornwall*, I, part 8, p. 53. His total for the Domesday county (excluding tenants and sub-tenants) came to 5,298.

[4] But see p. 368 below for the complication of serfs.   [5] See p. 335 below.

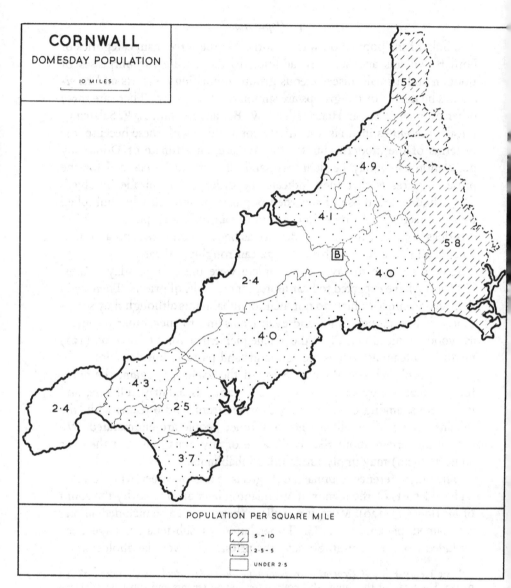

Fig. 72. Cornwall: Domesday population in 1086 (by densities).
B indicates the borough of Bodmin.

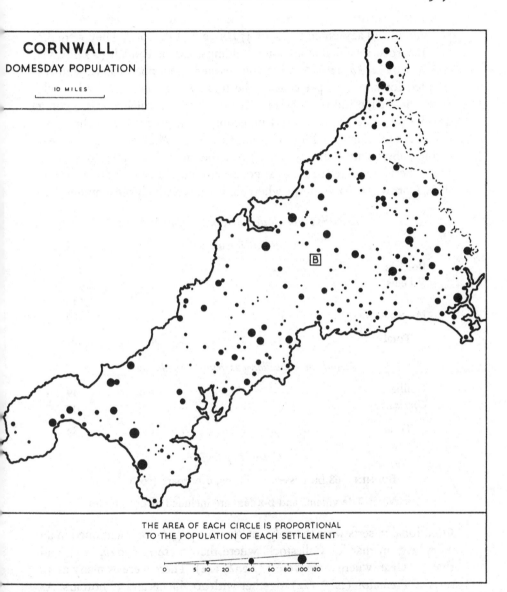

Fig. 73. Cornwall: Domesday population in 1086 (by settlements).
B indicates the borough of Bodmin.

Bordars constituted the most important element numerically in the population and amounted to about 45 per cent of the total. They were the sole recorded inhabitants on some holdings, e.g. at Penhalt (*243b*, 124) and Bosent (*228b*, 122). Next in numerical order came villeins, who accounted for about 32 per cent of the total. A common Exeter formula states the hidage and plough-lands occupied by the villeins before going on to enumerate the categories of people on a holding; and the word *villani* is sometimes used in this connection even when no villeins appear in the subsequent enumeration, e.g. in the entry for Trebarfoot (*239*, 122b). Serfs amounted to about 21 per cent of the total. Next to Gloucestershire, Cornwall was the county where they were relatively most numerous.[1]

Recorded Population of Cornwall in 1086

A. *Rural Population*

Bordars	2,421
Villeins	1,702
Serfs	1,148
Miscellaneous	89
Total	5,360

Details of Miscellaneous Rural Population

Coliberts	49
Cervisarii	40
Total	89

B. *Urban Population*

BODMIN   68 burgesses, 5 villeins, 6 bordars (see p. 335).

Note: The villeins and bordars are included in both tables.

Occasionally, serfs were the only category of population mentioned in an entry, e.g. in that for Callestock, where there were 12 (*202b*, 121), and that for Gear, where there were 3 (*225b*, 122). There were as many as 50 serfs at Trematon (*256*, 122) together with 20 villeins and 30 bordars. As

[1] A roughly contemporary document shows that serfs were by no means exclusively Celtic. Manumissions of slaves at the altar of St Petrock in Bodmin are entered on the blank pages and margins of the Bodmin Gospels. In 47 separate transactions at various dates between about 941 and 1050, some 10 per cent of the names were English (G. Oliver, *Monasticon Dioecesis Exoniensis*, Exeter, 1846, *passim*).

there were only 3 teams on the demesne and 7 with the peasantry, we may wonder how the serfs were employed, unless in connection with the castle there.

In the small miscellaneous group, coliberts amounted to about 1 per cent of the total recorded population, and they appear for only two manors —41 at Winnianton and 8 at Rinsey (*99*, *225b*; 120). Finally, there were 40 *cervisarii* on the royal manor of Helston (*100b*, 120). The only other Domesday reference to *cervisarii* occurs in the Little Domesday account of Bury St Edmunds in Suffolk (*372*). It has been variously translated, but may here refer to tenants who paid their dues in ale.[1]

(5) *Values*

The value of an estate is generally given for two dates—for 1086 and for some earlier year. Some entries give only one value, and a small number give no value at all. The following examples illustrate some of the variations to be found in the Exchequer text. The first two are the formulae most commonly encountered:

> Tretheake (*250*, 123): *Olim xx solidos. Modo valet xv solidos.*
> Treliever (*199*, 120b): *Olim et modo valet iiii libras.*
> Ashton (*258b*, 122): *Valet x solidos.*
> Binnerton (*111b*, 120b): *Reddit x libras.*
> Ellbridge (*245b*, 122): *Olim valebat xv denarios.*[2]

The Exchequer *valet* often appears as *reddit* in the Exeter text, e.g. in entries for Alverton (*255*, 121b) and Liskeard (*228*, 121b). The general use of the word *olim* in the Exchequer folios leaves the date of the earlier valuation in doubt, but a few entries are more explicit and speak of the day on which the existing holder entered into possession. There are seven such entries on folio 121, and that for Treknow will serve as an example: *Quando comes accepit xxv solidos. Modo valet xv solidos.*[3] In

---

[1] A charter of 1076 in the cartulary of the priory of St Stephen at Launceston tells of land on which were 'freemen, *cervisarii* and serfs' (C. Henderson, *Essays in Cornish History*, Oxford, 1935, p. 70). See p. 341 below for the render of a cask of ale in connection with the church of St German.

[2] Earlier values alone appear also for Skewes (*99*, 120) and Trembraze (*99b*, 120) but each of these was said to be unoccupied (*vacua*) in 1086. The Exeter text also contains other entries for these two Winnianton manors (see p. 309 above). Skewes is said to be *penitus vastata* (*225b*) and Trembraze to be *vastata* (*226*)—see p. 333 below.

[3] The seven entries, with their Exeter folios, are: Treknow (*203b*), St Keverne (*205b*), Crantock (*206*), Perranzabuloe (*Lanpiran*, *206b*), St Buryan (*Ecglosberria*, *207*), St Neot (*207*) and St Constantine's half-hide (*207*).

contrast with the Domesday Book, the Exeter text almost invariably gives the earlier date as *quando recepit*.[1] Here are the two versions of the valuation of Landulph:

> E.D. (*260*): *valet per annum x solidos et quando comes recepit xv solidos.*
> D.B. (122b): *Olim xv solidos. Modo valet x solidos.*

Occasionally, when the Domesday Book states a value for only one date, the Exeter text adds the earlier value, sometimes the same as, sometimes different from, that of 1086, as the following examples show:

> Truthwall in Ludgvan
> E.D. (*208b*): *valet per annum xx solidos et quando comes terram accepit.*
> D.B. (120b): *Valet xx solidos.*
>
> Rosebenault
> E.D. (*241*): *valet x solidos et quando recepit xx solidos.*
> D.B. (123b): *Valet x solidos.*

The various accounts of the Winnianton complex[2] state different values for some manors in 1086; Garah is valued at 5s. on one Exeter folio (*99b*) and at 10s. on another (*226*), Mawgan at 3s. (*99b*) and 5s. (*226*), and Trenance in Mullion at 30d. (*99*) and 2s. (*225b*).

The amounts are usually stated in round numbers of pounds and shillings, e.g. £2 or 30s. The majority of the valuations at both dates are multiples of 5s., although low values such as 1s., 3s., 4s., and 8s. can be encountered. The royal manor of Connerton (*111b*, 120) was said to render £12 'by tale', *numero* in the Exeter text, *ad numerum* in that of the Exchequer. The values of some other royal manors were said to be 'by weight' or 'by weight and assay',[3] e.g.:

> Climsom (*102*, 120): *Reddit vi libras ad pondus.*
> Helston (*100b*, 120): *Reddit viii libras ad pondus et arsuram.*

For manors valued in this way, only a 1086 figure is given. The silver mark (the equivalent of 13s. 4d.) is used to give the earlier values of Brannel (*249*, 121b), Trenowth (*249*, 121b), and Treglasta (*237*, 121b), each said to be worth 12 silver marks (*xii markae argenti*). The silver

---

[1] Among the Exchequer variations the following should be noted: (1) Crantock (*206*, 121): *Valet v libras. Quando comes terram saisivit valebat xl solidos.* (2) St Neot (*207*, 121): *valet v solidos; prius valebat xx solidos.*

[2] See p. 309 above.

[3] No non-royal manors are valued in this way.

mark is also used in the 1086 value of Rillaton (*264*, 121 b): *Olim xxx libras. Modo valet xv libras et unam markam argenti et v solidos.*

Values had occasionally risen; very often they had remained the same; but on about three quarters of the Cornish holdings they had fallen. The rise on a relatively few manors more than balanced the more widespread small reductions on other manors. The great increases on thirteen of the principal demesne manors of the count of Mortain are striking: they are described in the first thirteen entries of the count's fief in the Exchequer version, and are set out below. L. F. Salzman drew attention to the curious features of these valuations—the frequent occurrence of sums of £8 and of amounts ending in 18s. 4d. 'There is clearly some meaning in these remarkable valuations, but it appears at present impossible to

Table of the first thirteen manors entered in the Exchequer account of the count of Mortain's fief (121b)

Notes: (1) The Exeter folios are given in parentheses.

(2) The earlier values of Treglasta, Trenowth and Brannel were 12 marks each, i.e. £8.

(3) The 1086 value of Rillaton was £15 + 1 silver mark + 5s., i.e. £15. 18s. 4d.

(4) L. F. Salzman believed that the earlier £15. 18s. 4d. for Helstone and the £12 for Tybesta were probably mistakes for £8 or 12 marks (*V.C.H. Cornwall*, I, part 8, p. 52).

	Plough-lands	Plough-teams	Value when received			Value 1086		
			£	s.	d.	£	s.	d.
Fawton (*228*)	30	21	8	0	0	16	18	4
Liskeard (*228*)	60	16	8	0	0	25	18	4
Stratton (*237*)	30	19	30	0	0	35	18	4
Helstone (*237*)	15	12	15	18	4	15	18	4
Treglasta (*237*)	20	14	8	0	0	15	18	4
Tybesta (*247*)	30	13	12	0	0	15	18	4
Trenowth (*249*)	40	17	8	0	0	25	18	4
Brannel (*249*)	20	23	8	0	0	12	18	4
Moresk (*249*)	10	7	5	0	0	9	18	4
Trewirgie (*249b*)	16	13	5	0	0	8	0	0
Alverton (*255*)	60	15	8	0	0	20	0	0
Tehidy (*255*)	50	17	8	0	0	20	0	0
Rillaton (*264*)	15	10	30	0	0	15	18	4

unravel the mystery.'[1] Apart from the figures themselves, the amounts of the increases are striking. So are the increases on some of the bishop of Exeter's manors, e.g. from £10 to £24 at Pawton (*199b*, 120b) and from £8 to £17 at Lawhitton (*200b*, 120b). Some decreases in value are also striking, e.g. from £10 to £3 on St Petrock's manor of Cargoll (*203*, 121), from £2 to 3*s*. at Treworder (*100*), and from £2 to 30*d*. at Tregoose (*99b*).[2] Were such changes the results of farming vicissitudes? Or could they have arisen from changes in the composition of manors, changes about which we are told nothing? A hint of such changes appears in the account of the St Piran's manor of Perranzabuloe (*206b*, 121), the value of which had fallen from 40*s*. to 12*s*.; but we are told not only that the count had taken away two manors (*terrae* in the Exchequer version) from the original manor, but that from one of these two manors he had taken away all the livestock (*abstulit comes totam pecuniam*).

Generally speaking, the greater the number of teams and men on a holding, the higher its value; but there is much variation, and it is impossible to discern any consistent relationship, as the following figures for five holdings, each rendering £3 in 1086, show:

	Teams	Popula-tion	Other resources
Calstock (*256*, 122)	8	72	Wood, pasture
Cargoll (*203*, 121)	9	50	Mill, wood, pasture
Gulval (*200b*, 120b)	4	20	Meadow, pasture
Ludgvan (*260*, 122b)	12	63	Pasture
Sheviock (*180b*, 121)	5	27	Wood, pasture

These are but some of the perplexities presented by the Cornish folios.

Conclusion

For the purpose of calculating densities, Cornwall has been subdivided into eleven units, a division based partly upon the hundreds as deduced from the Geld Accounts and partly on the main physical and geological features. The boundaries of the density units have been drawn to coincide with those of the civil parishes, and this gives the units an artificial character because many parishes extend across more than one kind of soil

---

[1] L. F. Salzman in *V.C.H. Cornwall*, I, part 8, p. 52.
[2] Treworder and Tregoose were Winnianton manors, and their separate values are omitted in the Exchequer version—see p. 309 above.

and country, e.g. the parishes around Bodmin Moor. The division into eleven units is therefore not a perfect one, but it may provide a basis for distinguishing broad variations over the face of the countryside.

Of the five standard formulae, those relating to plough-teams and population are most likely to reflect something of the distribution of wealth and prosperity throughout the county. The main feature that stands out on both maps is the paucity of teams and population as compared with, say, Devon and Somerset, to say nothing of the rest of southern England. The density of teams never rose above 1·2 per square mile and that of population never above 5, except along the eastern border of the county where it adjoins Devonshire (Figs. 70 and 72). There were variations but not marked contrasts, and this reflected the somewhat uniform character of much of the peninsula.

Figs. 71 and 73 are supplementary to the density maps, but it is necessary to make reservations about them. As we have seen on p. 300, some Domesday names covered two or more settlements, and several of the symbols should thus appear as two or more smaller symbols. Even so, the small size of so many of the symbols is an outstanding feature of the map and indicates the dispersed nature of much of the settlement of the county. This limitation, although serious locally, does not affect the general impression conveyed by the maps, which confirm and amplify the information on the density maps.

WOODLAND

Types of entries

The Cornish folios give information both for wood itself and for underwood. The amount of each is expressed in terms either of acres or of linear dimensions. The amount of underwood is always expressed in terms of acres.

Woodland is called *silva* in the Exchequer folios and *nemus* in the Exeter folios. The great majority of entries state the amount in terms of acres. The figures range from one acre at, for example, Hammett (*261*, 125) to 100 acres at Glynn (*230*, 124) and to 200 acres at Moresk (*249*, 121b) and at Fawton (*228*, 121b). The figure for Liskeard is as much as 300 acres in the Exeter text (*228*), which is given as 400 acres in that of the Exchequer (121b).[1] But amounts are mostly small, usually less than 20 acres. The figures are often in round numbers that suggest estimates,

[1] The Exeter figure of 300 has been plotted on Fig. 74.

e.g. the 60 acres at Cosawes (*224b*, 122); but occasional entries seem to be precise, e.g. the 1 acre at Tinten (*200b*, 120b) and the 13 acres at Penheale (*102b*, 120). No attempt has been made to convert these figures into modern acres, and they have been plotted merely as conventional units on Fig. 74.

The remainder of the wood entries (less than 16 per cent) express amounts in terms of leagues and furlongs. The exact wording varies slightly, but the following entry for Ashton indicates the kind of phrasing that was used:

E.D. (*258b*): *vi quadragenariae nemoris in longitudine et iii in latitudine.*
D.B. (122): *Silva vi quarentenis longa et iii quarentenis lata.*

The six at Ashton is the highest number of furlongs recorded. The largest amount of wood entered under a single name is the 4 leagues by 2 leagues for the canons' share of the manor of St Germans (*199b*, 120b).[1] Large amounts entered for composite manors may have covered tracts of wood lying some distance away from the main manor,[2] and so may imply some process of addition whereby the dimensions of separate tracts of wood were consolidated into one sum. The exact significance of all these linear dimensions is far from clear, and we cannot hope to convert them into modern acreages.[3] All we can do is to plot them diagrammatically as on Fig. 74, and we have assumed that a league comprised twelve furlongs. There are no abnormal entries as there are in some of the south-west counties.

A small number of entries mention underwood—*nemusculus* in the Exeter text and *silva minuta* in that of the Exchequer. In the accounts of Froxton (*334b*, 125) and Thurlibeer (*233b*, 124), what is recorded as underwood in the Exchequer version appears as wood in the Exeter text, and has been so marked on Fig. 74. One of the entries for Draynes speaks of underwood (*231*, 124b) and the other of wood (*232*, 124b), but this is the only example of both types of wood appearing under one place-name. The amounts of underwood are always given in acres, and range

[1] This has been plotted at South Petherwin—see p. 301 above.
[2] It has, for example, been claimed that the 1 league by $\frac{1}{4}$ league of wood recorded for Winnianton (*99*, 120) really referred to Merthen Woods some 5 miles to the north-east on the steep south-facing slopes of the Helford River (C. Henderson: (1) *Essays in Cornish History*, Oxford, 1935, p. 136; (2) *A History of the Parish of Constantine in Cornwall* (Royal Institution of Cornwall, Long Compton, 1937), p. 90.
[3] See p. 372 below.

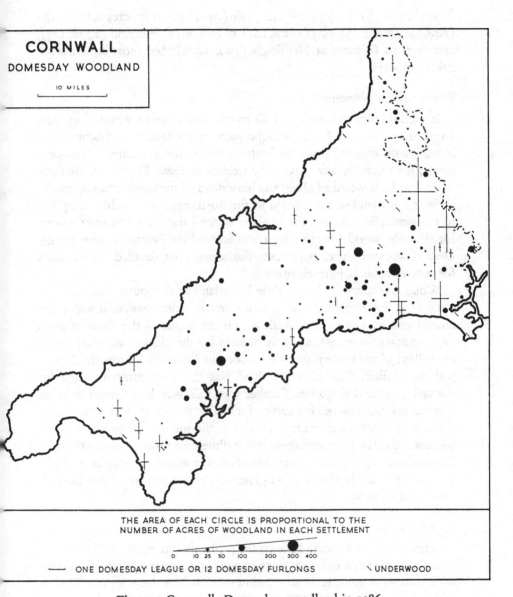

Fig. 74. Cornwall: Domesday woodland in 1086.

When wood and underwood are entered for the same settlement, both are shown.

from 1 acre at, for example, Worthyvale (*24 t*, 123) to 30 acres at Launcells (*244*, 124) and at Antony (*180 b*, 121), to 40 acres at Poughill (*233b*, 124), and even to 60 acres at Methleigh (*199*, 120b); but most amounts are below 10 acres.

Distribution of woodland

It would seem that much of Cornwall was without wood (Fig. 74). Exposure to the salt-laden Atlantic gales was effective in discouraging tree-growth, especially on the higher granitic bosses. Land's End peninsula, for example, was apparently treeless in 1086. There was, furthermore, but little wood all along the hinterland of the north coast up to the sheltered lowland around Bude in the north-east. A possible exception to the generalisation was the lowland around the Camel estuary where, for example, wood 2 leagues by 1 was entered for Pawton (*199b*, 120b). One must, moreover, remember the absence of detailed information for seven of the St Petrock manors.

Wood was more frequent in the hinterland of the south coast, and we can only suppose that in 1086, as today, much, if not most, of it was in the incised and so less exposed valleys. To the south of the Bodmin mass varying amounts were frequently entered for the villages associated with the valleys of the Fowey, the West Looe, the Looe, the Seaton, the Tiddy, and the Lynher. Further west, wood was also frequently recorded for the valley of the Fal system. Further west still, wood, measured in linear dimensions, was entered for some of the villages around the Carnmenellis mass. It is, however, difficult to be sure of the amounts of wood involved because we cannot convert these linear dimensions into Domesday acres. This, of course, is a general consideration that affects the map as a whole. Even so, it is likely that Fig. 74 gives us a broad picture of the Cornish woodland of 1086.

MEADOW

Types of entries

A feature of the Cornish folios is the relatively rare mention of meadow. It is entered under only 42 place-names, and many substantial manors with numerous teams have no meadow recorded for them, e.g. Callington with 18 teams (*102*, 120). The formula is always the same—'*n* acres of meadow' (*n acrae prati*). Quantities are very small. Seventeen places had but 1 acre of meadow apiece, and 14 places had but 2 acres apiece. The total number of acres entered for the county as a whole amounted only

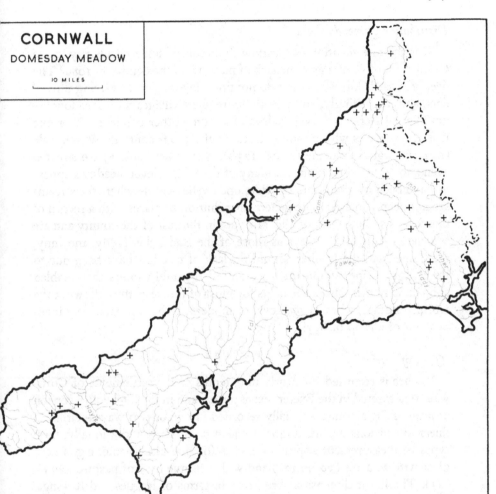

Fig. 75. Cornwall: Domesday meadow in 1086.

A cross indicates a settlement with recorded meadow.

to 139. The largest single amount was the 30 acres on the royal manor of
Kilkhampton (*101*, 120). A surprising entry is that for Pigsdon (*397b*,
125), where there was only 1 bordar with 1 team but as much as 10 acres
of meadow. In view of the small amounts involved, no attempt has been
made to indicate quantities on Fig. 75.

Distribution of meadowland

Was the rare mention of meadow the result of some peculiarity in the Cornish folios or of a genuine lack of meadow in the county in 1086? The deeply incised Cornish streams do not provide extensive low-lying alluvial flats, and, accordingly, some of the infrequent villages with meadow are not aligned along the river valleys as in some other counties. Over one half of the places with meadow lie above the 300 ft contour. Worthyvale (*241*, 123) and Tredaule (*242b*, 123b), with 1 acre apiece, are situated even above the 600 ft contour. Many of these high-level meadows appear to have been on the floors of wide, open valley-heads where the streams are not deeply incised. The general distribution of places with a record of meadow is somewhat sporadic. Most are in the east of the county and are associated with such valleys as those of the Bude, the Tiddy, the Inny, the Camel, and the Lynher. Could the lack of meadow have been due to the long growing season for grass and to the mild winters that enabled animals to remain out in the fields for most of the year, thus allowing the Cornish farmer, whether medieval or modern, to be comparatively independent of a crop of hay?

PASTURE

Types of entries

Pasture is recorded for nearly nine tenths of the settlements of Cornwall. It is entered in the Exeter text as *pascua* and in the Exchequer version as *pastura*. Its amount is usually recorded in one of two ways—either in linear dimensions or, much more frequently, in acres. Occasionally, both types of measurement appear in entries for one place-name, e.g. Ellenglaze with 20 acres (*202b*, 121) and with 1 league by ½ of pasture (*203b*, 121). The linear dimensions are given in terms of leagues and furlongs. The exact wording varies slightly, but the following entry for St Winnow shows the kind of phrasing that was used:

E.D. (*201*): *dimidia leuga pascuae in longitudine et tantundem in latitudine.*
D.B. (120b): *Ibi pastura dimidia leuga longa et tantundem lata.*

The exact significance of these linear dimensions is far from clear, and we cannot hope to convert them into modern acreages[1]. All we can do is to plot them diagrammatically as on Fig. 76, and we have assumed that a league comprised twelve furlongs. There are some large amounts, e.g. the

[1] See p. 372 below.

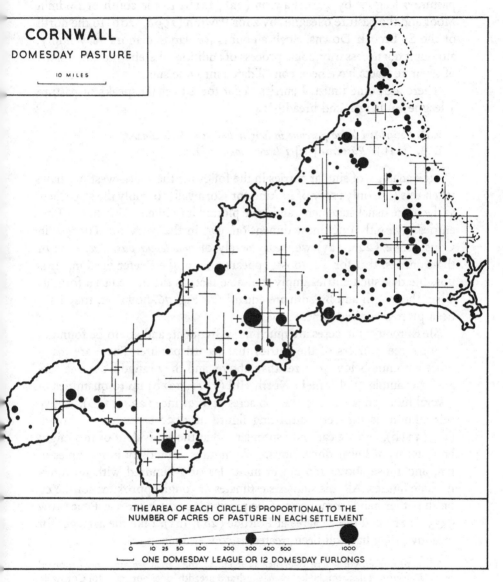

CORNWALL
DOMESDAY PASTURE

10 MILES

THE AREA OF EACH CIRCLE IS PROPORTIONAL TO THE
NUMBER OF ACRES OF PASTURE IN EACH SETTLEMENT

0   10 25   50   100   200 300 400 500   1000
└─────┘ ONE DOMESDAY LEAGUE OR 12 DOMESDAY FURLONGS

Fig. 76. Cornwall: Domesday pasture in 1086.

When the pasture of a settlement is entered partly in linear
dimensions and partly in acres, both are shown.

pasture 7 leagues by 4 for Fawton (*228*, 121 b) to the south of Bodmin Moor and the pasture 6 leagues by 2 for Pawton (*199 b*, 120 b) to the north of the St Breock Downs. Such amounts for large manors may imply, although not necessarily, some process of addition whereby the dimensions of separate tracts have been consolidated into one sum.

There are some unusual entries. That for Otterham speaks of pasture 1 league 'in length and breadth':

E.D. (*259 b*): *i leuga pascuae in longitudine et in latitudine.*
D.B. (122 b): *Pasturae* [sic] *1 leuua longa et lata.*

On the analogy of similar entries in the folios for the south-west, we have taken this (the only entry of its kind for Cornwall) to imply the same facts as the more usual statement, and have plotted it as intersecting lines.[1] Five entries give only one linear dimension, e.g. in the entry for Tucoyse in Constantine (*225*, 125) we hear merely of *una leuua pasturae.* One of these entries, that for Tregoose, appears only in the Exeter version.[2] It is possible that such entries imply the same facts as the more usual formula when the length and breadth are equal.[3] On Fig. 76, however, they have been plotted as single lines.

Measurements in acres are much more frequent, and are to be found in about three quarters of the entries that record pasture. There are many more amounts below 50 acres than above, and they range from one acre at, for example, Tolcarne in North Hill (*181 b*, 121 b) up to quantities of several hundred acres, e.g. the 500 acres at Trewince (*224*, 124b). Above this amount is the very surprising figure of 1,000 acres for Trenowth (*249*, 121 b), and we can only speculate what the equivalent of this might be in terms of linear dimensions.[4] The figures are mostly in multiples of ten, and those above 100 are in multiples of a hundred with no intermediate figures. All this suggests estimates or rough approximations. Yet, on the other hand, some seem to be precise, e.g. the 7 acres at Pencarrow (*233*, 122b) and the 17 acres at Polrode (*204*, 121). Like the acreages for meadow, they have all been treated as conventional units.

[1] See p. 100 above, where reference is made to the 'areal leagues and furlongs' of R. W. Eyton. The formula 'between length and breadth' does not occur for Cornwall.
[2] The five entries are for Arrallas (*249 b*, 123), Bodigga (*247*, 122b), Newton Ferrers (*258 b*, 122), Tregoose (*226*) and Tucoyse (*225*, 125).
[3] See p. 101 above.
[4] It could be an error. On the other hand Trenowth is situated on the south-west edge of the Hensbarrow granite mass where the pasture might have been.

Distribution of pasture

The fact that pasture was recorded for nine tenths of the settlements of Cornwall means, as Fig. 76 shows, that it was widely distributed throughout the county. The settlements most plentifully endowed with pasture were generally near upland areas, unsuitable for tillage, but how much of these upland areas was rough pasture and how much moorland, we cannot say (Fig. 78). Around Bodmin Moor were Fawton (*228*, 121b), for example, with 7 leagues by 4 leagues of pasture, Hamatethy (*259b*, 122b) with 5 leagues by 2, and Helstone (*237*, 121b), Rosecraddock (*234*, 123), and Trezance (*228b*, 122), each with 3 leagues by 2. To the west, near the Hensbarrow upland, were Tremoddrett (*250b*, 123) and Brannel (*249*, 121b), each with pasture 4 leagues by 2; also nearby was Trenowth (*249*, 121b), with its 1,000 acres of pasture. North of Hensbarrow was Pawton (*199b*, 120b), near the St Breock Downs, with 6 leagues by 2. Further west still were Helston (*100b*, 120) with 5 leagues by 3, Cosawes (*224b*, 122) with 5 leagues by 2, and Goodern (*248*, 122b) and Tywarnhayle (*202b*, 121), each with 5 leagues by 1.

Fairly large amounts were also entered for other districts. Near the uplands of the eastern border of the county were Climsom (*102*, 120), for example, with 4 leagues by 4 leagues, St Germans (*199b*, 120b) with 4 leagues by 2, St Stephens by Launceston (*206b*, 120b) with 3 leagues by 2, Calstock (*256*, 122) with 3 leagues by 1, and Callington (*102*, 120) with 3 leagues by half a league.

WASTE

Among the Winnianton manors described under *Terra Regis*,[1] the Exeter text (*99–99b*) says that two are unoccupied (*vacua*). The description of both manors under the count of Mortain's fief (*225b–6*) says that they were waste or wholly waste (*vastata, penitus vastata*). These are the relevant entries:

Skewes (*99*): hanc potest i carruca arare sed modo est vacua et quando comes accepit valebat xv solidos.

Skewes (*225b*): hanc potest arare una carruca...et penitus vastata est et quando comes accepit valebat xv solidos.

Trembraze (*99b*): hanc potest i carruca arare et quando comes accepit valebat xv solidos, et modo vacua est.

Trembraze (*226*): vastata est et quando comes accepit valebat x solidos. No plough-lands are mentioned.

[1] See p. 309 above.

The discrepancy between the two statements of the value of Trembraze is, presumably, the result of separate returns. Trembraze and Skewes were situated near one another in the Lizard peninsula, and the waste of these entries was, we can assume, not the waste of moor and heath but land that had gone out of cultivation. What local tragedy was responsible for their waste condition, we cannot say. We might also remember those other holdings without teams or people, yet for which no waste was specifically entered (Fig. 82).

MILLS

Domesday Cornwall is remarkable in that mills are recorded for only five places. The mills with their annual renders are indicated below:

> Cargoll (*203*, 121): 1 rendering 30*d*.
> Connerton (*111b*, 120): 1 rendering 30*d*.
> Launceston (*264b*, 121b): 2 rendering 40*s*.
> Liskeard (*228*, 121b): 1 rendering 12*s*.
> Trevisquite (*260*, 122b): 1 rendering 2*s*.

A total of 6 mills seems a strangely small number for Cornwall in 1086. Can we conclude that other mills were omitted by the Inquest? Or can we assume that water-power was but little used for grinding grain in this remote part of the realm? Cornwall had fewer mills than Devon, which in turn had fewer than Somerset and Dorset. All four counties were in the same circuit and so we might expect some uniformity of treatment. If the record reflected the reality, the absence of mills from Cornwall as compared with Devon, and from Devon as compared with Somerset and Dorset, would be understandable if the use of water-power was being introduced from the eastern parts of the realm. Beyond these speculations we cannot go.

CHURCHES

The Cornish folios do not enumerate churches in connection with settlements, but that churches were numerous in the county we know from two sets of evidence in the folios themselves. In the first place, we hear of at least twelve churches as holders of land.[1] It is probable that many were

---

[1] St Germans (*199b*, 120b); St Stephen by Launceston (*101b*, *206b*; 120, 120b); St Michael (*208b*, *258b*; 120b, 125); St Keverne (*205b*, 121); St Probus (*206*, 121); St Crantock (*206*, 121); St Piran (*206b*, 121); St Buryan (*207*, 121); St Neot (*207*, *230b*; 121, 124); St Constantine (*207*, 121); St Ewe in Trenance in St Austell

collegiate churches because we hear of their canons, e.g. at five different places on fo. 121 of the Exchequer text and on fos. *205b–7* of the Exeter text. We are not told of the existence of churches at Bodmin but there was obviously one there because 'Boia, the priest of Bodmin,' held Pendavy, and he is described variously as *sacerdos, presbyter* and *clericus* (*101b, 507*; 120). The 'priests of Bodmin' also held land at Hollacombe near Holsworthy (*481b*, 117b) and Newton St Petrock in Devon (*483*, 117b).

In the second place, the names of Celtic saints appear not only frequently in the dedications of churches but sometimes also in place-names themselves. Five Domesday place-names include 'Saint' followed by a personal name.[1] Three other place-names begin with *eglos*, which is the Celtic word for 'church'.[2] There is, furthermore, one place-name that incorporates the word 'church'.[3] Perhaps here also should be mentioned those 25 Domesday place-names beginning with 'Lan'; this Celtic word signified an enclosure but was usually used in the sense of an enclosure dedicated to a saint.

Two priests appear as landholders and so were not necessarily resident in the places they held. 'Brismar the priest' had held Truthwall in Ludgvan in 1066 (*258b*, 125), and 'Bernard the priest' held Tackbear of the count of Mortain in 1086 (*244b*, 125), but this latter place is now in Devonshire.

URBAN LIFE

The Cornish folios include an entry for the manor of Bodmin which, in the Exchequer version (120b), reads as follows:

The church of St Petrock holds Bodmin. There is one hide of land which never paid geld. There is land for 4 plough-teams. There, 5 villeins have 2 plough-teams, with 6 bordars. There [are] 30 acres of pasture and 6 acres of underwood. There, St Petrock has 68 houses and one market. The whole is worth 25s.

(*245b*, 125). A twelfth church was that of St Petrock which held land at a number of places (*202–5*, 120b–1). A thirteenth church, that of St Goran, was also mentioned in the Geld Accounts (*72*).

[1] The forms in the Exeter text are: *Sanguinas* (*238b*, 123b) or St Gennys; *San Winnuc* (*201*, 120b) or St Winnow; *Sanctus Constantinus* (*207*, 120b) or St Constantine; *Santmawant* (*99b, 226b*; 120) or St Mawgan; *Sainguilant* (*238b*, 122b) or St Juliot.

[2] The forms in the Exeter text are: *Hecglosenuda* (*203*, 121) or St Enoder; *Eglostudic* (*204b*) or St Tudy; *Eglossos* (*250b*, 123) or Eglosros *alias* Philleigh.

[3] The form in the Exeter text is *Maronacirca* (*232b*, 123) or Marhamchurch.

In this, as in the Exeter version (202), we are not specifically told that Bodmin was a borough, nor are burgesses mentioned. The 'priests of Bodmin' are mentioned in various entries in both the Cornish and the Devonshire folios.[1] Towards the end of the Exeter Domesday (529 b) there is a Summary of the St Petrock fief which includes in its enumeration an item of 68 burgesses. Although the Summary does not mention Bodmin by name, we may assume that the 68 burgesses refer to the 68 houses of the main entry. Elsewhere (101 b, 507; 120) we read that 'Boia, the priest of Bodmin,' held Pendavy.

We may not be far wrong in envisaging Bodmin as being a small market centre with an agricultural flavour and also with a well-established collegiate church. Its recorded population of 79 may imply a total population of at least 400.

LIVESTOCK

The entries of the Exeter text record information about livestock—not about total stock but only about that on the demesne. We cannot say how complete even these restricted figures were, and it may well be that some demesne animals were omitted. A number of entries for holdings with demesne land make no reference to livestock, e.g. those for Torwell (235 b) and for Treverbyn (248). Conversely, stock is entered occasionally for holdings without recorded demesne, e.g. for Westcott (239 b) and for Wilsworthy (239). Other entries mention customary dues of sheep and oxen (only for 1066, it is true) and they would seem to imply the existence of unrecorded stock.[2] From one hide at Perranzabuloe (206 b, 121), the count of Mortain had taken away all the stock (totam pecuniam), but we are given no details.

The entry for Caradon (245) seems to involve a scribal error:

...Ibi habet Turstinus dimidium ferlingum in dominio et iii oves et villani habent aliam terram et iii boves. Ibi habet Turstinus vi bordarios et iiii servos et vi oves et v capras...

The first *oves* looks like a mistake for *boves*, and has been counted as such; a total of 6 oxen, however, is not mentioned in the Exchequer version (123), which makes no reference to ploughing oxen or teams. The table on p. 337 shows the variety of livestock mentioned, and also the number in each category.

---

[1] See p. 335 above.　　　[2] See p. 341 below.

Horses are mentioned in a variety of ways. Rounceys (*roncini* or *runcini*) were most probably pack-horses. The 12 mares (*equae*) were at Cargoll (*203*). A distinction is made between *equae indomitae* and *equae silvestres*, but we may wonder whether there was any difference between the two categories. The king himself seems to have been a substantial horse-breeder with, for example, 45 *equae indomitae* at Binnerton (*111b*) and 40 *equae silvestres* at Connerton (*111b*). The count of Mortain is also credited with fairly large numbers, e.g. 27 *equae indomitae* at Ludgvan (*260*) and 20 at Cosawes (*224b*).

Animals (*animalia*) presumably included all non-ploughing beasts. The entry for Rillaton (*264*) does not apparently specify a number, and we read merely of *xv equae indomitae et animalia*.[1] Cows (*vaccae*) appear infrequently and never on the same holding as *animalia*; the maximum on any manor was 4 and often only 1 was entered. Salzman suggested that the milk 'was probably reserved for the lord of the manor'.[2] The solitary bull (*taurus*) at Bodardle (*249b*) is the only one mentioned in the Domesday Book.

Demesne Livestock in Cornwall in 1086

Sheep (*oves*)	13,003
Wethers (*berbices*)	240
Swine (*porci*)	505
She-goats (*caprae*)	906
Animals (*animalia*)	1,063
Cows (*vaccae*)	55
Bull (*taurus*)	1
Rounceys (*roncini* or *runcini*)	21
Unbroken mares (*equae indomitae*)	352
Forest mares (*equae silvestres*)	58
Mares (*equae*)	12

She-goats (*caprae*) are much fewer than in Devon. The flocks were much smaller than those of sheep, and the number entered for the demesne of a manor was usually less than 50 and quite often less than ten. The 50 recorded for Launcells (*244*) and the 40 each for Kilkhampton (*101*), Thurlibeer (*233b*) and Whitstone (*255b*) are exceptionally large numbers. No goats at all are entered for the Penwith district.

[1] We have counted 15 unbroken mares on the assumption that the figure for animals was omitted.
[2] *V.C.H. Cornwall*, I, part 8, p. 56.

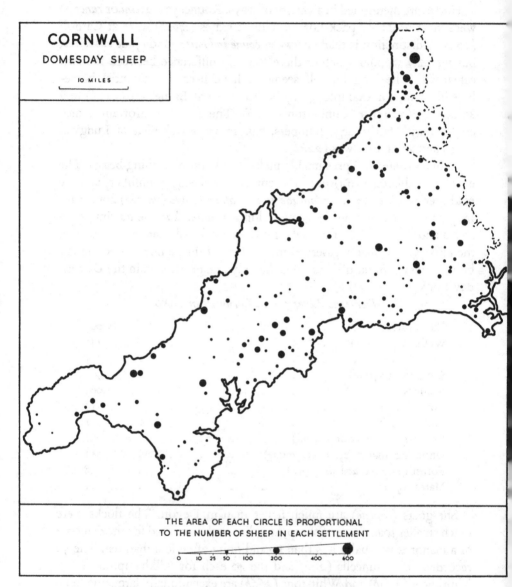

CORNWALL

DOMESDAY SHEEP

10 MILES

THE AREA OF EACH CIRCLE IS PROPORTIONAL
TO THE NUMBER OF SHEEP IN EACH SETTLEMENT

0   10 25 50   100   200   400   600

Fig. 77. Cornwall: Domesday sheep on the demesne in 1086.
Wethers (*berbices*) have been included with sheep.

Swine were recorded usually in smaller numbers. There were as many as 20 each at Kilkhampton (*101*) and Lizard (*226*), but these were unusually large numbers. Swine were often entered for holdings for which no wood was recorded, e.g. for Lizard with the 20 just mentioned and for Ludgvan (*260*) with 17.

Sheep were far and away the most numerous of all the recorded livestock, and flocks of many hundreds were frequently encountered. It is true that the large figures for some complex manors may be deceptive in the sense that they covered separate unnamed places, e.g. the 600 sheep at Kilkhampton with 35 plough-teams (*101*), the 300 at Stratton with 19 teams (*237*), and the 200 at, say, Callington with 18 teams (*102*); but a manor such as that of Crantock with only 1½ teams had as many as 110 sheep (*206*). Most figures, like these, are round numbers, but the count sometimes seems to have been made with precision, e.g. there were 287 sheep at Fawton (*228*) and 57 at Pelynt (*230b*). Wethers (*berbices*) are mentioned only once, for Cosawes (*224b*), where there were 240. Fig. 77 shows that demesne sheep, at any rate, were fairly well distributed throughout the county, maybe even more so when one remembers the absence of information for some of the St Petrock manors.[1]

MISCELLANEOUS INFORMATION

Markets

Markets are recorded for five places and an annual fair for one place. The count of Mortain had gained control over much of this commercial activity at the expense of the Church. He had seized the market at St Stephens by Launceston from the canons there, and had 'placed it within his castle'. He had established a market near another castle, at Trematon, with the result that the bishop of Exeter's market at St Germans nearby (some three or so miles away) had been rendered worthless; reference to this is also made under *Terrae Occupatae* (*507*). Then he acquired the annual fair at Methleigh, which, again, had belonged to the bishop of Exeter. Finally, he held the market at Liskeard. Only that at Bodmin seems to have remained to the Church (to the canons of St Petrock), but we are not told how much it was worth. The relevant Exeter entries read as follows:

St Stephens by Launceston (*206b*, 120b): *i mercatum quod ibi iacebat ea die qua rex E. fuit vivus et mortuus, quod abstulit inde comes de Moritonio et*

---

[1] See p. 302 above.

posuit in castro suo, et reddit per annum xx solidos. The castle is mentioned again under *Dunhevet* or Launceston (*264b*, 121 b).

Trematon (*256*, 122): *i mercatum quod reddit per annum iii solidos...In ea mansione habet comes i castrum.*

St Germans (*200*, 120b): *In ea mansione erat i mercatum ea die qua rex E. fuit vivus et mortuus, in dominica die, et modo adnichilatur per mercatum quod ibi prope constituit comes de Moritonio in quodam suo castro in eadem die.* Virtually repeated under *Terrae Occupatae* (*507*).

Liskeard (*228*, 121 b): *i mercatum quod reddit iiii solidos.*

Bodmin (*202*, 120b): *i mercatum.*

Methleigh (*199*, 120b): *de hac mansione habet comes de Moritonio annuale forum quod tenebat Leuricus episcopus ea die qua rex E. fuit vivus et mortuus.* Repeated under *Terrae Occupatae* (*507b*), which speaks of *quaedam feria* and says that the count holds it unjustly.

There were probably other eleventh-century markets that went unmentioned in the Domesday folios. There is certainly non-Domesday evidence of a fair and of a Thursday market at or near St Michael's Mount.[1]

Castles

Two castles are mentioned, both belonging to the count of Mortain. The Exchequer version of the relevant entries reads as follows:

Dunhevet (or Launceston) (*264b*, 121 b): *ibi est castrum comitis.*
Trematon (*256*, 122): *In ea mansione habet comes i castrum.*

Both, as we have seen, had markets associated with them. There is another reference to a Cornish castle in the Devonshire folios, where we read that the count of Mortain had exchanged the Devonshire manors of Benton and Haxton for the 'castle of Cornwall': *istas ii predictas mansiones dedit comes de Moritonio episcopo propter escanbium castell Cornugallie* (*118*, 101 b). It is practically certain that this refers to the castle at *Dunhevet* or Launceston.[2]

[1] See: (a) P. L. Hull, *The Cartulary of St Michael's Mount* (Hatfield House MS no. 315), Devon and Cornwall Record Society, new series v (Torquay, 1962); (b) J. H. Round, *Calendar of Documents preserved in France illustrative of the history of Great Britain and Ireland A.D. 918–1206*, I (H.M.S.O., London, 1899), p. 256; (c) D. Matthew, *The Norman Monasteries and their English Possessions* (Oxford, 1962), pp. 22–4, 35–8.
[2] H. P. R. Finberg, 'The castle of Cornwall', *Devon and Cornwall Notes and Queries*, XXIII (Exeter, 1949), p. 123.

Salt-pans

There is but one reference to salt-pans, of which there were 10 entered for the manor of Stratton, not much more than a mile from Bude and its estuary on the north coast: *x salinae quae reddunt x solidos* (237, 121 b).

Mints

There is no Domesday record of mints in Cornwall, but there is evidence of a pre-Conquest mint at Castle Gotha and earlier at Launceston.[1]

Customary dues in Cornwall

Unless otherwise stated the entries are taken from the
Exchequer text (120–1), the alternative versions
being on fos. *204b–6b* of the Exeter Domesday.

To St Petrock
Coswarth: *pro consuetudinem i bos et vii oves.**
Tregona: *xv denarii de consuetudine.*
Trevornick: *xii oves et xv denarii.*
Trenhale: *vi oves et viii denarii.*
Tolcarne in Newquay: *unus bos.*
Tremore: *unus bos et xv denarii et xii oves.*
Lancarffe: *xv denarii.*
Treninnick: *xv denarii et v oves.*
St Tudy: *pro consuetudinem xxx denarii.*†
Carworgie (*112, 507b*; 120b): *pro consuetudinem viii denarii.*

To St Germans
Anonymous holding (*201, 507*; 120b): *pro consuetudinem una cupa cervisiae et xxx denarii.*

To St Piran
Perranzabuloe: *firma iiii septimanorum et decano xx solidi pro consuetudinem.*

* Another entry on fo. 120b says: *pro consuetudinem xxx denarii aut i bos*; the Exeter equivalent (*111b*) reads: *xxx denarii et i bos et vii oves.*
† Omitted from the Exchequer version.

Customary dues

Ten holdings are said to have rendered, at some time previous to 1086, customary dues of money, sheep or oxen to the church of St Petrock, but

[1] See R. H. M. Dolley (ed.), *Anglo-Saxon Coins* (London, 1960), pp. 146, 148.

nine of these seem to have been appropriated by the count of Mortain; the tenth due had come from the royal manor of Carworgie. Moreover, the count of Mortain had appropriated from the church of St German a holding which had once rendered a cask of ale and 30*d*. Finally, the men of the count had taken from the canons of St Piran four weeks' *firma* and from their dean 20*s*. The details of these dues are set out on p. 341.

The interior of the peninsula of Cornwall is occupied by a series of uplands associated largely but not entirely with four main granite masses which have resisted erosion and so rise above the general surface. In the east of the county is Bodmin Moor, the largest of these masses, for the most part over 800 ft above sea-level and reaching 1,375 ft in Brown Willy. This high ground continues northward, beyond the limit of the granite, on to the Upper Devonian sandstone in the neighbourhood of Davidstow and so to the coast; much of this countryside is over 900 ft O.D., and it even reaches 1,009 ft. Further west lie the Hensbarrow Downs, mostly above 500 ft O.D. and rising to 1,062 ft. This upland also continues northward, beyond the granite, into the St Breock Downs, which reach 700 ft. West of Truro is the third granite mass—that of Carnmenellis— mostly above 400 ft O.D. and rising to 819 ft. The fourth granite area is that of Land's End peninsula, again over 400 ft O.D. and rising to 827 ft. There is also a fifth area of granite, but this has been dissected by erosion and drowned by submergence to produce the 140 or so Isles of Scilly, which are not mentioned in the Domesday Book.

Today, much of the surface of these upland areas, especially that of Bodmin Moor, is occupied by moorland, but there is also a great deal of improved land, more particularly in the granite area of Carnmenellis and in the non-granitic areas of Davidstow and St Breock Downs. There is very little evidence of the condition of these uplands in 1086. There were settlements on the non-granitic areas, but the almost complete absence of settlements from the granite areas seems to indicate that they were occupied by stretches of moorland much more extensive than those of today. The ill-drained soils made cultivation difficult, and exposure to the strong westerly winds hindered the growth of wood. These open moorlands seem not to have been entirely without value, because the large amounts of pasture entered for the surrounding settlements may well

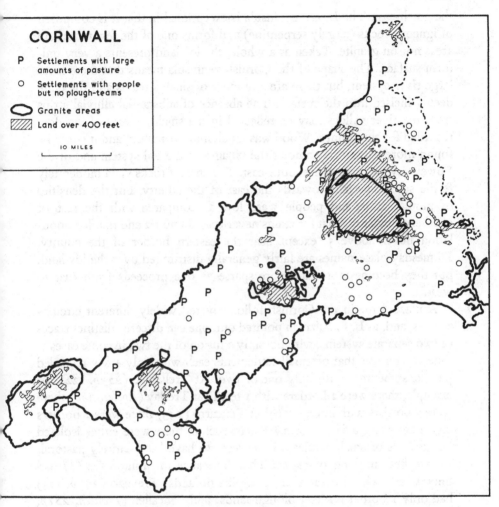

CORNWALL

P   Settlements with large amounts of pasture

O   Settlements with people but no plough-teams

⬭   Granite areas

▨   Land over 400 feet

10 MILES

Fig. 78. Cornwall: Pastoral features.

Settlements with large amounts of pasture have been taken to be *either* those with one league by one league and over *or* those with 200 acres and over. See Figs. 76 and 92.

refer to stretches of rough grazing on the moorlands (Fig. 78). To what extent the turf and peat were cut, we cannot say, because the Domesday Book makes no reference to such activities.

The lowland that surrounded these higher areas is composed mainly of Devonian slates, shales and grits and Carboniferous grits and shales in the north-east. The Lizard peninsula is exceptional in that it is composed of igneous rocks (largely serpentine) and forms one of the few moorland areas not on granite. Taken as a whole, the lowland presents a very uniform surface. The shape of the Cornish peninsula means that there is no large river system, but there are a number of small streams and these are deeply incised into the surface. The absence of substantial alluvial strips from the river valleys may be reflected in the small amount of meadow recorded for the county. Wood was much more common, and it could be found more particularly among the villages of the Fal system and of the Fowey–Lynher area in the south-east. The ancient rocks yield moderately fertile soils, especially towards the east of the county, but the densities of plough-teams and people were low as compared with the rest of southern England. That for teams never rose above 1·2 and that for population never above 5, except near the eastern border of the county. Domesday place-names are fairly generally distributed over the lowland, but they become more and more sparse as one proceeds from east to west.

A place-name in the Cornish folios covered widely different circumstances, and, as L. F. Salzman pointed out, one can discern 'distinct traces of two separate systems, with possibly a fusion of the two in some cases'.[1] One system was that of small settlements each with only a few recorded people, sometimes with only one or two. At Trehawke (*257b*, 122), for example, there were 2 bordars with 1 team; at Treslay (*Roslet*, *240*) there were 3 bordars with but 3 oxen;[2] at Pencarro (*233*) there were 2 bordars and 2 serfs and again 3 oxen.[3] Places such as these were either isolated homesteads or small hamlets. Some seem to have been entirely pastoral, for we hear nothing of teams. Thus Trenance in Mullion (*225b*) had only 2 serfs with no teams on its 4 plough-lands; Trevego (*231b*, 125) had only 1 bordar on its 3 plough-lands; and Lancallen (*Lantien*, *251b*, 122) had only 3 bordars on its 2 plough-lands. Fig. 78 shows how frequent were these settlements with inhabitants but without teams. Their names are characteristically Celtic, incorporating such elements as *pen*, *lan*, or, most frequently, *tre*; but there were occasional English names such as Callestock (*202b*, 121), to the north-west of Truro, with 12 serfs but no teams on its 4 plough-lands.

---

[1] L. F. Salzman, in *V.C.H. Cornwall*, I, part 8, p. 54.
[2] The Exchequer text says 3 bordars with half a team (123b).
[3] The Exchequer text omits the 3 oxen (122b).

The other system was that of moderately substantial villages each with a recorded population of 20 or more and with at least 3 or 4 teams at work. These were similar to the villages of much of the English plain, and they often, but not exclusively, had English names. Such was Rame (*181*, 121) with 23 people and 4 teams, or Hilton (*233b*, 124) with 21 people and 5 teams, or Milton (*232*, 123) with 40 people and 8 teams. A number of these place-names had quite large figures entered for them, e.g. Pawton (*199b*, 120b) with 86 people and 43 teams or Alverton (*255*, 121 b) with 71 people and 15 teams. It seems as if these were large manors each covering not only a village cluster but also unnamed hamlets and isolated farmsteads. This is especially true of those large royal manors described in fos. 120–120b of the Exchequer text. It may not be irrelevant to note that many of their names end in the English 'ton'—Helston, Kilkhampton (*Chilchetone*), Blisland (*Glustone*), Caradon (*Carneton*), Climsom (*Clismestone*), Callington, Connerton and Binnerton.

The landscape of Cornwall, in its human aspect, was made up not only of nucleated villages but also of hamlets and isolated farmsteads. While we may not associate English names exclusively with the former and Celtic names with the latter, there is a degree of correspondence. The presence of so many small settlements accounts for the fact that so few Cornish Domesday place-names have become names of modern parishes. These small settlements, with the Celtic names, gave to the human geography of Domesday Cornwall a character quite unlike that of the rest of southern England.

BIBLIOGRAPHICAL NOTE

(1) There were a number of nineteenth-century translations of the Exchequer text and also a literal extension:

JONATHAN COUCH, 'Translations from Domesday', *Trans. Natural History and Antiquarian Society of Penzance*, II (Penzance, 1852–4), pp. 110–25, 167–85, 244–65.

W. B. SANDERS, *A literal extension of the text of Domesday Book in relation to . . . Cornwall. To accompany the facsimile copy photo-zincographed etc.* (London, 1861).

W. POLSUE LAKE, *A Complete Parochial History of the County of Cornwall*, IV (Truro, 1872), pp. 27–65.

W. B. SANDERS, *A literal translation of the text of Domesday Book in relation to . . . Cornwall* (London, 1875).

(2) *The Victoria County History of Cornwall*, vol. I, part 8 (London, 1924) includes a translation of the Exeter text by T. Taylor (pp. 61–103) together with an introduction by L. F. Salzman (pp. 45–59).

(3) The following is a selection of papers in *Devon and Cornwall Notes and Queries* (Exeter) that deal with various Domesday problems. They are set out in chronological order.

J. H. ROWE, 'The forty mythical brewers of Cornwall', VI (1910), pp. 39–41.

J. H. ROWE, 'The boundaries of hundreds, ancient and modern' (in Devon and Cornwall), XI (1921), pp. 256–7, and XII (1923), pp. 325–6.

R. P. CHOPE, 'Domesday mills in Devon and Cornwall', XII (1922), pp. 21–3.

J. J. ALEXANDER, 'The hundreds of Cornwall', XVIII (1934), pp. 177–82.

H. P. R. FINBERG, 'The castle of Cornwall', XXIII (1949), p. 123.

H. P. R. FINBERG, 'Boieton, Elent, Trebicen, Trewant', XXII (1946), p. 95.

(4) Some other works of interest in the Domesday study of the county are as follows (arranged in chronological order):

J. CARNE, 'An attempt to identify the Domesday manors in Cornwall', *Jour. Roy. Inst. Cornwall*, I (Truro, 1865), pp. 11–19; II (Truro, 1867), pp. 219–22.

H. M. WHITLEY, 'The Cornish Domesday and the geld inquest', *Jour. Roy. Inst. Cornwall*, XIII (Truro, 1898), pp. 548–75.

T. TAYLOR, 'St Michael's Mount and the Domesday Survey', *Jour. Roy. Inst. Cornwall*, XVII (Truro, 1908), pp. 230–5.

J. A. RUTTER, 'Cornish acres in Domesday', *Notes and Queries*, 12th ser., VII (London, 1920), pp. 392, 437, 471–2.

F. W. MORGAN, 'Domesday woodland in south-west England', *Antiquity*, X (Gloucester, 1936), pp. 306–24. This contains a map of the Domesday woodland of Cornwall.

W. M. M. PICKEN, 'The Domesday Book and East Cornwall', *Old Cornwall*, II (St Ives, 1936), pp. 24–7.

N. J. G. POUNDS, 'The ancient woodland of Cornwall', *Old Cornwall*, III (St Ives, 1942), pp. 523–8. This contains a map of places with Domesday wood.

N. J. G. POUNDS, 'The Domesday geography of Cornwall', *109th Ann. Report Roy. Cornwall Polytechnic Soc.* N.S., X, pt I (Falmouth, 1942), pp. 68–82.

R. LENNARD, 'Domesday plough-teams: the south-western evidence', *Eng. Hist. Rev.* LX (London, 1945), pp. 217–33.

S. VIVIAN, 'The Domesday geography of Cornwall', *113th Ann. Report Roy. Cornwall Polytechnic Soc.* (Falmouth, 1946), pp. 27–32.

H. P. R. FINBERG, 'The Domesday plough-team', *Eng. Hist. Rev.* LXVI (London, 1951), pp. 67–71.

R. S. HOYT, 'The *Terrae Occupatae* of Cornwall and the Exon Domesday',
Traditio, IX (New York, 1953), pp. 155–75.
R. LENNARD, 'The composition of the Domesday caruca', *Eng. Hist. Rev.*
LXXXI (1966), pp. 770–5.

(5) A valuable aid to the Domesday study of the county is the large collec-
tion of parish histories and other material made by Charles Henderson between
1920 and 1933, and now housed in the Royal Institution of Cornwall Museum
at Truro. See also C. Henderson, *Essays in Cornish History* (Oxford, 1935).

(6) For the Exeter Domesday and Geld Accounts, see p. 393 below.

(7) The English Place-Name Society has not yet published a volume dealing
with the county, but there is an unpublished manuscript by J. E. B. Gover on
'The Place-Names of Cornwall' which is housed in the Royal Institution of
Cornwall Museum at Truro. See also:

H. JENNER, 'Cornish place names', *Jour. Roy. Inst. Cornwall*, XVIII (Truro,
1910), pp. 140–9.
T. F. G. DEXTER, *Cornish Names* (London, 1926).
J. E. B. GOVER, 'Cornish Names', *Antiquity*, II (Gloucester, 1928), pp. 319–27.

CHAPTER VI

THE SOUTH-WESTERN COUNTIES

BY H. C. DARBY, LITT. D., F.B.A.

The Exchequer folios for the south-western counties are of especial interest because of the existence of another version for Cornwall and Somerset, for almost the whole of Devonshire, for about one third of Dorset, and for a solitary Wiltshire manor. It is now clear that this was an early version and that it was the source of the corresponding text in the Exchequer Domesday Book itself, but the latter omitted any mention of, for example, demesne livestock. Clearly, a comparison of the two versions raises many points of interest, and the differences, in so far as they affect a geographical analysis, are set out in Appendix II. This earlier version, conveniently called the Exeter Domesday, is part of the Document known as the *Liber Exoniensis* which, so far as we can tell, has been in the custody of the authorities of Exeter Cathedral since the eleventh century. The *Liber Exoniensis* also includes a number of documents associated with the Domesday Inquest. One is *Terrae Occupatae*, a record of special points for Cornwall, Devonshire and Somerset; and, in some ways, it resembles the *Clamores* for Huntingdonshire, Yorkshire and Lincolnshire that appear in the Exchequer Domesday Book. Another is the so-called Geld Rolls or Geld Accounts, which are a record of the geld levied from each hundred at a time near that of the Inquest, and which are of help in the identification of the place-names of all five counties. A third document comprises two lists of hundreds for Cornwall, Devonshire and Somerset, which agree neither with one another nor with the hundreds of the Geld Accounts. Finally, there are Summaries of the details for a few fiefs and also a list of 26 fiefs. The contents of the *Liber Exoniensis* are set out in Appendix I.

The area was assessed in terms of hides, virgates and ferlings and of what appear to be geld-acres. The 5-hide unit may be discerned in at least a fifth of the holdings of each county except Devon and Cornwall. Some people have believed that the hides of all five counties were small and that they comprised only 40 or 48 acres, not the usual 120, but the evidence for this is doubtful. Cornwall, however, was exceptional in that its hide comprised only 12 geld-acres, but there is no evidence to show whether this implies a small hide or a different system of enumeration. Some manors or groups of manors are said never to have paid geld, but, except in Devon and Cornwall, the majority of these were royal manors belonging to the Ancient Demesne of the Crown and they had often been responsible for the ancient render of a night's *firma*. The liability of some manors for geld had been reduced, and for Cornwall a distinction is usually made between the full assessment and the smaller liability for geld when the fiscal demesnes were exempt. For Dorset and Somerset, we occasionally hear not of hides but of *carucatae nunquam geldantes*, and these may represent land brought into cultivation or added to a manor since the time when its assessment had been made or last revised.[1] Carucates also appear very occasionally in the folios for Wiltshire and Devon, but we are not told whether or not they were non-gelding.

One striking fact about the assessment is its varying incidence as between one county and another. Somerset was under-rated as compared with Dorset and Wiltshire, while Cornwall and Devonshire were very much under-rated, as the table below shows.

Relative Assessments of the South-western Counties

	Hidage	Plough-lands	Plough-teams
Wiltshire	3,893	3,360	2,926
Dorset	2,351	2,275	1,842
Somerset	2,902	4,775	3,886
Devonshire	1,159	7,932	5,759
Cornwall	428	2,563	1,218

To what extent this extreme under-rating of Devonshire and Cornwall was, as Maitland suggested, 'perhaps as old as the subjection of West Wales', we cannot say.[2]

The statement about plough-lands is fairly uniform. The Exchequer Domesday Book records *Terra n carucis* in entry after entry, and the Exeter version runs: *hanc terram possunt arare n carucae*. There is an occasional blank space where a figure was never inserted, and a few entries, strangely enough, record teams but not plough-lands. Taking the Domesday formulae at their face value, they seem to indicate land fit for ploughing

[1] R. Welldon Finn, *The Domesday Inquest* (London, 1961), p. 132.
[2] F. W. Maitland, *Domesday Book and Beyond* (Cambridge, 1897), p. 467.

in 1086, but many of the figures look suspiciously artificial, e.g. North Molton in Devon (94b, 100b) had 100 plough-lands for 47 teams and Bruton (90b, 86b) in Somerset had 50 for 21 teams. The relation between plough-lands and teams varies a great deal from entry to entry, as the following table shows.

Domesday Entries: Relation between Plough-lands and Teams

The percentages in each category refer to entries that record both.

	Equal	Deficient in teams	With excess teams
Wiltshire	57	37	6
Dorset	46	46	8
Somerset	35	51	14
Devonshire	23	71	6
Cornwall	5½	94	½

When plough-lands and teams were equal, we are occasionally told so, as in the entry for Huish in Burnham on Sea in Somerset (355, 95 b): *Terra est i caruca quae ibi est.* When the number of teams is smaller than that of plough-lands, we can only infer that a holding was not being tilled up to capacity. But was this due to land that had gone out of cultivation since 1066 or to potential arable as yet untilled? If the former, we should expect the decline to be reflected in a decrease in value. Yet while some under-stocked holdings had fallen in value, others had remained constant, and others had even increased. No general correlation between plough-team deficiency and decrease in value is possible. It would seem, therefore, that the plough-land figures indicate the total arable land, used and unused. The overwhelming number of apparently 'understocked' settlements in Cornwall and Devon is very striking, and yet these areas had not been devastated, as had the northern counties of England. We are immediately faced with the possibility that this feature of Cornwall and Devon was the result of the infield-outfield system whereby small numbers of teams successively ploughed various parts of large tracts of territory over a number of years.[1] Certainly, the agrarian arrangements of the south-west were very different from those over most of the rest of England.

The presence of settlements with excess teams raises other problems. They were not to be found in Cornwall, apart from the solitary instance

[1] F. W. Maitland, *op. cit.* p. 425.

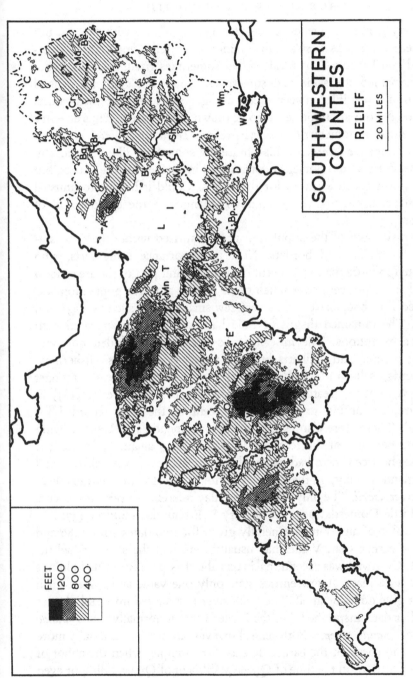

Fig. 79. South-western Counties: Relief.

Domesday boroughs are indicated by initials: A, Axbridge; B, Barnstaple; Ba, Bath; Be, Bedwyn; Bo, Bodmin; Bt, Bradford on Avon; Bp, Bridport; Bt, Bruton; C, Calne; Cr, Cricklade; D, Dorchester; E, Exeter; F, Frome; I, Ilchester; L, Langport; Ly, Lydford; M, Malmesbury; Ma, Marlborough; Mi, Milborne Port; Mn, Milverton; O, Okehampton; S, Salisbury; Sh, Shaftesbury; T, Taunton; Ti, Tilshead; To, Totnes; W, Wareham; Wa, Warminster; Wi, Wilton; Wm, Wimborne Minster.

FEET
1200
800
400

of Antony (*180b*, 121) where there were 6 plough-lands with 7 teams. But elsewhere there was a substantial minority of such apparently 'over-stocked' settlements. Their values had sometimes risen, but had some-times remained the same and sometimes dropped. Again, no general correlation is possible between excess plough-teams and increased values. Occasionally our attention seems to be drawn to the excess. Thus an entry for Coleford in Somerset (*427*, 93b) reads: *Terra est dimidia caruca. In dominio tamen est i caruca.* This phrase, or something like it, occurs in 8 Dorset entries, in 6 Somerset entries, and in 15 Devonshire entries, but we are never given a reason for the excess. Could it be due to unusual agrarian arrangements or to unknown changes in the composition of manors?

The main bulk of the population was comprised within the two cate-gories of villeins and bordars. Next in importance came serfs, who amounted to more than 15 per cent over the greater part of the area and to more than 20 per cent over much of it. Hardly any free population was recorded for 1086, apart from a very few small groups such as *Angli* and *Franci*. The unnamed thegns of 1066 had apparently disappeared with but rare exceptions. The miscellaneous population included not only coscets, cottars, and coliberts but also a variety of others—fishermen, swineherds, salt-workers, bee-keepers, smiths and potters, amongst others. An unusual feature was the large number of coscets (1,385) in Wiltshire, amounting to about 80 per cent of the total recorded for England. There were also as many as 370 swineherds in Devon, being about 70 per cent of the total for England. Then again, Wiltshire and Somerset between them had 440 coliberts, or just over one half the total for England. Finally, on the Cornish manor of Helston (*100b*, 120) there were 40 *cervisarii*. The only other Domesday reference to *cervisarii* occurs in the Little Domesday account of Bury St Edmunds in Suffolk (372).

The value of an estate is generally given for two dates—for 1086 and for some earlier year. Values had usually remained the same or had in-creased; Cornwall was exceptional in that about 75 per cent of its holdings showed a decrease. A few entries state only one value and a very small number give no value at all. The Exchequer text almost invariably leaves the earlier date unspecified, but the Exeter version normally states *quando recepit* or *quando accepit*. Both texts, however, are very occasionally more specific and speak of the earlier date as, for example, when the abbot of Athelney died, or in the time of Queen Matilda or of Queen Edith, or even

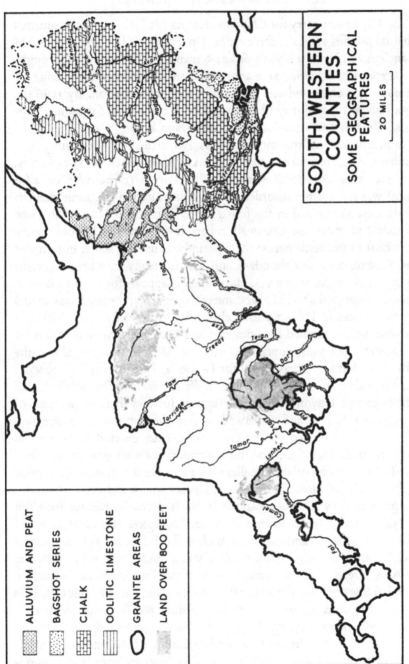

Fig. 80. South-western Counties: Some geographical features.
Rivers flowing across blown sand, alluvium and lowland peat are not marked.

T.R.E. The Exeter entry for Chaldon Herring (*59b*), in Dorset, is unusual in that it specifies values at three dates. On a few royal manors we hear of the ancient render of a night's *firma*: 6 manors in Wiltshire, 3 boroughs and 5 groups comprising 21 manors in Dorset, 5 groups comprising 12 manors in Somerset, and one group of 3 manors in Devon. By 1086 the obligations of the majority of these had been commuted for money payments. We also hear occasionally of renders in kind.

Prominent among the miscellaneous resources of each county was woodland. It was measured either in terms of linear dimensions or of acres, but there were occasional variants. Underwood (*nemusculus* or *silva minuta*) was quite often recorded and was measured in the same manner. Forests were mentioned in the folios for Wiltshire and Dorset. Meadow was abundant except in Cornwall, where it was entered for only some 13 per cent of the settlements; the amount was usually stated in terms of acres. Pasture was also abundant and was normally measured in terms of linear dimensions or of acres. Marsh (*mora*) appears for twelve places in Somerset, measured also in linear dimensions or acres; the *morae* associated with two places in Devon presumably refer to upland moors. Fisheries are mentioned or implied for Somerset and Devon and very occasionally for Dorset. Salt-pans appear mostly for Devon (mainly along the south coast) but also occasionally for Dorset and Cornwall; their absence from the Somerset record is strange. Mills are entered frequently for all counties except Devon, where they appear for less than 10 per cent of the recorded places, and Cornwall, where they appear in connection with only 5 out of 330 places. No churches are entered for Cornwall except as landholders; and for the remaining counties they appear only sporadically. Among other miscellaneous items are the vineyards recorded for about a dozen places in Somerset, Dorset and Wiltshire.

A general view of these various items is given in tabular form on pp. 434-9. These tables do not provide a complete statement of every variation in content and language within the folios for the five south-western counties. They are intended only as a guide to the salient features that have already been discussed in the preceding chapters. While the general framework of the Inquest was the same for all counties, it is clear that the detailed recording of information was far from uniform.

Composite maps have not been drawn for all items of the Inquest, but only for those most relevant to an understanding of the landscape and its economic geography—for settlements, plough-teams, population, woodland, meadow, pasture, fisheries and salt-pans, and also for that unusual element in the English scene, the vineyard. These maps must now be discussed separately.

SETTLEMENTS

A number of reservations must be borne in mind when looking at a map of Domesday place-names in the south-western counties. One arises from the fact that some Domesday place-names remain unidentified and so cannot be plotted on a map. Another reservation derives from the fact that no place-names appear in a number of entries, and it is possible that some of these refer to holdings at places not named elsewhere in the text; some entries, for example, speak of *terra* or *mansio* or *alibi*. With these may be grouped those other unnamed holdings added to, or taken from, named manors such as the Somerset Littleton in Compton Dundon (*433b*, 94) to which two unnamed manors had been added (*Huic mansioni sunt additae ii mansiones*).

A much more important reservation springs from the fact that a number of Domesday place-names covered more than one settlement. Many Domesday names, for example, are represented in later times by groups of two or more adjoining place-names with distinguishing appellations such as Great and Little, East and West, or with some more distinctive epithets. A number of these groups may have come into being as a result of post-Domesday colonisation, but we cannot always be sure whether or not a Domesday name covered only one vill; whether, say, the single name of *Fontel* (65b, 72b) in Wiltshire covered two settlements in the eleventh century, as it certainly did by the fourteenth—Fonthill Bishop and Fonthill Gifford. Only occasionally does the Domesday text itself distinguish between the units of such groups, e.g. when it speaks of *Litelbride* and *Langebride*, that is, Littlebredy and Long Bredy in Dorset (*37b*, 78); or when it distinguishes South Cadbury (*Sut Cadeberia*) from North Cadbury (merely *Cadeberia*) in Somerset (*383, 383b*; 97b).

The problem is more acute when groups of adjoining settlements take their respective names from the streams along which they were aligned. Thus, in the Dorset folios there are 15 entries for Tarrant, and today there are 8 settlements along the river of that name. In the same way there are 10 entries for Frome and 7 modern settlements; and there are 18 entries for Piddle and 11 modern settlements. A similar feature is true of other Dorset rivers. Or, again, in the Devonshire folios there are 14 entries for

Otri and 10 modern settlements along the River Otter, 10 entries for
Clyst and 10 modern settlements. Sometimes the name of the river has
changed. Thus there are 9 Wiltshire entries for Deverill (the old name
of the upper Wylye) and 5 adjoining places with that basic name
along the river; 7 entries for Winterbourne (the present River Bourne)
and 4 places; 12 entries for another Winterbourne (the present River Till)
and 7 places. Clearly in each of these groups there was very likely more
than one settlement in the eleventh century, but how many we cannot
say.

Yet another reservation arises from the fact that the constituent
members of some large manors are not named. Thus Brokenborough (67)
was a manor with 64 teams, 102 recorded people and 8 mills. Separate
details are given for one of its sub-tenancies, Corston; reference is also
made to 3 other sub-tenancies, but whether they were at places named
elsewhere in the Wiltshire folios, we cannot say. The descriptions of a
number of manors refer to *appendicii*, but without telling us where they
were. The Somerset manor of Taunton (*173b*, 87b) had 73 teams and
265 recorded inhabitants apart from its burgesses, and we know from
non-Domesday sources that it included other places in existence in the
eleventh century but unnamed in the Domesday Book. The Dorset
manor of Sherborne (77) had 69½ teams and 239 recorded inhabitants.
The manor of Crediton (*117*, 101b) in Devonshire had as many as 185
teams and 407 recorded inhabitants. Or, again, we can hardly suppose
that Pawton (*199b*, 120b), the largest of the Cornish manors, and the
only one mentioned in the hundred of that name, was the sole inhabited
place in the hundred, with its 43 teams and its 86 recorded people. There
are other places with recorded populations of over, say, 60; the entries for
them do not mention the existence of sub-tenancies, but the high figures
involved make it likely that such existed, and maybe at places that are
not named in the Domesday Book.

Furthermore, the entries for many smaller Domesday manors may
also have covered a number of separate unnamed settlements. The south-
western peninsula, especially that part to the west of the Parrett, is very
largely a land of isolated farms. Some estimates have placed the number
of such settlements in Domesday Devonshire, for example, at many
times that of the recorded place-names. Attempts have been made by
W. G. Hoskins to apportion the details of some Devonshire manors
among the isolated farms of modern parishes, but, as he made clear,

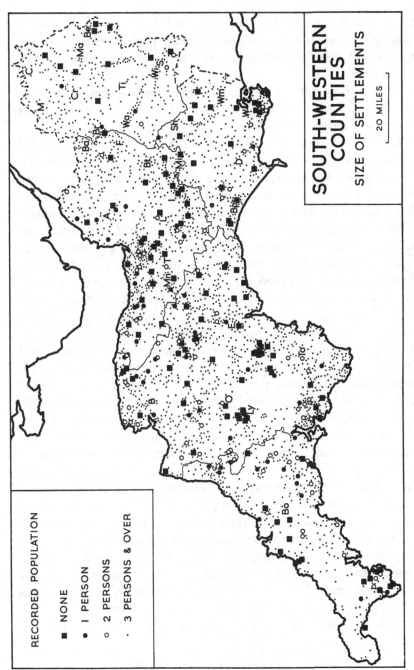

Fig. 81. South-western Counties: Size of settlements.

For the key to the initials of the boroughs, see Fig. 79.

such attempts are fraught with uncertainties.[1] Even so, we are safe in envisaging the county, and much of the south-west in general, to have been a land of hamlets and isolated farms in the eleventh century as it very largely is today.

Moreover, many of the named places were themselves very small. Some had no recorded population and no teams. Their people and resources may have been included with those of other places, or their land may have gone out of cultivation by 1086. Others had but one or two recorded inhabitants. We cannot, for example, regard as vills such Cornish settlements as Pigsdon (*397b*, 125) with but 1 bordar and 1 team, or Trehawke (*257b*, 122) with 2 bordars and 1 team, such Devonshire settlements as Willsworthy (115b) with 4 serfs and 1 team, or such Somerset settlements as Downscombe (*430*, 94) with only 1 bordar and half a team. Some of these small settlements were without recorded teams, e.g. the Cornish Trenance in Mullion (*225b*) with only 2 serfs on its 4 plough-lands, the Devonshire Speccott (*407b*, 115) with only one villein on its 3 plough-lands, or the Somerset Woodwick (*186b*, 89b) with 5 bordars on its 3 plough-lands; perhaps we can envisage such places as being entirely pastoral. L. F. Salzman discerned in Cornwall 'distinct traces' of two contrasting types of settlement.[2] On the one hand, there were those small settlements, often, but not always, with Celtic names incorporating such elements as *pen*, *lan* or, most frequently, *tre*. At the other extreme, on the other hand, there were substantial villages each with a recorded population of 20 or more and with at least 3 or 4 teams at work. They were similar to the villages of much of the English plain and they usually, but not exclusively, had English names, e.g. Hilton (*233b*, 124) with 21 recorded people and 5 teams, or Milton (*232*, 123) with 40 people and 8 teams. The same contrast may well have been true of at least Devonshire and Somerset west of the Parrett, although, in these latter counties, the smaller settlements, like the larger, had English names. Figs. 81 and 82 show how widespread were these places with but few people or none, and often with no teams. These maps give only a very incomplete picture of the widespread nature of the isolated settlements in the south-west. As we have seen, many isolated settlements, indeed most, may have been described anonymously under the names of other settlements.

In view of all these reservations, Fig. 83 gives a very imperfect and

---

[1] W. G. Hoskins, *Provincial England* (London, 1963), pp. 20 ff.
[2] L. F. Salzman in *V.C.H. Cornwall*, I, part 8, p. 54.

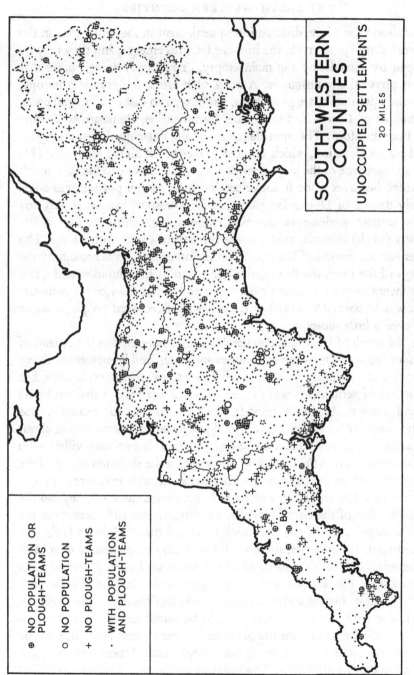

SOUTH-WESTERN
COUNTIES

UNOCCUPIED SETTLEMENTS

20 MILES

⊕ NO POPULATION OR
 PLOUGH-TEAMS

○ NO POPULATION

+ NO PLOUGH-TEAMS

· WITH POPULATION
 AND PLOUGH-TEAMS

Fig. 82. South-western Counties: Unoccupied settlements.
For the key to the initials of the boroughs, see Fig. 79.

generalised view of the distribution of settlement in the south-west in the eleventh century. Even so, the broader features emerge, and they may be grasped by considering the main empty, or relatively empty, areas that reflect physical circumstances. Among such areas, the granite outcrops, with their poor soils, are prominent. From west to east in Cornwall come the masses of Land's End peninsula, Carnmenellis, Hensbarrow Downs and Bodmin Moor. The infertile nature of the last-named area is emphasised by its elevation, which is mostly over 800 ft above sea level. This is even more true of the large granitic mass of Dartmoor to the east, much of which lies over 1,200 ft above sea-level. Its central portion was completely devoid of Domesday names, and its western margin largely so; a few western settlements, however, were at high altitudes, e.g. Willsworthy (115 b) at about 900 ft, where there were 4 serfs with a team. The lower eastern margin of Dartmoor, on the other hand, was broken by the valleys of the Dart, the Bovey, the Teign and their tributaries, and a few settlements were to be found even up to and above the 1,000 ft contour; such was Natsworthy (113 b) at 1,200 ft with 5 recorded people, 2 teams and even a little meadow.

In the north of Devon, and extending into Somerset, was the upland of Exmoor, built of Devonian sandstones and slates, and lying mostly above 1,000 ft O.D. Its peaty soils provided little inducement to settlement, but a number of settlements were to be found in the valleys that broke its upland surface. At about 1,200 ft O.D. in Devon, for example, was Radworthy (*415*, 111) with 4 recorded people and 1¼ teams, and at about the same height was Lank Combe (*337*, 114) with a solitary villein who had no team. Across the border, in Somerset, in the sheltered valley of the Exe, was Almsworthy (*430*, 94), at about 1,000 ft, with 17 recorded people and 4 teams. The empty area of Exmoor continued eastward, beyond the Exe, into that of the Brendon Hills, and further east still there were the smaller empty areas of the Quantock Hills and the Blackdown Hills.

Eastwards in Somerset, there were two areas conspicuously devoid of settlements. One was the lowland of the Somerset Levels. It would seem that most of this alluvial and peaty region was marsh in the eleventh century; but its surface was broken by hillocks and knolls of higher ground, and it was here that settlements were to be found and also on the large island of Wedmore and on the peninsula of the Polden Hills. But on the silt and clay coastal strip fronting the Bristol Channel there were frequent and prosperous settlements. The other empty area of Somerset was that

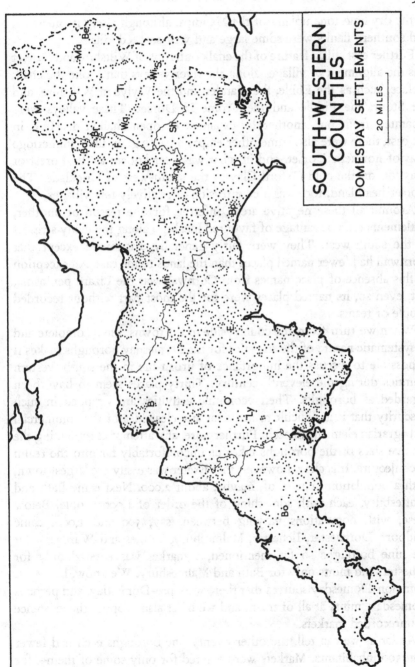

Fig. 83. South-western Counties: Domesday place-names.

For the key to the initials of the boroughs, see Fig. 79.

of the dry limestone upland of the Mendips, although along its northern and southern flanks were some large and well-stocked vills.

Further east still, a feature of the chalk outcrops of Wiltshire and Dorset was the alignment of villages along the river valleys that broke the chalk surface, such as the Ebble, the Nadder and the Wylye in Wiltshire, and the Stour, the Piddle and the Frome in Dorset. These valleys were separated from one another by expanses of open chalkland. Here, in the east, there were also some other empty areas, e.g. on the Kimmeridge Clay of northern Dorset, and in what was later known as the Forest of Braydon on the cold Oxford Clay to the south-west of Cricklade. The Dorset heathland, too, was a countryside of relatively few villages.

Around all these negative areas, infertile for one reason or another, settlements took advantage of favoured localities along the many streams of the south-west. They were fairly generally distributed except that Cornwall had fewer named places than the lands to the east. An exception to this absence of place-names from Cornwall was the Lizard peninsula, but, even so, its named places were for the most part without recorded people or teams.

When we turn from the countryside to the towns, the incomplete and unsystematic nature of the statistics for the Domesday boroughs makes it impossible to discuss the importance of urban life in the south-western counties during the eleventh century. Thirty places seem to have been regarded as boroughs. Their economic activities are wrapped in such obscurity that it is difficult to arrive at any clear idea of the commercial and agrarian elements in each. Furthermore, any attempt at estimating the relative sizes of the boroughs takes us uncomfortably far into the realm of conjecture. It is clear, however, that Exeter was easily the largest town, with a population in 1086 of, maybe, about 2,000. Next came Bath and Shaftesbury, each with something of the order of 1,000 people. Below these, with populations varying between, say, 500 and 1,000, came Bridport, Dorchester, Ilchester, Malmesbury, Totnes and Wareham. Of the nine boroughs so far mentioned, a market was entered only for Ilchester, and mints only for Bath and Malmesbury. We know, however, from non-Domesday sources that there were pre-Domesday, and perhaps Domesday, mints at all of them, and we must also suppose the existence of unrecorded markets.

As far as we can tell, the other twenty-one boroughs each had fewer than 500 inhabitants. Markets were entered for only some of them—for

Bradford on Avon, Bodmin, Frome, Milborne Port, Milverton and Okehampton, and both a market and a mint for Taunton. On the other hand, markets were recorded for a number of places that were not boroughs, and even an annual fair was entered for the village of Methleigh in Cornwall (*199*, 120b). Clearly the Domesday record is incomplete. It is very difficult to form any idea of the relative sizes of these smaller boroughs because many seem to have formed parts of large manors, and we cannot apportion the recorded population between a borough and the manor on which it stood. The Wiltshire manor of Bedwyn, for example, had a recorded population of 172 together with 25 burgesses. The Devonshire manor of Okehampton had a recorded population of but 66 together with only 4 burgesses, although it had a market and a castle. Can we regard it as an incipient borough in a rural setting? Frome and Milverton each had a market and their 'third penny' is mentioned, but no burgesses are specifically recorded for either place. As Tait wrote: 'In this land of petty boroughs, burghal status was precarious.'[1] Some commentators have thought there is a case for counting Yeovil and Watchet in Somerset as Domesday boroughs, but they have not been so regarded in this analysis.[2] In any case, if we guess aright, some of the smaller boroughs were not very different from the larger villages around.

Whatever the uncertainties, one thing is clear. Some boroughs may have been smaller in 1086 than they had once been. We frequently hear of wasted houses. Thus Exeter had nearly 400 houses but another 50 lay waste, presumably to make room for the castle, although we cannot assume that the presence of wasted houses necessarily implied the building of a castle. Over one half of the houses of Dorchester lay waste, nearly one half of those of Wareham, over a third of those of Lydford, about a third of those of Barnstaple and of Shaftesbury and one sixth of those of Bridport. There were also a few waste houses at Bath and at Malmesbury. But the Domesday text is silent about other changes that had taken place in the urban geography of the south-western towns since the Conquest.

[1] J. Tait, *The Medieval English Borough* (Manchester, 1936), p. 55.
[2] See pp. 196–9 above.

POPULATION AND PLOUGH-TEAMS

The density maps in the earlier volumes, dealing with eastern and midland England, did not take account of the figures for the boroughs. But in the present volume, as in those dealing with south-eastern and northern England, the plough-lands, the plough-teams and the rural population of the boroughs have been taken into consideration in the calculation both of county totals and of densities. The urban element itself (e.g. burgesses and houses) has been disregarded, and therefore, in estimating the areas of the density units, a quarter of a square mile has been deducted for each borough. For the purpose of comparing these maps with those of eastern and midland England, these adjustments make no appreciable difference.[1] One further point must be mentioned. As in all four preceding volumes, the boundaries between the various density units inevitably have an artificial appearance because they are based upon those of nineteenth-century civil parishes.

Taken as a whole, densities of population in the south-west were somewhat similar to those of much of the rest of England, ranging mostly between 5 and 10 people per square mile (Fig. 84). They rose above 10 only in a limited number of favoured areas: in Devonshire, in the lower Exe basin and in the coastal district of the south between the Teign and the Dart; in Somerset, only in the Vale of Taunton and the oolitic scarplands and some neighbouring areas in the east and south of the county; in Dorset, only on the small plain to the west of the Wey; in Wiltshire, only in the valleys of the Bristol Avon and the Salisbury Avon and in the Vale of Pewsey. There were, on the other hand, some areas where the densities fell below 5, and in places even below 2·5, people per square mile. This was true of the greater part of Cornwall except for a strip along the border with Devon. It was also true of the uplands of Dartmoor and Exmoor and their surrounding districts, and of much of the Somerset Levels, the hilly region of south-western Somerset, and the Mendip area. The heathlands of Dorset were also sparsely occupied, with under 5 people per square mile. So were some districts in Wiltshire—the forest regions of the east (Savernake, Chute, Clarendon and Melchet) and the south-west (Selwood), and parts of the chalk upland (the district in the north-west of Salisbury plain and the Grovely–Great Ridge area), and

[1] In the summary volume, the density maps for all counties will be calculated on a uniform basis.

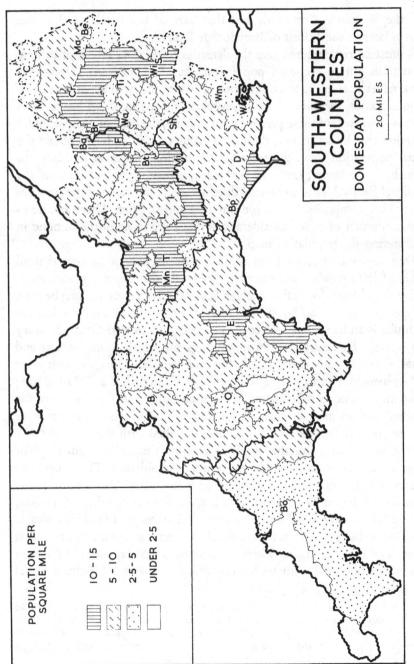

Fig. 84. South-western Counties: Domesday population in 1086.

For the key to the initials of the boroughs, see Fig. 79.

also the Wiltshire Cotswolds and that part of the northern clay vale known later as the Forest of Braydon or Braden.

It must be remembered that the densities on Fig. 84 refer not to total population but to recorded population. As Maitland said: 'Domesday Book never enables us to count heads. It states the number of tenants of various classes, *sochemanni*, *villani*, *bordarii* and the like, and leaves us to suppose that each of these persons is, or may be, the head of a household.'[1] Whether this be so or not, the fact remains that, in order to obtain the actual population from the recorded population, we must multiply the latter by some factor, say 4, or perhaps 5, according to our ideas of the medieval family.[2] This, of course, does not affect the value of the statistics for making comparisons between one area and another. There are, however, a number of other considerations that have to be borne in mind in interpreting the population map.

One reservation arises from the fact that we cannot be sure that all heads of households were counted. There are those many places about which very little information is given beyond their names and maybe their hidages or plough-lands; all we hear of Sidmouth in Devonshire, for example, is an incidental reference in an entry relating to Ottery St Mary (*195*, 104). There are also those holdings for which plough-teams and other resources are enumerated, but not people. Thus the entry for St Stephens by Launceston (*206b*, 120b), in Cornwall, is a fairly full one; it mentions, among other items, 3 demesne teams and 6 teams 'with the villeins' and an annual value of £4, but neither villeins nor any other people are enumerated. Who, we may ask, tilled with the 9 teams? We cannot be certain about the significance of such omissions, but they (or some of them) seem to imply unrecorded inhabitants. Then there are entries which bear witness to their own imperfections. An entry for Overton (65b), in Wiltshire, leaves a space for the number of villeins, and the letters *rq* (*require*) are to be found in the margin in red ink; what is more, no other people were recorded, but there were as many as 7 teams there. There is also a blank where the number of coscets should have been inserted in one of the entries for the Wiltshire manor of Tollard Royal

---

[1] F. W. Maitland, *op. cit.* p. 17.

[2] Maitland suggested 5 'for the sake of argument', *op. cit.* p. 437. J. C. Russell more recently suggested 3·5 (*British Medieval Population*, University of New Mexico Press, Albuquerque, U.S.A., 1948, pp. 38, 52). But for evidence in support of the traditional multiplier of 5, or something near it, see J. Krause, 'The medieval household: large or small?', *Econ. Hist. Rev.* 2nd ser. IX (1957), pp. 420–32.

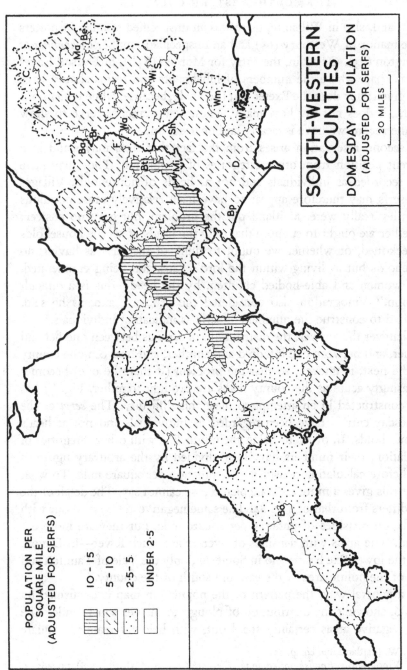

POPULATION PER
SQUARE MILE
(ADJUSTED FOR SERFS)

10 - 15

5 - 10

2·5 - 5

UNDER 2·5

SOUTH-WESTERN
COUNTIES

DOMESDAY POPULATION
(ADJUSTED FOR SERFS)

20 MILES

Fig. 85. South-western Counties: Domesday population in 1086 (adjusted for serfs).

On this map, the serfs have been regarded as individuals and not as heads of households. Their numbers have been divided by the arbitrary figure of four before calculating densities of population. For the key to the initials of the boroughs, see Fig. 79.

(71 b); and, also in Wiltshire, there was an unspecified number of potters in the manor of Westbury (65) and an unspecified number of *homines* at Lavington (73). Or again, the entry for Morden (83 b), in Dorset, leaves a blank space where the numbers of inhabitants and teams should have been given; nor does the Exeter text (*56*) help, because, while it mentions villeins, it does not say how many. These few examples serve to show that the record we have is clearly an imperfect one.

A second reservation arises from the fact that serfs may stand in a different position from other categories of population. They may have been recorded as individuals and not as heads of households. Villages with serfs may therefore appear to have been relatively more populous than they really were. Maitland put the problem, but gave no answer: 'Whether we ought to suppose that only the heads of servile households are reckoned, or whether we ought to think of the *servi* as having no households but as living within the lord's gates and being enumerated, men, women and able-bodied children, by the head—this is a difficult question.'[1] Vinogradoff also considered the problem, and, as he said, hesitated to construe the numbers of serfs as indicating individuals.[2]

Whatever the answer, the distribution of *servi* as between one fief and another, and between one hundred and another, and between one county and the next, is uneven, so that the problem (if there be one) becomes increasingly acute as the county maps are brought together. Fig. 85 has been constructed in an attempt to meet this difficulty. The *servi* of the Domesday entries have been regarded as individuals and not as heads of households. In order to bring them into line with other categories of population, their numbers have been divided by the arbitrary figure of four before calculating densities of population per square mile. To what extent this gives a more accurate picture, we cannot say. The detail of the map differs from that of Fig. 84. The same negative areas stand out with population densities of under 5 per square mile, but they are more extended. The areas with densities of over 10 are much fewer—in Devon, only the lower Exe basin; and in Somerset, only the Vale of Taunton and parts of the oolitic area of the east and south of the county.

In a general way, the pattern of the population map is confirmed by Fig. 86, showing the distribution of plough-teams per square mile. The main negative areas certainly stand out, with below 1 team per square

[1] F. W. Maitland, *op. cit.* p. 17.
[2] P. Vinogradoff, *English Society in the Eleventh Century* (Oxford, 1908), pp. 463–4.

369

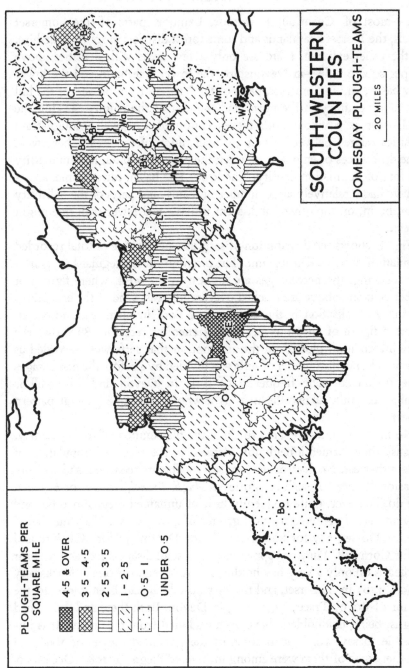

Fig. 86. South-western Counties: Domesday plough-teams in 1086.

For the key to the initials of the boroughs, see Fig. 79.

mile—most of Cornwall, Dartmoor, Exmoor, parts of the Somerset Levels, the Dorset heathlands and some forest districts in south Wiltshire. At the other extreme, there are only a few areas with densities above 3·5 per square mile—in Devonshire, the lower Exe basin and the lower Taw basin; and in Somerset, parts of the Vale of Taunton and of the hill and valley country in the north and the east of the county. Taking the south-west as a whole, with the exception of Cornwall, plough-team densities ranged between 1 and 3·5 over much of the area. There were some differences of emphasis between the pattern of plough-team density and that of population density. The lower Exe basin, for example, seems to have had a relatively large number of teams, and parts of the Salisbury Avon basin, on the other hand, seem to have had fewer than one might expect.

Fig. 87 shows the distribution of serfs as a percentage of total recorded population for each density unit. The distribution is irregular but, generally speaking, the percentages are highest in the west, where figures of 20 per cent and above are reached over large areas. One of the exceptions to this generalisation is the area around Bath in north-east Somerset, where a figure of above 25 per cent is encountered. Fig. 88 shows the same information plotted on the assumption that serfs were recorded as individuals, i.e. the number of serfs was divided by 4 before the percentages were calculated. The frequency of serfs in Devon and Cornwall is emphasised, but, taking the map as a whole, the same general pattern appears.

No map has been prepared to show the distribution of free peasantry, because they formed a negligible element in the recorded population of the south-west. No *liberi homines* or *sochemanni* are recorded, and we only occasionally hear of such categories as *Angli, Franci, censores* or *homines*. We do, however, quite frequently hear of unnamed thegns in 1066, and they are usually entered in small groups of 4 or 5 or so. They numbered 340 for Dorset, 307 for Somerset, 141 for Devon, 54 for Wiltshire and 17 for Cornwall. They had apparently almost all disappeared, and we hear of thegns on only a very few holdings in 1086, e.g. the one at Marksbury (*169b*, 90b) in Somerset, and the 15 whose land had been added to the manor of Bovey Tracey (*135*, 102) in Devon. Both these and the named thegns, being landholders, have been excluded from our count. It is impossible to say what had become of the unnamed thegns of 1066 and whether some of them were among the named thegns of 1086. Occasional

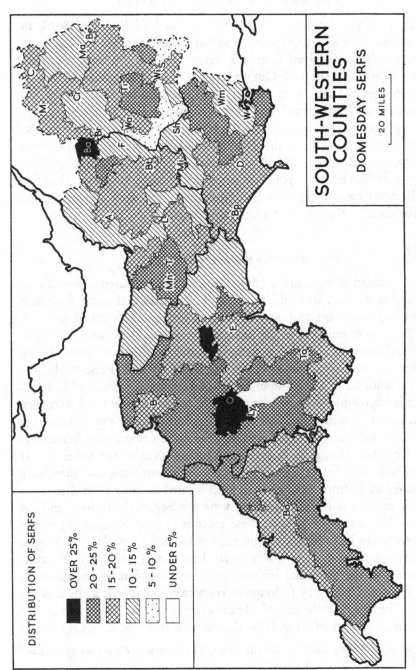

Fig. 87. South-western Counties: Distribution of serfs in 1086.
For the key to the initials of the boroughs, see Fig. 79.

figures prompt speculation. Seven thegns had held Poorton (80b), in Dorset, in 1066 and there were 7 villeins there in 1086. Or, again, in Dorset, 5 thegns had held Higher Kingcombe (80b) in 1066, and there were 5 villeins there in 1086. Can we suppose that at both places the villeins of 1086 were the thegns of 1066 or their descendants? But, usually, the figures do not lend themselves so conveniently to such suggestions.

Supplementary to the maps are the tables on p. 439 which summarise the relative importance of the different categories of population in each county as a whole for 1086. Any discussion of the significance of these maps and tables lies far beyond the scope of this study. Moreover, their full implications cannot begin to be appreciated until they are set against similar data for the whole of Domesday England.

WOODLAND AND FOREST

The woodland of the south-western counties was measured in one of two ways, that is, apart from the miscellaneous entries that occur for every county—in linear dimensions and in acres. It is normally called *nemus* in the Exeter text and *silva* in the Exchequer version. Underwood was also recorded and measured in the same way, and is called *nemusculus* in the Exeter text and *silva minuta* in the Exchequer version. Presumably, the smaller amounts of wood and underwood were usually measured in acres, but it is impossible to equate the two units of measurement and so reduce them to a common denominator. We certainly cannot say whether the relative visual impression conveyed by the symbols for linear dimensions and for acres on Fig. 89 is correct. Another point to remember is that wood was recorded in connection with settlement sites, and may have stretched away from them or even been located in some other district.

The units of linear measurement were the league, the furlong and the perch. The length of each of these units is open to doubt.[1] When the quantities are in furlongs, the number is normally less than the twelve usually supposed to make a Domesday league, and assumed in the construction of the map; but higher figures are occasionally encountered, e.g. the wood 16 by 13 furlongs at Halberton (*110b*, 101b) in Devonshire. The exact significance of a linear entry is far from clear. Is it giving extreme diameters of irregularly shaped woods, or is it making rough

[1] See H. C. Darby and I. B. Terrett (eds.), *The Domesday Geography of Midland England* (Cambridge, 1954), p. 433.

373

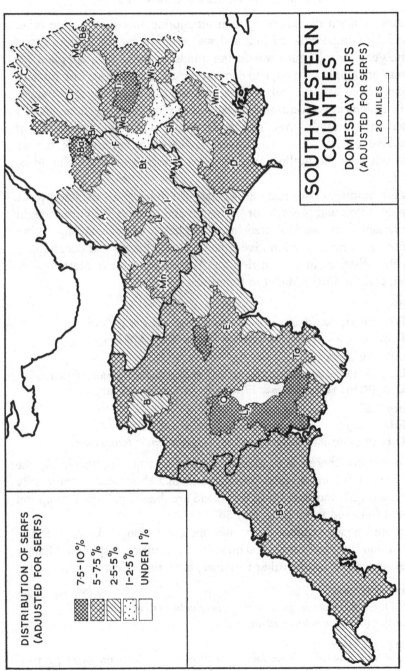

Fig. 88. South-western Counties: Distribution of serfs in 1086 (adjusted for serfs).

On this map, the serfs have been regarded as individuals and not as heads of households. Their numbers have been divided by the arbitrary figure of four before calculating percentages. For the key to the initials of the boroughs, see Fig. 79.

estimates of mean diameters, or is it attempting to convey some other notion? It is impossible to tell, and we certainly cannot assume that a definite geometrical figure was in the minds of those who supplied the information. Nor can we hope to convert these measurements into modern acreages by an arithmetical process. All we can safely do is to regard the dimensions as conventional units, and to plot them diagrammatically as intersecting straight lines. The resulting map will give us some idea of the relative distribution of wood over the face of the countryside, but we cannot tell how the outlines of individual stretches of wood should be filled in.

As for pasture, occasional entries state the dimensions of wood to be 'between length and breadth' or 'in length and breadth'.[1] Eyton thought these variant formulae indicated 'areal leagues' or 'areal furlongs',[2] but when the Exchequer version gives one form, the Exeter text occasionally gives the other, as in the entries for Adber and Worth Matravers in Dorset, and for Curry Mallet in Somerset:

Adber
> E.D. (*493*): *i quadragenaria nemoris in longitudine et in latitudine.*
> D.B. (99): *una quarentena silvae in longitudine et latitudine.*

Worth Matravers
> E.D. (*51 b*): *vii quadragenariae nemoris in longitudine et totidem in latitudine.*
> D.B. (82b): *vii quarentenae silvae inter longitudinem et latitudinem.*

Curry Mallet
> E.D. (*429*): *dimidia leuga nemoris in longitudine et latitudine.*
> D.B. (93): *dimidia leuua silvae inter longitudinem et latitudinem.*

One is tempted therefore to assume that the variant formulae imply the same facts as the more usual formula, and that they occur occasionally when the length and breadth of woodland are the same. This assumption has been followed in plotting Fig. 89.

A more frequent variant gives only one dimension, and, again, Eyton thought that these indicated areal quantities, but the corresponding Exeter entries sometimes follow other formulae, as in the following examples:

Clawton (Devon)
> E.D. (*318b*): *una leuga nemoris in longitudine et latitudine.*
> D.B. (108b): *una leuua silvae.*

---

[1] See p. 382 below.
[2] R. W. Eyton, *A key to Domesday... the Dorset survey* (London, 1878), pp. 31-5.

Hurpston (Dorset)
> E.D. (*60 b*): *i quadragenaria nemoris in longitudine et tantundem in lati-*
> *tudine.*
> D.B. (84): *i quarentena silvae.*

Pitcott (Somerset)
> E.D. (*146*): *i quadragenaria nemoris inter longitudinem et latitudinem.*
> D.B. (88 b): *una quarentena silvae.*

It would seem that single dimensions were also occasionally used when length and breadth were the same. On Fig. 89, however, such entries have been plotted as single lines, that is in the absence of an alternative version which mentions length and breadth.

Occasionally, one set of measurements is found in a linked entry covering a number of widely separated places. This seems to imply, although not necessarily, some process of addition whereby the dimensions of separate tracts of wood were consolidated into one sum. On Fig. 89 such measurements have been plotted for the first-named settlement only and the other places in the entry have each been indicated by the symbol for 'miscellaneous'.[1] The amount of wood entered for large manors may likewise have been at separate places. The largest amount entered under a single name is the 6 leagues by 4 for Amesbury in Wiltshire (64b). The Exeter text sometimes gives separate quantities for manorial components that are combined into single amounts in the Exchequer version. With one exception, the amounts involved are in acres, so that only simple addition is involved. The exception is the Somerset entry for Hornblotton, Alhampton and Lamyatt, where wood is measured in linear dimensions. Here are the details:

> E.D. (*169 b*)
> Hornblotton: Wood 4 furlongs by 1 furlong.
> Alhampton: Wood 5 acres.
> Lamyatt: Wood 3 furlongs by ½ furlong.
> D.B. (90b)
> All three holdings: Wood 9 furlongs by 1½ furlongs.

It seems impossible to reconcile the sum of the Exeter components with the Exchequer total, and this impossibility defeats any attempt to discover

---

[1] If a settlement (other than the first-named) in a composite entry also has wood entered for it in a separate entry, that wood has been plotted instead of the symbol for 'miscellaneous'.

a clue to the arithmetic that lay behind the ubiquitous formula '*m* furlongs by *n* furlongs'.[1] The Exeter dimensions have been plotted on Fig. 89.

When wood is measured in acres, the amounts range from half an acre to several hundred, e.g. to 500 acres at Winkleigh (*109*, 101 b) in Devon, but the great majority do not rise above 30 acres. The round figures in many entries (e.g. 20, 40 or 60) may indicate that they are estimates rather than precise amounts, but the detailed figures in other entries (e.g. 1½, 9 or 31) suggest precision. No attempt has been made to convert these Domesday figures into modern acreages, and they have been plotted merely as conventional units on Fig. 89. A few entries appear to use acres as linear measurements or to combine furlongs and acres in a curious manner. Here are two examples from Wiltshire:

> Calne (64 b): *Silva ii quarentenis longa et una quarentena et xxiiii acris lata.*
> Woodhill (66): *Silva una quarentena longa et iii acris lata.*

Eyton explained these as references to 'lineal acres' comprising 4 perches,[2] but on Fig. 89 they have been indicated under the category of 'miscellaneous'.

There is a variety of miscellaneous entries. We hear of a hide, of a virgate, and of arpents of wood, and of spinney (*spinetum*), thorn-wood (*runcetum*), brushwood (*broca*), alder-wood (*alnetum*) and of a grove (*lucus*). The Dorset entry for Renscombe (*38 b*, 78) mentions *silva infructuosa*; and in the Exeter text for Nettlecombe (*38*, 78), also in Dorset, there is a reference to *nemus nullum fructum fert*; and there are yet other minor variants.

A relatively small number of entries speak of underwood. The *silva minuta* of the Exchequer text has variants of *silva parva, parva silva* and *silva modica*; the last of these, at any rate, is rendered in the Exeter text as *nemusculus*, the usual Exeter rendering of *silva minuta*. Underwood, like wood, is measured sometimes in linear dimensions and sometimes in acres. The former have variants similar to those for wood. The latter are usually in small amounts of under 10 acres, but quantities of up to 300 acres can occasionally be encountered. There were, for example, 300 acres of underwood entered for Shirwell (*298*, 106b) in Devon.

---

[1] For similar pasture entries for manorial components, see p. 384 below.
[2] R. W. Eyton, *op. cit.* pp. 25–7.

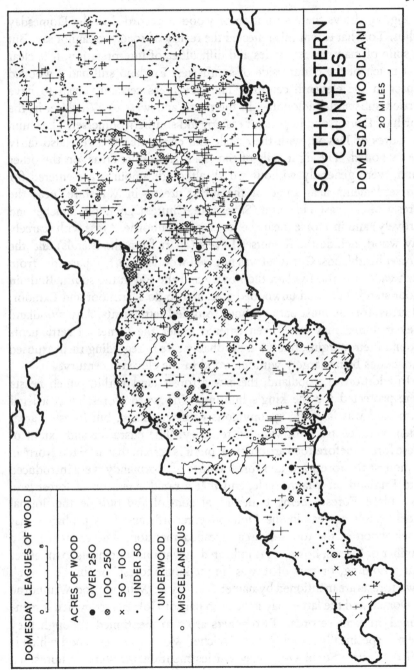

DOMESDAY LEAGUES OF WOOD

0   8

ACRES OF WOOD

●   OVER 250

⊗   100 - 250

×   50 – 100

·   UNDER 50

/   UNDERWOOD

○   MISCELLANEOUS

SOUTH-WESTERN COUNTIES

DOMESDAY WOODLAND

20 MILES

Fig. 89. South-western Counties: Domesday woodland in 1086.

Fig. 89 shows the distribution of wood as recorded in the Domesday folios. To what extent this reflected the reality of 1086, we cannot say. But in spite of many uncertainties and difficulties of interpretation, it is clear that much of the countryside of south-west England still had a wooded aspect in the eleventh century. The clay areas certainly seem to have carried substantial amounts—in the north and west of Wiltshire, in the north of Dorset, and in parts of eastern and southern Somerset. The Culm Measures of Devon, with their shale and clay outcrops, were also fairly well wooded. The light soils of the chalk and oolite areas, on the other hand, were generally without wood, although amounts are entered for the settlements near some of the forest areas—in Wiltshire, near the forests of the east and south (Savernake, Chute, Clarendon, Melchet and Grovely) and in Dorset near the forest of Wimborne. Areas with scarcely any wood include the Somerset Levels (apart from the islands) and the Dorset heathlands. Cornwall was not well wooded apart, apparently, from such valleys as the Lynher, the Fowey and the Fal in the south. Bodmin Moor stands out as an unwooded area, and so do Dartmoor and Exmoor, all areas, for the most part, without recorded settlements. Any woodland they included may have been entered under the names of settlements around their margins, but we are probably safe in regarding their exposed surfaces as being largely without wood in the eleventh century.

In addition to woodland, there was forest land within which beasts were protected for the king's hunting. The word 'forest' was a legal term, and was far from synonymous with woodland, but forested areas often included tracts of wooded territory. The character and extent of these forests before 1066 are obscure, but it is certain that after the Norman Conquest the forest law and forest courts of Normandy were introduced into England on a large scale, and that a rapid extension of forest land took place. Forests, being under royal control and outside the normal order, rarely appear in the Domesday record, and it is probable that much wood in the forested areas escaped mention. The existence of a number of forests, however, is indicated by various statements to the effect that a holding, or part of it, was 'in the king's forest' or 'in the forest'. Two forests are mentioned by name: Grovely in Wiltshire and Wimborne in Dorset, and the latter may have been part of Holt Forest, which is mentioned in later records. Two others are also mentioned although they are not specifically called forests, Melchet Wood and 'the wood which is called Chute'. Some Domesday entries mention foresters or huntsmen

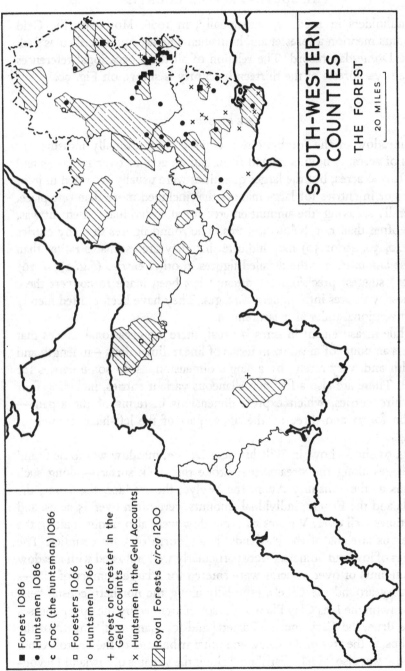

Fig. 90. South-western Counties: Domesday and later Forests.

The extent of the royal forests is based mainly on: (1) M. L. Bazeley, 'The extent of the English forest in the thirteenth century', *Trans. Roy. Hist. Soc.* 4th ser. IV (London, 1921), p. 148; (2) *V.C.H. Wilts.* IV (London, 1959), pp. 331–460.

Legend:

- ■ Forest 1086
- ● Huntsmen 1086
- c Croc (the huntsman) 1086
- □ Foresters 1066
- ○ Huntsmen 1066
- + Forest or forester in the Geld Accounts
- × Huntsmen in the Geld Accounts
- ▨ Royal Forests *circa* 1200

as landholders in 1086 or, occasionally, in 1066. Moreover, the Geld
Accounts mention a forester and huntsmen who are not described as such
in the Domesday record. The relation of all these occasional references
to the royal forests of the thirteenth century is shown on Fig. 90.

MEADOW

The meadow of the south-western counties was normally measured in
terms of acres. Amounts ranged from half an acre to over 100 acres and
even to 300 acres, but the larger quantities were usually recorded in joint
entries or in entries for large manors that included more than one place.
Generally speaking, the amount entered for an individual settlement was,
more often than not, below 30 acres. The round figures in many entries
(e.g. 20, 30, 40 or 50) may indicate that they are estimates rather than
precise amounts, but the detailed figures in other entries (e.g. 1½, 9, 26¼
or 107) suggest precision. No attempt has been made to convert these
Domesday figures into modern acreages. They have been plotted merely
as conventional units of measurement.

While measurement in acres is usual, there are occasional entries that
express amounts of meadow in terms of linear dimensions—in length and
breadth and, very rarely, by a single dimension, as do some entries for
wood. There are also a few miscellaneous variant entries, including four
Wiltshire entries, which express dimensions in terms of the arpent—
one on fo. 73 and three on the upper part of the left-hand column of
fo. 74b.

Fig. 91 shows how, in Wiltshire and Dorset, meadow was to be found
in villages along the streams that broke the chalk surface—along such
streams as the Salisbury Avon, the Wylye, the Nadder, the Stour, the
Piddle and the Frome; individual amounts were often over 25 acres, and
sometimes well over. Villages with meadow were also frequent along the
numerous streams of the claylands in the north of both counties. The
villages of lowland Somerset were particularly well endowed with meadow
and amounts of over 25 acres were entered for a large number of vills—
for many around the Levels, especially along the northern coastal belt;
for many on the Lias Clay Plain to the south, drained by the upper courses
of the Brue, the Cary and the Parrett; and for many to the north of the
Mendips, in the area of the Chew and other tributaries of the Bristol Avon.
Meadow was also widely distributed along the streams of lowland Devon-

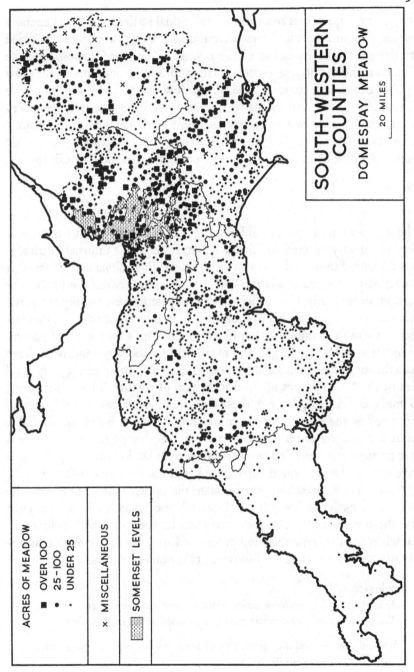

Fig. 91. South-western Counties: Domesday meadow in 1086.

shire, but the quantities recorded for individual settlements were generally smaller than in Somerset. Even so, amounts above 25 acres were recorded for settlements in the basins of the Exe, the Taw, the Torridge and the Tamar; some of these settlements with appreciable amounts of meadow lay above the 400 ft contour. Finally, meadow was recorded for only some 13 per cent of the settlements of Cornwall, and the amounts were very small, usually only one or two acres; some settlements with these small amounts lay above 400 ft. The infrequent mention of meadow may result from a genuine absence or from some peculiarity in the Cornish folios.

PASTURE

The south-western counties differed from most others in that pasture was very frequently entered for them. Apart from a few unusual entries, it was measured in one of two ways—in linear dimensions and in acres. It is normally called *pascua* in the Exeter text and *pastura* in the Exchequer version. Presumably, the smaller amounts of pasture were usually measured in acres, but it is impossible to equate the two units of measurement and so reduce them to a common denominator.[1] On Fig. 92 amounts of pasture of one league by one league and more, and of 200 acres and over, have been distinguished from smaller amounts in an attempt to convey a general idea of the main features of the distribution of pasture. This is not meant to imply any close equation of these two amounts, and we certainly cannot say whether the relative visual impression conveyed by the symbols for linear dimensions and for acres is correct. Another point to remember is that pasture was recorded in connection with settlement sites, and may have stretched away from them or even been located in some other district.

As for wood, occasional entries state the dimensions of pasture to be 'between length and breadth' or 'in length and breadth'. Eyton thought that these variant formulae indicated areal leagues' or 'areal furlongs',[2] but when the Exchequer version gives one form, the Exeter text occasionally gives another, as in the following two Somerset entries:

Pendomer
    E.D. (*275*): *iiii quadragenariae pascuae inter longum et latum.*
    D.B. (*92*b): *iiii quarentenae pasturae in longitudine et latitudine.*

[1] For the league and the significance of linear entries, see pp. 372–4 above.
[2] R. W. Eyton, *op. cit.* pp. 31–5.

Winscombe
    E.D. (*161*): *i leuga pascuae in longitudine et in latitudine.*
    D.B. (90): *una leuua pasturae in longitudine et latitudine.*

One is tempted, therefore, to assume that the variant formulae imply the same facts as the more usual formula, and that they occur occasionally when the length and breadth of pasture were the same. This assumption has been adopted in plotting Fig. 92.

Another variant gives only one dimension, and, again, Eyton thought that these indicated areal quantities, but the corresponding versions sometimes follow other formulae, as in these three Devonshire entries, one of which (that for King's Nympton) is unique:

Clawton
    E.D. (*318b*): *i leuga pascuae in longitudine et latitudine.*
    D.B. (108b): *una leuua pasturae.*

Holne
    E.D. (*367b*): *i leuga pascuae inter longitudinem et latitudinem.*
    D.B. (111): *i leuua pasturae.*

King's Nympton
    E.D. (*98*): *i leuga pascuae ab omni parte.*
    D.B. (101): *una leuua pasturae in longitudine et latitudine.*

It would seem that single dimensions were also occasionally used when length and breadth were the same. On Fig. 92, however, such entries have been plotted under the category of 'miscellaneous', that is, in the absence of an alternative version which mentions length and breadth.

Occasionally, one set of measurements is found in a linked entry covering a number of widely separated places. This seems to imply, although not necessarily, some process of addition whereby the dimensions of separate tracts of pasture were consolidated into one sum. On Fig. 92 the larger measurements have been plotted for the first-named settlement only, and other places in the entry have each been indicated by the symbol for 'miscellaneous'.[1] The amount of pasture entered for large manors may likewise have been at separate places. The largest amount entered under a single name is the 7 leagues by 7 for Melksham in Wiltshire (65). The Exeter text for two Dorset entries gives separate quantities for

---

[1] If a settlement (other than the first-named) in a composite entry also has a large amount of pasture entered for it in a separate entry, that pasture has been plotted instead of the symbol for 'miscellaneous'.

manorial components that are combined into single amounts in the Exchequer version. The details are given below; the figures for wood and meadow are also given in order to make clear the division into two equal portions:

Frome
> E.D. (*48b*): (*a*) 4½ acres of wood; 10 acres of meadow; pasture 8½ furlongs by 8½ furlongs.
> (*b*) 4½ acres of wood; 10 acres of meadow; pasture 8½ furlongs by 8½ furlongs.
> D.B. (81b): 9 acres of wood; 20 acres of meadow; pasture 17 furlongs by 17 furlongs.

West Stafford
> E.D. (*55b*): (*a*) 12 acres of meadow: 8 furlongs of pasture and 4 acres.
> (*b*) 12 acres of meadow; 8 furlongs of pasture and 4 acres.
> D.B. (83b): 24 acres of meadow; 16 furlongs of pasture and 8 acres.

It seems impossible (certainly in the Frome entry) to reconcile the sum of the Exeter components of pasture with the Exchequer total, and this impossibility defeats any attempt to discover a clue to the arithmetic that lay behind the ubiquitous formula '*m* furlongs by *n* furlongs'.[1]

When pasture is measured in acres, the amounts range from one acre to 550 and even to 1,000 acres in one Cornish and one Somerset entry, but the great majority do not rise above 50 acres. The round figures in many entries (e.g. 30, 40, 60 or 100) may indicate that they are estimates rather than precise amounts, but the detailed figures in other entries (e.g. 13, 26, 31 or 53) suggest precision. No attempt has been made to convert these Domesday figures into modern acreages, and they have been regarded merely as conventional units in the preparation of Fig. 92. A few entries appear to use acres as linear measurements. Here are two examples from Devonshire:

> Abbotskerswell (*184*, 104): *v quadragenariae pascuae in longitudine et xxx agri in latitudine.*
> Throwleigh (*458b*, 113b): *dimidia leuga pascuae in longitudine et iiii agri in latitudine.*

Eyton explained these as references to 'lineal acres' comprising 4 perches,[2] but on Fig. 92 they have been indicated under the category of 'miscellaneous'.

---

[1] For a similar wood entry for manorial components, see p. 375 above.
[2] R. W. Eyton, *op. cit.* pp. 25–7.

385

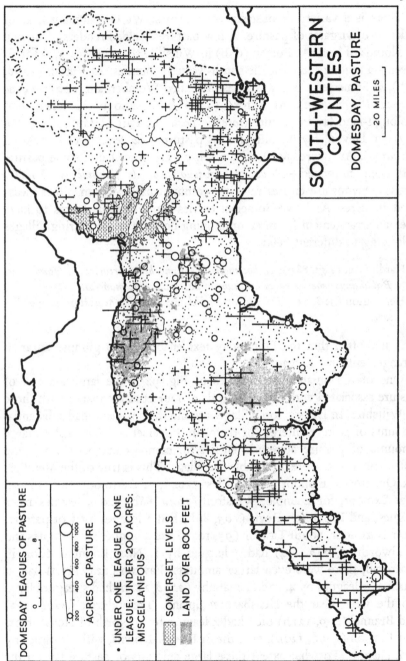

Fig. 92. South-western Counties: Domesday pasture in 1086.

There is a variety of miscellaneous entries. We occasionally hear of hides or of virgates of pasture. A few pastures yielded renders of money or blooms of iron. At Porton (69b) in Wiltshire there was pasture for 50 sheep—a unique entry for the south-western counties. At Yatton (*159b*, 89b), in Somerset, there was a pasture called Wemberham (*i pascua quae vocatur Weimorham*). At Exford (*359b*, 95b), also in Somerset, half a plough-land lay in pasture (*iacet in pastura*). Part of Mullacott (*469*, 117), in Devon, lay uncultivated as pasture (*alia terra iacet vastata ad pasturam*). At Swyre (80b), in Dorset, on the other hand, some pasture had been cultivated—*Prius erat pascualis, modo seminabilis*. Finally, twelve Devonshire entries refer to 'common pasture', sometimes measured in acres. And two Somerset entries show us what seems to have been an arrangement for intercommoning between neighbouring villages belonging to different fiefs:

> Hardington (*147*, 88b): *In hoc manerio est una hida pertinens ad Hamintone. Balduuinus tenet et habet communem pasturam huic manerio.*
> Hemington (*315*, 93): *De hac terra i hida est in communi pastura in Hardintone.*

It is not often that the Domesday text enables us to glimpse agrarian arrangements in this way.

One of the outstanding features of Fig. 92 is the large amount of pasture associated with the chalklands of Dorset as compared with those of Wiltshire. In Dorset, too, villages around the heathland had substantial amounts of pasture entered for them. In Somerset and Devonshire large amounts of pasture appear in entries for complex manors close to the upland areas that were unsuitable for tillage. This is true of the Mendips, the Quantocks, Exmoor and, to a less extent, of Dartmoor. In Somerset near Exmoor, for example, Winsford (*104b*, 86b) had 4 leagues by 2 leagues, and Brompton Regis (*103*, 86b) had 3 leagues by 1 of pasture. Near Dartmoor, South Tawton (*93*, 100b) had 4 leagues by 4 leagues and Bradworthy (*335b*, 114) had 3 leagues by 1. Near Bodmin Moor, in Cornwall, there were even larger amounts. Fawton (*228*, 121b) had as much as 7 leagues by 4, and Hamatethy (*259b*, 122b) had 5 leagues by 2. To the west, near the Hensbarrow upland, Tremoddrett (*250b*, 123) and Brannel (*249*, 121b) each had 4 leagues by 2. North of Hensbarrow was Pawton (*199b*, 120b), near the St Breock Downs, with 6 leagues by 2 of pasture. To what extent these large amounts of pasture lay on the moorlands themselves, we cannot say. It is clear, however, that large amounts of pasture were not limited to settlements close to uplands. They were also to be found on the lower-lying lands of Somerset, more so on those of Devon, and especially on those of Cornwall. In the west of Cornwall there were Helston (*100b*, 120) with pasture 5 leagues by 3, Cosawes (*224b*, 122) with 5 leagues by 2, and a number of other places also with large amounts. The landscape of the south-west was characterised by its pastoral character, and nowhere was this more true than in Cornwall.

FISHERIES

Fisheries are specifically mentioned, or implied, in connection with relatively few places in the south-west. Some fisheries yielded renders— in money (from 1*s.* to 25*s.*), in eels or in salmon. Normally, the presence of only one fishery is mentioned except for some settlements around the Somerset Levels where we sometimes hear of two or three. Only half a fishery is recorded for Weare Gifford (*412b*, 115) in Devon, and there is no clue to the other half.[1] A holding at Lyme Regis (75 b), in Dorset, had only an unspecified number of fishermen who paid a due of 15*s.* in lieu of fish for the monks. There was another teamless community of 1 villein and 4 fishermen at Bridge (*57*; 83, 83b, 84) which may have been somewhere near the mouth of the Wey in Dorset. At the Devonshire settlement of Hollowcombe in Fremington (*408*, 115b) three salt-workers rendered, amongst other things, a load of fish from which we must suppose the presence of a local fishery.

Fig. 93 shows that there were three main groups of fisheries: (1) those in and around the Somerset Levels; (2) those along the lower courses of the Taw and the Torridge in northern Devonshire; and (3) those along the lower courses of various rivers that flowed to the south coast, mainly in Devonshire. The fishery in the east of Dorset was at a mill on the small River Tarrant, which is a tributary of the Stour. To what extent all these were river-fisheries we cannot say. We can, however, say that it is most unlikely that the 30 symbols on Fig. 93 provide anything like a full picture of the fishing activity along the rivers of the south-west in the eleventh century.

[1] But non-Domesday evidence shows it to have been at Monkton—see p. 269 above.

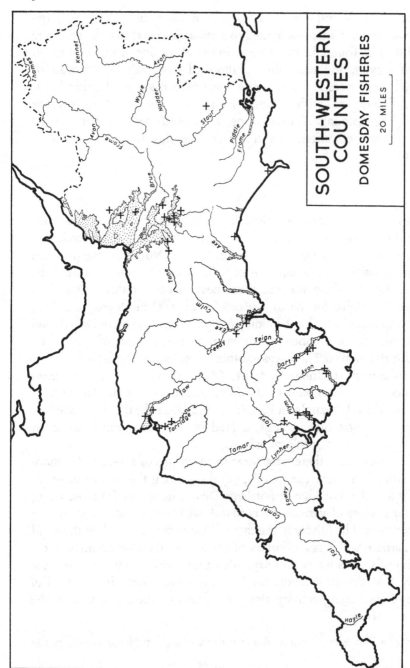

SOUTH-WESTERN
COUNTIES

DOMESDAY FISHERIES

20 MILES

Fig. 93. South-western Counties: Domesday fisheries in 1086.

Not only *piscariae*, but places with fishermen and with eel and other renders of fish are shown. Areas of blown sand, alluvium and lowland peat are shown; rivers flowing across these areas are not marked.

SALT-PANS

Salt-pans are specifically mentioned or implied in connection with 34 places, mainly in Devonshire, with a few in Dorset and Cornwall. The number of pans on a holding is usually stated and also their render, almost always in money but sometimes in loads of salt or fish. But there were a number of much larger salt-making centres, e.g., in Dorset, Studland (79b) had 32 pans and *Waia* (79) had 12; in Devonshire, Bishopsteignton (*117*, 101b) had 24 pans; and in Cornwall, Stratton (*237*, 121b) had 10 pans. We are told nothing of the circumstances of salt manufacture, but on some manors we hear not of salt-pans but of salt-workers, from which we must assume the presence of pans. In Dorset, there were 27 salt-workers at Lyme Regis (77b, 85), 16 at Charmouth (80) and 13 at Ower (*44b*, 78). In Devonshire, there were 33 at Otterton (*194b*, 104) and smaller groups at a number of other places.

Fig. 94 shows that the majority of salt-making places were on the south coast, mainly along the estuaries of the Devonshire rivers and in the neighbourhood of Poole Harbour in Dorset. On the north coast, the record of salt-making was limited to Stratton in Cornwall and to a few places on the estuaries of the Taw and Torridge in Devon. We may well suppose that the Domesday folios do not give a complete picture of salt manufacture in the south-west in the eleventh century. The entry for Salcombe Regis (*118b*, 102) in Devonshire, for example, makes no reference to salt-pans, although the name means 'salt valley', and is mentioned in a pre-Domesday charter. The other Salcombe (west of Dartmouth) is not mentioned in either Domesday text.

VINEYARDS

As Fig. 95 shows, vineyards are recorded in connection with thirteen places in the south-west, although three of these places appear together in a joint entry. Most were measured in terms of the French unit of the arpent, the exact size of which is uncertain. All were on estates held by tenants-in-chief. Lacock and Wilcot belonged to Edward the sheriff of Wiltshire; Durweston, Wootton and Tollard Royal to Aiulf the sheriff of Dorset, although the last-named place was just over the border in Wiltshire; Glastonbury, Meare and Panborough to Glastonbury Abbey; Muchelney,

390

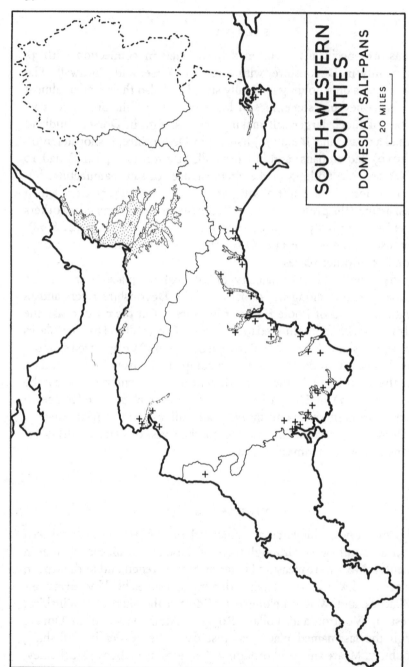

Fig. 94. South-western Counties: Domesday salt-pans in 1086.

Not only *salinae*, but places with *salinarii* are shown. Areas of blown sand, alluvium and lowland peat are shown.

391

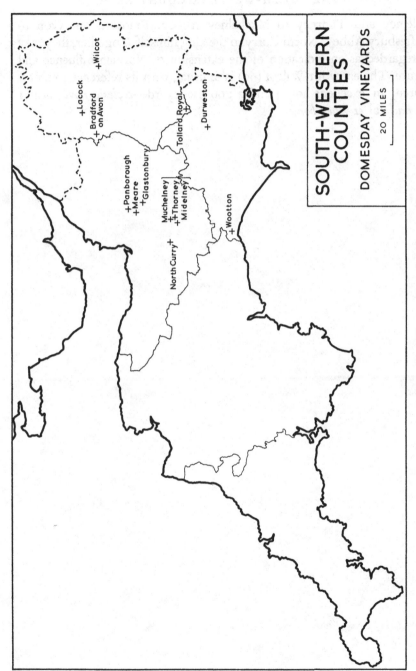

Fig. 95. South-western Counties: Domesday vineyards in 1086.

The three islands of Muchelney, Midelney and Thorney had one arpent of vineyard between them.

Midelney and Thorney to Muchelney Abbey; Bradford on Avon to Shaftesbury Abbey; North Curry to the king himself. Together, they may be regarded as an indication of the extension of Norman influence and culture. The entry for Wilcot (69) is unusual, with its reference to a new church, an excellent house and a good vineyard—*ecclesia nova, domus obtima* [sic] *et vinea bona.*

APPENDIX I

NOTE ON THE 'LIBER EXONIENSIS'

The *Liber Exoniensis*, preserved in the Library of Exeter Cathedral, includes the sole surviving example of a first draft, based upon individual fiefs, of the material elicited by the Domesday Inquest. The text for Somerset and Cornwall is all but complete; that for Devon lacks some entries involving six fiefs; that for Dorset lacks all the entries for as many as 46 fiefs; that for Wiltshire is entirely lacking except for a solitary entry. In addition to the Domesday text, the *Liber Exoniensis* also includes: (1) the Geld Accounts, a record of geld levied at a time near that of the Inquest; (2) *Terrae Occupatae*, a condensed record of illegalities noted during the Inquest for the counties of Somerset, Devon and Cornwall; (3) two lists of the hundreds of the same three counties; and (4) Summaries of the holdings of certain tenants-in-chief, together with a list of 26 fiefs. The folios of the original manuscript were unnumbered, and some were left blank. A printed version, not always accurate, was issued in 1816, together with indexes of place-names and proper names.[1] The arrangement of the printed text is artificial and was designed to accord as far as possible with that of the Exchequer version. Its contents are set out on p. 394.

A selected list of books and articles dealing with the general nature and purpose of the *Liber Exoniensis* is set out below:

Sir H. ELLIS, *A General Introduction to Domesday Book*, 2 vols. (London, 1816).

T. W. WHALE and O. J. REICHEL, 'Analysis of the Exon Domesday', *Trans. Dev. Ass.* xxviii (Ashburton, 1896), pp. 391–463; xxxv (Sidmouth, 1903), pp. 662–712; xxxvi (Teignmouth, 1904), pp. 156–72.

T. W. WHALE, 'Principles of Domesday and the Feudal Aids', *Trans. Dev. Ass.* xxxii (Totnes, 1900), pp. 521–51.

T. W. WHALE, 'Index to the Exon Domesday', *Trans. Dev. Ass.* xxxiv (Bideford, 1902), pp. 289–324.

[1] This is the fourth volume of the Record Commission's *Libri Censualis vocati Domesday Book*. It omits a leaf (*347–347 b*) which contains six Devon manors, and which was accidentally discovered after 1816—*V.C.H. Devonshire*, I (London, 1905), pp. 375–6.

T. W. WHALE, 'Analysis of the Exon Domesday', *Trans. Dev. Ass.* xxxvii (Princetown, 1905), pp. 391–463.

F. H. BARING, 'The Exeter Domesday', *Eng. Hist. Rev.* xxvii (London, 1912), pp. 309–18.

R. WELLDON FINN, 'The Evolution of Successive Versions of Domesday Book', *Eng. Hist. Rev.* lxvi (London, 1951), pp. 561–4.

P. H. SAWYER, 'The Place-Names of the Domesday Manuscripts', *Bull. John Rylands Lib.* xxxviii (Manchester, 1956), pp. 289–324.

R. WELLDON FINN, 'The Immediate Sources of the Exchequer Domesday', *Bull. John Rylands Lib.* xl (Manchester, 1957), pp. 47–78.

R. WELLDON FINN, 'The Exeter Domesday and its Construction', *Bull. John Rylands Lib.* xli (Manchester, 1959), pp. 360–87.

V. H. GALBRAITH, *The Making of Domesday Book* (Oxford, 1961).

R. WELLDON FINN, *Domesday Studies: The Liber Exoniensis* (London, 1964).

CONTENTS OF THE 'LIBER EXONIENSIS'

APPENDIX II

COMPARISON OF EXETER AND
EXCHEQUER VERSIONS

BY R. WELLDON FINN, M.A., H. C. DARBY, LITT.D., F.B.A.
AND G. R. VERSEY, M.A.

The Exchequer version is a condensation and simplification of the Exeter text. Much was abbreviated and much omitted. One striking omission is that of demesne livestock, and there is, moreover, only infrequent reference to *terra villanorum*, while, for the majority of fiefs, the hidage of manorial demesne is not stated. Demesne teams and villagers' teams are not always kept separate, especially when only small quantities are involved, and fractions of teams are often disregarded. Manorial components are rarely distinguished, and their separate details are often combined into single statements with the result that the place-names of sub-tenancies are sometimes omitted. There are also frequent arithmetical mistakes. Furthermore, there are many verbal differences, and the Exchequer formulae are generally shorter, and rents and values are seldom stated to be *per annum*. The names of the majority of sub-tenants and of pre-Conquest landholders are reduced to Christian names.

We might well, however, be doing the Exchequer clerks an injustice always to blame them for the omission of some of the material which appears in the Exeter text. The surviving copy of the Exeter text cannot be the same as that from which the Exchequer clerks worked, and some discrepancies may be due to Exeter copyists and not to Exchequer compilers. What is more interesting is the fact that a few entries of the Exchequer text include information that does not appear in the surviving Exeter text. These can only have been subsequent additions to a text which has not survived. Furthermore, examination of the manuscripts shows that the surviving Exeter text itself must have been corrected after the copy used by the Exchequer clerks had been made; these late insertions were for the most part never added to the Exchequer version.

The table of differences does not set out every variation in content and language but only those of a statistical character that concern totals, maps and general geographical information.

TABLE OF DIFFERENCES BETWEEN EXETER
AND EXCHEQUER VERSIONS

Note: (1) Information omitted from the Exeter text but appearing in the Exchequer version is set out separately on pp. 427–8.

(2) Exeter details are given in the second column, Exchequer details in the third.

(3) The folio reference is that on which an entry begins, and the relevant detail may therefore be found on the succeeding folio.

(4) The arrangement under counties and sub-headings is that of this volume itself.

(5) Omissions are indicated by —.

(6) An asterisk (*) indicates that the condition of the MS does not enable one to say with certainty that a text is in error.

(7) No reference is made to those Exeter entries which have been defaced, thus preventing comparison with the corresponding Exchequer versions. A number of Exeter entries on folios *404–405b*, for example, are obliterated by ink blots.

(8) Where two or more holdings at one place are described on the same folio, the names of tenants are given in order to assist identification.

(9) When the Exchequer gives merely one figure for the sum of the Exeter components (including demesne and villagers' teams) this has not been noted unless there is an arithmetical error. But all equations of oxen and teams are given, including those in which 8 oxen equal one team.

(10) The omission of livestock from the Exchequer entries is not noted.

(11) Abbreviations (for plural as well as for singular nouns) are used as follows:

ac	acre	fg	ferling	pc	perch	t	team
b	breadth, broad	h	hide	pr	priest	u	underwood
bo	bordar	hd	hundred	*q.r.*	*quando recepit,*	v	villein
cos	coscet	l	length, long		*accepit*	vg	virgate
cot	cottar	lg	league	s	serf	vill	villagers'
dem	demesne	m	meadow	sw	swineherd	w	wood
f	furlong	p	pasture				

WILTSHIRE

Pasture

Sutton Veny (*47, 72*)	½ lg × 1 f	½ lg × 1

DORSET

Assessment

Warmwell (*60, 83b*)	1½ h 1 vg	1½ h

Plough-land

Wareham (*28b, 78b*)	1	—

Demesne teams

*Charborough (*25b, 75*)	1½	1
Nutford (*31b, 75b*)	1	—

Demesne teams (cont.)

Little Puddle (*36*, 77b)	1½	I
Nettlecombe (*38*, 78), *unus miles*	2	—
Graston (*57b*, 83b)	1½	I

Villagers' teams

Nettlecombe (*38*, 78)	2½	2
West Milton (*38*, 78)	3	5
Poorton (*42*, 78b)	1	—
Compton Abbas (*43*, 78)	2	—
Frome in Tollerford hd. (*48b*, 81b), Robert	1	—
Tarrant (*58*, 83b)	½	—
Durweston (*58b*, 83b)	½	—
Ringstead (*60*, 83b), Hugh	½	—
Wilkswood (*60b*, 84)	1 t, 2 ox	I

Population

Child Okeford (*25*, 75)	9 bo	8 bo
Frome St Quinton (*29*, 75b)	8 cot	—
*Hampreston (*30*, 75b)	2 v	5 v
Affpuddle (*36b*, 77b)	5 cot	—
Cerne in Modbury hd. (*45*, 78)	7 bo	5 bo
Winterborne, Eastern Group (*47b*, 82)	1 v	—
Frome in Tollerford hd. (*48b*, 81b), Dodeman	2 cot	—
Hampreston (*56b*, 83b)	1 bo	2 bo
Uploders (*58*, 83b)	*i rusticus*	1 v
Tarrant (*58*, 83b)	2 v	1 v
Ringstead (*60*, 83b), Hugh	2 bo	1 bo
Seaborough (*154*, 87b)	1 s	2 s

Values

Puddletown (*25*, 75)	q.r. £73	—
*Loders (*26b*, 75b)	reddit £34 et q.r. reddebat tantundem	*Valuit et valet* £33
Cranborne (*29*, 75b)	see p. 96	
Ashmore (*29b*, 75b)	see p. 96	
Bockhampton (*33*, 85)	q.r. £3	—
Winterborne, Western Group (*33*, 85)	q.r. £6	—
Cerne Abbas (*36*, 77b)	*Et iste tegnus reddit per annum xxx solidos ecclesiae quominus servitium*	—
Affpuddle (*36b*, 77b)	*haec mansio reddit per annum vii libras et x solidos*	—

Values (cont.)

Affpuddle (*cont.*)	*et quando abbas recepit valebant c solidos plus predictae duae mansiones quia per Hugo filio Grip fuerunt depredati* (the other manor was Bloxworth)	
Littlebredy (*37*, 78)	see p. 96	
Nettlecombe (*38*, 78)	*q.r.* £11. 15s. od.	*Valuit* £12
Askerswell (*42*, 78b)	*q.r.* 100s.	—
Littlewindsor (*49b*, 81b)	*q.r.* 40s.	—
Mapperton near Beaminster (*49b*, 81b)	*q.r.* 60s.	—
Poorton (*51*, 82b)	*q.r.* 20s.	—
Farnham (*57b*, 83b)	*q.r.* 10s.	—
Tarrant (*58*, 83b)	*q.r.* 10s.	—
Winterborne, Eastern Group (*58b*, 83b), Robert 1st entry	*q.r.* 30s.	—
Winterborne, Eastern Group (*58b*, 83b), Robert 2nd entry	*valet* 25s. *et q.r.* 20s.	*Valuit* 25s. *Modo* 20s.
Chaldon Herring (*59b*, 83b)	see p. 96	
Ringstead (*60*, 83b), Ralph	*q.r.* 40s.	—
Goathill (*278*, 92b)	*q.r.* 20s.	—
Poyntington (*279*, 93)	*q.r.* 30s.	—

Wood

Puddletown (*25*, 75)	1 lg 8 f × 1 lg w	—
Cranborne (*29*, 75b)	2½ lg × 2 lg w	2 lg × 2 lg w
Nettlecombe (*38*, 78)	*quod nemus nullum fructum fert*	—
Chelborough (*49*, 81b)	1 lg × ½ lg w	—
Poorton (*51*, 82b)	15 ac w	—
Worth Matravers (*51b*, 82b)	7 f × 7 f w	7 f *inter* l & b w
Tarrant (*59*, 83b), Ralph	6 f × 2 f w	—
Hurpston (*60b*, 84)	1 f × 1 f w	1 f w
Wilkswood (*60b*, 84)	4 f *in* l & b w	4 f w
Goathill (*278b*, 92b)	25 ac w	15 ac w
Adber (*493b*, 99)	1 f × 1 f w	1 f *in* l & b w

Pasture

Cranborne (*29*, 75b)	2½ lg × 1 lg 1 f	2 lg 1 f × 1 lg
Frome in Tollerford hd. (*48b*, 81b)	see pp. 108–9	
Tarrant (*59*, 83b), Ralph	8 f *inter* l & b	8 f
Ringstead (*60*, 83b), Hugh	4 f *inter* l & b	4 f
Creech (*60b*, 84)	7 f × 4 f	7 f × 3 f
Afflington (*62*, 82b)	1 f *inter* l & b	1 f *in* l & b

Waste

Hurpston (*60b*, 84)	*haec terra omnino devastata est*	—

Mills

*Cerne in Modbury hd. (*45*, 78)	1 *reddit* 25*d.*	1 *reddens* 20*d.*
Puddle in Bere hd. (*56b*, 83b)	1 *de* 10*s.*	—

Urban life

East Pulham (*48*, 81 b)	*i ortus in Wareham qui reddit iii denarios*	—
Goathill (*278b*, 92b)	*ii masurae in Meleborna*	—

Miscellaneous

Winterborne, Eastern Group (*28*)	*Aiulfus vicecomes habet i virgatam terrae in Wintreborna de revelanda et reddit v solidos per annum*	—
Puddle in Bere hd. (*56b*, 83b)	*ibi est dimidia hida et quattuor agri et i ortus quae nunquam gildavit sed celatum est*	—

SOMERSET

Settlement

Burnett (*114*, 87)	named place held with Keynsham	place-name omitted

Assessment

Brompton Regis (*103*, 86b)	3 h in dem	4 h in dem
Congresbury (*106*, 87), 3 thegns	3 h	3 h 3 vg
Martock (*113*, 87)	4 h added	3 h added
*Farmborough (*141*, 88), William	2 h	5 h
Monkton Combe (*185b*, 89b)	6 h *minus* 1 v in dem	6 h in dem
Ashbrittle (*269*, 92)	1½ h added	1 h added
Wincanton (*352*, 95)	*Supradictae iiii hidae reddiderunt gildum pro iii hidis*	—
Elworthy (*361*, 96)	see p. 132	
Poleshill (*363*, 96)	½ h added	1 h added
Brewham (*364b*, 96b)	2 h taken away	3 h taken away
Woodspring (*369b*, 96b)	4 h 3 vg in dem	blank in MS.
Ston Easton (*446*, 98)	1½ h	1 h

Plough-lands

Brompton Regis (*103*, 86b), priest	2	—
Keynsham (*113b*, 87), Ælfric	1	—
Stanton Drew (*113b*, 87)	10	—
Burnett (*114*, 87)	4	
Belluton (*114*, 87)	4	— (but in parallel entry on fo. 91b)
Chewton Mendip (*114b*, 87), manor church	3	—
*Farmborough (*141*, 88), William	2	5
Winford (*145*, 88b), Colsuain	2	—
Banwell (*157*, 89b), Serlo	8	—
Winscombe (*161*, 90), Roger and Ralph	5, total for 2 holdings	—
Eckweek (*276b*, 92b)	1	—
Broadwood (*358b*, 95b)	1½	1
Vexford (*427b*, 93b), Alric	1½	1
Blagdon (*452*, 97b), Lambert	1	—
Moreton (*453b*, 98)	5, (details add to 5½)	5

Demesne teams

Rode (*148*, 88b)	6½, total for all 6 holdings	7
Ashcott (*164b*, 90)	2 ox	—
Bleadon (*173b*, 87b), Saulf	1½	1
Taunton (*174b*, 87b), Godwine *et al.*	7½, total for 5 holdings	7
Steart (*272b*, 92)	2	3
Aller in Sampford Brett (*286b*, 91b)	1	—
Appley (*315b*, 93)	½	—
Oaktrow (*359b*, 96)	½	—
Luxborough (*360b*, 96), Nigel	1	—
Compton Pauncefoot (*383b*, 97b)	4 ox	½
Hadworthy (*422*, 93)	1	—
Otterhampton (*424*, 93)	1	—
Terra Teodrici (*425b*, 93b)	2 ox	—
Withiel (*426*, 93b)	½	—
Monksilver (*427*, 93b), Alric	2 ox	—
Vexford (*427b*, 93b), Alric	½	—
*Curry Mallet (*429*, 93)	1	2
Puckington (*429b*, 93b), manor added	½	—

Demesne teams (cont.)

Bagley in Stoke Pero	½	1
(430b, 94)		
Long Sutton (435b, 94b)	2, total for 2 holdings	3
Stone in Mudford (454, 98)	½	—
Draycott in Rodney Stoke	2 ox	—
(492, 99)		
Eastrip (493b, 99)	4 ox	½

Villagers' teams

Somerton (89b, 86)	3	4
Milborne Port (91, 86b)	½	—
Mudford (116, 87)	3	—
Barrow Gurney (143, 88)	12	—
Pitcott (146, 88b)	½	—
West Camel (189, 91),	1½	1
Dodeman		
Carhampton church	1½	—
(196b, 91b)		
Broadway (268b, 92)	½	—
Houndstone (280b, 93)	½	—
Sampford Brett (286b, 91b)	2	1
Appley (315b, 93)	1½	—
Stretcholt (350, 95),	½	—
Walscin, 1st entry		
Tarnock (351b, 95), Richard	1	—
Luxborough (360b, 96),	1	—
Nigel		
Combe Sydenham (362b,96)	*iiii animalia in carrucam*	½
Cheriton (364b, 96b)	3 ox	½
Spaxton (372, 97), *quidam*	2	—
miles		
Radlet (372b, 97)	1	—
Hadworthy (422, 93)	½	—
Otterhampton (424, 93)	½	—
Stockland (425, 93b)	1	—
Puckington (429b, 93b),	2 ox	—
manor added		
Littleton in Compton	½ t, 2 ox	1
Dundon (433b, 94)		
Skilgate (442, 94b)	1	½
Upper Cheddon (444,	1	—
94b), Robert		
Cannington (478, 98b)	2 ox	—
Stocklinch (491, 99)	½	—
Discove (493, 99)	½	—

Demesne and villagers' teams

Street (357, 95)	4 dem ox + 4 vill ox	1

Population

Brompton Regis (*103*, 86b)	15 sw who render 32*s*.	—
Coker (*107*, 87)	1 sw who renders 10 swine	—
Mudford (*116*, 87)	4 v, 8 bo, 2 s	—
Clewer (*141b*, 88)	4 v	6 v
Bishopsworth (*141b*, 88), Azelin	3 v	4 v
Winford (*145*, 88b), Roger and Fulcran	17 v, total for 2 holdings	19 v
Radstock (*146b*, 88b)	4 cot	3 cot
Babington (*147*, 88b)	3 cot	—
Emborough (*150*, 89)	3 cot	—
Banwell (*157*, 89b), Bishop Giso	23 v	24 v
Banwell (*157*, 89b), Serlo *et al.*	23 v, total for 6 holdings	25 v
Wells (*157b*, 89), Fastrad, Richard and Erneis	25 v, total for 3 holdings	17 v
Winsham (*158b*, 89b)	1 sw who renders 12 swine	—
Chew Magna (*159*, 89b), Bishop Giso	2 sw who render 24 swine	—
Chew Magna (*159*, 89b), Richard *et al.*	18 bo, total for 5 holdings	27 bo
High Ham (*165*, 90), Robert, Girard and Serlo	15 bo, total for 3 holdings	14 bo
*Pylle (*166b*, 90b)	11 cot	—
Hornblotton, Alhampton and Lamyatt (*169b*, 90b)	13 bo, total for 3 places	12 bo
Brent (*170b*, 90b), Ælfric, Ralph and Godwine	4 v, 13 cot, total for 3 holdings	3 v, 10 cot
Pitminster (*173b*, 87b)	4 s, 1 sw	—
Bleadon (*173b*, 87b)	7 s	8 s
Lydeard St Lawrence (*175*, 87b)	3 s	—
Bathampton (*186*, 89b), Hugh and Colgrim	9 bo, total for 2 holdings	6 bo
*Stoke-sub-Hamdon (*267b*, 92), Robert, 1st entry	9 v	2 v
Beer Crocombe (*271*, 92)	8 bo	7 bo
Charlton near Somerton (*273b*, 92b)	2 cot	—
Odcombe (*274*, 92b)	6 bo	16 bo
Cucklington (*277*, 92b)	4 cot	—
Weston Bampfylde (*278*, 92b)	2 s	—
Milborne Port (*278b*, 93)	4 s	2 s
Lexworthy (*282*, 91b)	4 bo	3 bo
Bratton Seymour (*352*, 95)	8 bo	—

Population (cont.)

Castle Cary (*352b*, 95)	8 sw who render 50 swine	—
Wembdon (*353*, 95)	6 cot	—
Burnham (*354*, 95)	3 cot	—
Luxborough (*360b*, 96), Nigel	6 v, 3 bo	—
Combe Sydenham (*362b*, 96)	1 v	—
Brewham (*364b*, 96b)	25 cot	—
Witham (*382b*, 97b)	3 v	—
Pardlestone (*428*, 93b)	3 bo	4 bo
Enmore (*432*, 94)	2 s *et* 2 s	2 s
Raddington (*442*, 94b)	2 s	3 s
Whitelackington (*443*, 94b)	25 bo, 6 sw	30 bo, 7 sw
Stoke near Chew Magna (*452b*, 97b)	1 v	—
Moreton (*453b*, 98)	9 bo, 2 cot	11 bo
Stoke near Chew Magna (*492*, 99)	3 v	—

Rents of swineherds

North Petherton (*88b*, 86)	20 sw who render 100s.	20 sw
Bruton (*91*, 86b)	1 sw who renders 5 swine	1 sw

Values

Old Cleeve (*103b*, 86b)	q.r. £23 *de albo argento*	—
Nettlecombe (*104*, 86b)	q.r. £ (blank) and 12s. *de albo argento*	—
Capton (*104*, 86b)	q.r. 46s. *de albo argento*	—
Langford Budville (*104b*, 86b)	q.r. £4. 12s. 0d.	—
Winsford (*104b*, 86b)	q.r. £10. 10s. 0d. *de albo argento*	—
Winsford (*104b*, 86b), half-hide added	q.r. 20s.	—
Creech St Michael (*105*, 86b)	q.r. £9. 4s. 0d. *de albo argento*	—
North Curry (*105*, 86b)	q.r. £23 *de albo argento*	—
Eastham (*105b*, 87)	q.r. 50s.	—
Congresbury (*106*, 87), the king	q.r. £28. 15s. 0d. *de albo argento*	—
Congresbury (*106*, 87), Bishop Giso	q.r. £4	—
Congresbury (*106*, 87), Serlo and Gilbert	q.r. 40s., total for 2 holdings	—
Queen Camel (*106b*, 87)	q.r. £23 *de albo argento*	—
Coker (*107*, 87)	q.r. £19. 1s. 0d. *de albo argento*	—

Values (cont.)

Hardington Mandeville (*107*, 87)	*q.r.* £12. 14s. od. *de albo argento*	—
Henstridge (*107*, 87)	*q.r.* £23 *de albo argento*	—
*Martock (*113b*, 87), Ælfric	*q.r.* 40s.	—
Stanton Drew (*113b*, 87)	*q.r.* £4	—
Bath (*114b*, 87)	*reddit* £11 *de xx in hora*	*reddit* £11
Dowlish (*136b*, 87b)	*valet* 23s. *et q.r. tantundem*	*Valuit et valet* 24s.
Chaffcombe (*136b*, 87b)	*q.r.* 40s.	—
Chaffcombe (*136b*, 87b), manors added	*q.r.* 20s.	—
Exton (*139*, 87b)	*q.r.* 40s.	—
Kitnor (*139*, 88)	*q.r.* 5s.	—
Long Ashton (*143b*, 88b), Roger	*q.r.* 100s.	—
Long Ashton (*143b*, 88b), Guy	*q.r.* 20s.	—
Stratton-on-the-Fosse (*145b*, 88b)	*valet* £4. 10s. od.	*Modo* £4
Tellisford (*148*, 88b)	*q.r.* 20s.	—
Rode (*148*, 88b)	*valent* £9. 3s. od. *et q.r.* £7. 13s. od., totals for all 6 holdings	*Totum valuit* £7. 10s. od. *Modo* £8. 5s. od.
Stony Littleton (*149*, 89)	*q.r.* 30s.	—
Kingston Seymour (*150b*, 89), Fulcran	*q.r.* 3s.	—
Thorent (*153b*, 87b)	*Valuit et valet* 14s.†	*Valuit et valet* 13s.
Yatton (*159b*, 89b), Fastrad and Ilbert	*q.r.* £5, total for 2 holdings	—
Wanstrow (*160*, 89b)	*q.r.* £4	—
Shapwick (*161b*, 90)	*q.r.* £7	—
Sutton Mallet, Edington, Chilton-upon-Polden and Catcott (*162*, 90)	*valet* £15 *et q.r.* £19, totals for 4 places	*valet* £19
Woolavington (*162*, 90)	*q.r.* £7	—
Cossington (*162b*, 90)	*valet* £5 *et q.r. tantundem*	*Valuit et valet* £6
Walton near Glastonbury (*163b*, 90)	*q.r.* 100s.	—
Compton Dundon, Ashcott and Pedwell (*163b*, 90)	*valet* £10 *et q.r.* £8, totals for 3 places	*valet* £8
Bagley in Wedmore (*164*, 90)	*q.r.* 6s.	—
Dundon (*164*, 90)	*q.r.* £6	—
Overleigh in Street (*164b*, 90)	*q.r.* 60s.	—
High Ham (*165*, 90), Glastonbury Abbey	*q.r.* £4	—

† This Exeter entry is in Exchequer script. A parallel Exeter entry on fo. *467* follows the usual Exeter formula.

Values (cont.)

High Ham (*165*, 90), Robert, Girard and Serlo	*q.r.* 115*s.*, total for 3 holdings	—
Butleigh (*165b*, 90), Glastonbury Abbey	*q.r.* 60*s.*	—
Butleigh (*165b*, 90), Thurstan and Roger	*q.r.* £6. 10*s.* 0*d.*, total for 2 holdings	*Valuit* £7
Butleigh (*165b*, 90), Ælward	*q.r.* 5*s.*	—
Shepton Mallet and Croscombe (*166*, 90)	*q.r.* £8. 10*s.* 0*d.*, total for 2 places	—
North Wootton, Pylle and Pilton (*166b*, 90b)	*q.r.* £8, total for 3 places	—
East Pennard (*166b*, 90b)	*q.r.* £4	—
Doulting (*167*, 90b)	*q.r.* £6	—
Charlton in Shepton Mallet and *una mansio* (*167b*, 90b)	*q.r.* 50*s.*, total for 2 holdings	—
Batcombe (*167b*, 90b), Glastonbury Abbey	*q.r.* 40*s.*	—
Batcombe (*167b*, 90b), Roger	*q.r.* 40*s.*	—
Westcombe (*168*, 90b)	*q.r.* 20*s.*	—
Mells (*168*, 90b)	*q.r.* 100*s.*	—
Whatley near Frome (*168b*, 90b), Walter	*q.r.* 30*s.*	—
Whatley near Frome (*168b*, 90b), John	*q.r.* 15*s.*	—
Ditcheat (*170*, 90b), count of Mortain	*q.r.* £7	—
Camerton (*170*, 90b), Glastonbury Abbey	*q.r.* £6	—
Camerton (*170b*, 90b), Roger	*q.r.* 10*s.*	—
Brent (*170b*, 90b), Glastonbury Abbey	*q.r.* £15	—
Brent (*171*, 90b), Roger	*q.r.* 20*s.*	—
Edingworth (*171*, 90b)	*q.r.* 40*s.*	—
Church of St Andrew in Ilchester (*171b*, 91)	*valet* 100*s. et q.r. tantundem*	— (but given in parallel entry in adjacent column)
Glastonbury (*172*, 90)	*q.r.* £10	—
Meare (*172*, 90)	*q.r.* 20*s.*	—
Panborough (*172*, 90)	*q.r.* 4*s.*	—
Andersey (*172*, 90)	*q.r.* 15*s.*	—
Stone in East Pennard (*172b*, 91)	*q.r.* £9	—
Rimpton (*173*, 87b)	*q.r.* £6	*Valuit* £7
Bleadon (*173b*, 87b), Saulf	*q.r.* 10*s.*	—

Values (cont.)

Taunton (*174*, 87b), Godwine *et al.*	q.r. 67s., total for 5 holdings	—
Corston (*186b*, 89b)	q.r. £7	—
Isle Abbots (*188b*, 91)	*quando abbas obiit valebat* 16s.	—
Cathanger (*188b*, 91)	*quando abbas obiit valuit* 27s.	—
West Camel (*189*, 91)	*valet* £9...*valet* 20s. *et q.r.* 10s.	*Totum valet* £10. 10s. 0d.
Cannington church (*196b*, 91b)	q.r. 30s.	—
Carhampton church (*196b*, 91b)	*quando episcopus fuit mortuus valebat* 40s.	—
Huntscott (*196b*, 91b)	q.r. 10s.	—
Church of St Mary in Stogumber (*197*, 91)	q.r. 30s.	—
Crewkerne church (*197*, 91), Caen Abbey	q.r. £7	—
Crewkerne church (*197*, 91), *unus miles*	q.r. £4	—
Compton Durville (*265b*, 91b)	q.r. 40s.	—
Whitestaunton (*265b*, 91b)	q.r. 30s.	—
Lopen (*266*, 91b)	q.r. 10s.	—
Isle Brewers (*266*, 91b)	q.r. 60s.	—
Tintinhull (*266b*, 91b)	q.r. £10	—
Stoke-sub-Hamdon (*267*, 92), Robert, 2nd entry	q.r. 60s.	—
Draycott in Limington (*267*, 92)	q.r. 20s.	—
Swell (*268*, 92)	q.r. 40s.	—
Brushford (*268*, 92)	q.r. £8	—
North Bradon (*268b*, 92)	q.r. 20s.	—
Ashill (*268b*, 92)	q.r. £4	—
Greenham (*269*, 92)	q.r. 10s.	—
Appley (*269b*, 92)	q.r. 10s.	—
South Bradon (*269b*, 92)	*valet* 40s. *et q.r.* 20s.	*Valet* 40s. *et valuit* (sic)
North Bradon (*269b*, 92)	q.r. 20s.	—
Charlton near Somerton (*273b*, 92b)	*valet* £6 *et q.r. tantundem*	—
Chinnock (*274*, 92b), Count Eustace	q.r. £12	*Valuit* 100s.
Chinnock (*274b*, 92b), Alured	q.r. £3	—
Stoney Stoke (*276*, 92b)	q.r. £4	—
Redlynch (*276*, 92b)	q.r. £4	—
Keinton Mandeville (*276b*, 92b)	q.r. £5	—
Carlingcott (*276b*, 92b)	q.r. 60s.	—

Values (cont.)

Clapton in Cucklington (278, 92b)	q.r. £3	—
Lufton (280b, 93)	q.r. 12s.	—
Yeovil (281, 93), count of Mortain	q.r. 20s.	—
Yeovil (281, 93), Hamond	q.r. 10s.	—
Chelwood (282b, 91b)	q.r. 40s.	—
Compton Pauncefoot (283, 91b)	q.r. £4	—
Henstridge (286b, 91b)	q.r. £5	—
Stretcholt (350, 95), Walscin, 2nd entry	q.r. 40s.	—
Crosse (350, 95)	q.r. 10s.	—
Chapel Allerton (351, 95)	valet 100s. (et 5 interlined)	Modo 100s.
Tarnock (351b, 95), Ludo	q.r. 15s.	
Bradney (353b, 95)	q.r. 15s.	—
Horsey (353b, 95)	q.r. 60s.	—
Burnham (354, 95), Reinewal	q.r. £6	—
Brean (354, 95)	q.r. £8	—
Wick in Brent Knoll (354b, 95)	q.r. 10s.	—
Cutcombe (357b, 95b)	valet £6. 10s. 0d.	Modo £6
Quarme (358b, 95b)	q.r. 7s. 6d.	olim 7s.
Staunton (359, 95b)	q.r. 7s. 6d.	—
Staunton (359, 95b), virgate added	q.r. 3s.	—
Exford (359b, 95b), William, 2nd entry	q.r. 12d.	—
East Myne (360, 96)	q.r. 10s.	—
Luxborough (360b, 96), Ranulf	q.r. 15s.	—
Coleford (361b, 96)	q.r. 3s.	—
Watchet (361b, 96)	q.r. 5s.	—
Poleshill (363, 96)	q.r. 10s.	—
Leigh in Milverton (363, 96)	valet 12s. 6d.	Modo 12s.
Stringston (372, 97)	q.r. 40s.	—
Nether Stowey (373, 97), Alured	valet £10 et q.r. £8	—
Woolston in North Cadbury (383b, 97b)	q.r. 15s.	—
Blackford near Wincanton (383b, 97b)	q.r. 20s.	—
Keyford (384, 97b)	q.r. 5s.	—
Newton in North Petherton (422, 93)	q.r. 10s.	—
Perry (422, 93)	q.r. 30s.	—

Values (cont.)

Waldron (*422*, 93)	*valet* 20s. (*et* 15d. interlined) *et q.r. tantundem*	*Valuit et valet* 22s.
Rima (*423*, 93)	q.r. 30d.	—
Chilton Trinity (*423*, 93)	*valet* 15s. *et q.r. tantundem*	*Valuit et valet* 20s.
Terra Colgrini (*423b*, 93)	q.r. 7s.	—
Terra Teodrici (*425b*, 93b)	q.r. 20s.	—
Edstock (*426*, 93b)	q.r. 10s.	—
Knowle St Giles (*426b*, 93b)	q.r. 10s.	—
Lopen (*427*, 93b)	q.r. 10s.	—
Halsway (*427*, 93b)	q.r. 20s.	—
Coleford (*427*, 93b)	q.r. 4s.	—
Huish in Nettlecombe (*427b*, 93b)	q.r. 20s.	—
Vexford (*427b*, 93b), Alric	q.r. 10s.	—
Emble (*428*, 93b)	q.r. 5s.	—
Weacombe (*428b*, 93b)	1 virgate *reddit* 7s. 6d. and 3 virgates *reddunt* 25s. *et q.r. tantundem*	*Totum valet* 32s.
Ashway (*428b*, 93b)	q.r. 20s.	—
Broford (*429*, 93b), William, 1st entry	q.r. 5s.	—
Broford (*429*, 93b), William, 2nd entry	*valet* 2s. 6d. *et q.r. tantundem*	*Valet* 26d.
Exford (*430*, 94), Ednoth	q.r. 2s.	—
Combe in Withycombe (*430b*, 94)	q.r. 5s.	—
Aller in Carhampton (*430b*, 94)	*valet* 7s. 6d. *et q.r.* 5s.	*Valet* 8s.
Golsoncott (*431*, 94)	q.r. 3s.	—
Holnicote in Selworthy (*431*, 94)	*valet* 22s. *et q.r. tantundem*	*Valet* 21s.
Doverhay (*431*, 94)	*valet* 7s. 6d. *et q.r.* 10s.	*Valet* 8s.
Hone (*431*, 94)	q.r. 30d.	—
Earnshill (*431b*, 94)	q.r. 20s.	—
Holford (*433b*, 94), William	*reddit* 17s. 6d. *et q.r.* 10s.	*Valet* 18s.
Littleton in Compton Dundon (*433b*, 94)	*valet* 60s. *et q.r. tantundem*, total for 2 holdings	*Valuit et valet* 40s.
Whatley in Winsham (*438*, 96b)	q.r. 10s.	—
Laverton (*438b*, 96b)	*valet* £7 *et q.r.* £8	q.r. £7. *Modo* £8
Newton in North Petherton (*441*, 94b)	q.r. 5s.	—
Skilgate (*442*, 94b)	q.r. 40s.	—
Timberscombe (*442b*, 94b), ferling added	q.r. 6s.	—
Sydenham (*443*, 94b)	q.r. 5d.	—

Values (cont.)

Halswell (*443*, 94b)	q.r. 7s. 6d.	—
Ash Priors (*443b*, 94b)	q.r. 20s.	—
Cheddon Fitzpaine (*444*, 94b)	q.r. 50s.	—
Marston Bigot (*444b*, 94b)	q.r. £6	—
Tickenham (*448b*, 98)	q.r. 60s.	—
Stoke near Chew Magna (*452b*, 97b)	q.r. 20s.	—
Ridgehill (*452b*, 98)	q.r. 20s.	—
Wheathill (*453*, 98)	valet 30s.	*Modo* 40s.
*Bossington (*463b*, 97)	q.r. 15s.	*Valuit* 20s.
Treborough (*463b*, 97)	q.r. 7s.	—
Tadwick (*464b*, 99)	q.r. 15s.	*Olim* 10s.
Knowle in Shepton Montague (*465b*, 99)	valet 40s. et q.r. £4	*Olim* 40s. *Modo* £4
Perry (*477*, 98b)	q.r. 10s.	*Olim* 11s.
Capland (*491b*, 98b), half-hide added	valet 6s. et q.r. *fuit vastata*	—
Stoke near Chew Magna (*492*, 99)	valet 25s.	—

Wood and underwood

Pitcott (*146*, 88b)	1 f *inter* l & b w	1 f w
Rode (*148*, 88b)	21 ac w + 12 ac u	33 ac w
Bishop's Lydeard (*157*, 89)	1 lg × 2 f w	1 lg × 3 f w
Banwell (*157*, 89b)	2½ lg *in* l & *in* b w	2½ lg *in* l & b w
Litton (*160*, 89b)	3 f *in* l & *in* b w	3 f *in* l & b w
Walton near Glastonbury (*163b*, 90)	7 f × 4 f w	7 f × 3 f w
Pilton (*165b*, 90)	1½ lg × ½ lg w	1 lg × ½ lg w
North Wootton, Pylle and Pilton (*166b*, 90b)	2 ac w + 2 ac u	4 ac w
Hornblotton, Alhampton and Lamyatt (*169b*, 90b)	see p. 174	
Brent (*171*, 90b), Roger de Courseulles	6 ac u	—
Butleigh (*173*, 91)	2 f × 1 f u	2 f × 1 f w
Taunton (*174b*, 87b), Godwine *et al.*	60 ac w + 1 ac u, total for 5 holdings	61 ac w
Taunton (*175*, 87b), Alured	15 ac w	20 ac w
Lydeard St Lawrence and Leigh in Lydeard St Lawrence (*175*, 87b)	15 ac w + 34 ac w & 2 ac w	49 ac w
Bathford (*185b*, 98b)	1 lg *in* l & b u	1 lg *inter* 1 & b u
Seavington (*265b*, 91b)	10 ac w	10 ac u
Donyatt (*270*, 92)	50 ac w	50 ac m

Wood and underwood (cont.)

Clapton in Cucklington (*278*, 92b)	2 f × 1 f w	—
Weston Bampfylde (*278*, 92b)	4 ac w	—
Belluton (*282b*, 91b)	3 f × 2 f w	4 f × 2 f w
Henstridge (*286b*, 91b)	3 f × 1 f w	4 f × 1 w
Bratton Seymour (*352*, 95)	6 f *in* l & *in* b w	6 f *in* l & b w
Holford St Mary (*362*, 96)	104 ac w	4 ac w
Broomfield (*363b*, 96)	1 lg *in* 1 & *in* b w	1 lg *in* l & b w
Nether Stowey (*373*, 97), Alured, 2nd entry	1 lg w	—
Eastrip (*382b*, 97b)	1 f *in* l & *in* b w	1 f l & b w
Cheriton (*386*, 96b)	1 lg 7 f × 7 f w	7 f × 7 f w
Curry Mallet (*429*, 93), Roger, 1st entry	½ lg *in* l & b w	½ lg *inter* l & b w
Curry Mallet (*429*, 93), Roger, 2nd entry	½ lg *in* l & *in* b w	½ lg *in* l & b w
Timberscombe (*442b*, 94b), ferling added	4 ac u	4 ac w
Kittisford (*443*, 94b)	12 ac u	12 ac w
Halswell (*443*, 94b)	14 ac u	14 ac w
Charlton near Somerton (*443b*, 94b)	2 ac u	2 ac w
Sutton Bingham (*444*, 94b)	3 f × 2 f w	3 f × 2 f p
Marston Bigot (*444b*, 94b)	1 f × 1 f w	—
Penselwood (*445*, 94b)	4 f × 13 pc w	4 f × 12 pc w
Lyde (*445*, 94b)	2 ac u	2 ac w
Milton Clevedon (*450*, 98)	10 f *in* l & *in* b w	10 f *in* l & b w
Ridgehill (*452b*, 98)	3 f × 3 f w	3 f *in* l & b w
Moreton (*453b*, 98)	30 ac w	15 ac w
Walton in Kilmersdon (*480*, 98b)	1 f *in* l & *in* b w	1 f *in* l & b w
West Lydford (*493*, 99)	1 lg *in* l & *in* b w	1 lg *in* l & b w

Meadow

Chard (*156*, 89)	15 ac	20 ac
Sutton Mallet, Edington, Chilton-upon-Polden and Catcott (*162*, 90)	119 ac, total for 4 places	99 ac
Brent (*171*, 90b), Roger de Courseulles	5 ac	—
Stoke Trister (*277*, 92b)	16 ac	15 ac
Weston Bampfylde (*278*, 92b)	8 ac	—
Bradney (*353b*, 95)	20 ac	25 ac
Staunton (*359*, 95b)	2 ac	3 ac
Woolston in Bicknoller (*361*, 96)	6 ac	7 ac

Meadow (cont.)

Stocklinch (*363b*, 96)	9 ac	8 ac
Perry (*422*, 93)	23 ac	33 ac

Pasture

Dowlish (*136b*, 87b)	4 f × 3 f	4 f × 4 f
Winford (*145*, 88b), Roger	4 ac	—
Pitcott (*146*, 88b)	2 f *inter* l & b	2 f
Banwell (*157*, 89b)	1 lg × 1 lg	1 lg l & b
Wedmore (*159b*, 89b)	1 lg *in* l & b	1 lg *inter* l & b
Winscombe (*161*, 90)	1 lg *in* l & *in* b	1 lg *in* l & b
una mansio, i.e. Nunney (*198*, 91)	20 ac & 20 ac	20 ac
Merriott (*271b*, 92)	½ lg *in* l & *in* b	½ lg *in* l & b
Pendomer (*275*, 92b)	4 f *inter* l & b	4 f *in* l & b
West Harptree (*354b*, 95b)	1 lg *in* l & *in* b	1 lg *in* l & b
Newton in Bicknoller (*361*, 96)	1 lg *in* l & *in* b	1 lg *in* l & b
Combe in Bruton (*383*, 97b)	1 lg *in* l & *in* b	1 lg *in* l & b
Haselbury Plucknett (*492*, 99)	½ lg *in* l & *in* b	½ lg *in* l & b
Discove (*493*, 99)	3 f *in* l & *in* b	3 f *in* l & b

Marsh

Wedmore (*159b*, 89b)	*ibi morae quae nichil reddunt*	—

Waste

Almsworthy (*430*, 94)	*q.r. erat penitus vastata*	—
Downscombe (*430*, 94)	*q.r. erat vastata*	—
Exford (*430*, 94), Roger de Courseulles	*q.r. erat penitus vastata*	—
Stone in Exford (*431b*, 94)	*semper fuit vasta postquam Rogerus recepit*	*sed vasta est*
Capland (*491b*, 98b), half-hide added	*q.r. fuit vastata*	—

Mills

*Henstridge (*107*, 87)	1 *reddit* 20d.	1 *reddens* 30d.
Woolley (*144b*, 88b)	2 *reddunt* 10s.	2 *reddentes* 2s.
Radstock (*146b*, 88b)	1 *reddit* 12s.	1 *reddens* 13s.
Rode (*148*, 88b)	see p. 191	
Chew Magna (*159*, 89b), Rohard and Stephen	2 *reddunt* 9s. 4d., total for 2 holdings	2 *reddentes* 10s.
Litton (*160*, 89b)	3 *reddunt* 10s. 10d.	3 *reddentes* 10s.
Camerton (*170*, 90b), Glastonbury Abbey	1 *reddit* 5s.	2 *reddentes* 5s.
Pitminster (*173b*, 87b)	1 *reddit* 16d.	—
una mansio, i.e. Nunney (*198*, 91)	½ *reddit* 30d.	1 *reddens* 30d.

Mills (cont.)

Stoke-sub-Hamdon (*267b*, 92)	2 *reddunt 9s. 8d.*	2 *reddentes 9s.*
Chilton Trivett (*425*, 93 b)	1 *reddit 20s.*	½ *reddens 20s.*
Huish in Nettlecombe (*464*, 96 b)	1 *reddit 4s.*	1 *reddens 3s.*

Churches

Kilmersdon (*198b*, 91 b)	*In ecclesia Chinemeredonae*	*In Chenemeresdone*
Horsey Pignes (*477*, 98 b)	*sacerdos de ecclesia istius villae*	*Ibi presbyter*

Urban life

Bruton (*107b*, 87)	*De Briuetona x solidi*	*De Bravetone xx solidi*
Keynsham (*113b*, 87)	*vii burgenses in burgo Badae*	*viii burgenses in Bade*
Backwell (*143*, 88)	*i borgisum qui manet ad Badam*	—
Goathill (*278b*, 92 b)	*ii masurae in Meleborna*	—
Milborne Port (*278b*, 93)	*v burgenses in Meleborna qui reddunt iii solidos et ix denarios*	*v burgenses reddunt iii solidos*
Oake (*433*, 94)	*in burgo Milvertonae*	*In Milvertone*
Hinton Charterhouse (*437*, 98)	*i domus in Bada quae reddit per annum vii denarios et i obolum et alia mansura in eodem burgo vacua*	*In Bade ii domus; una reddit vii denarios et obolum*
Yeovil (*439*, 96 b)	*xxii mansurae terrae*	*xxii masurae*

Miscellaneous

Over Stratton (*88b*, 86)	*reddit firmam in predicta mansione* (South Petherton) *scilicet lx solidos et xxiiii oves*	*Reddit modo lx solidos in firma regis*
Bruton (*91*, 86 b), 9 acres taken away	held in Redlynch	—
una mansio (*116*, 87)	separate from Pitney	with Pitney
Newton St Loe (*149*, 89)	9 thegns *T.R.E.*	2 thegns *T.R.E.*
Cricket St Thomas (*265*, 91 b)	*reddebat pro consuetudinem unoquoque anno in Sutpetret mansione regis vi oves cum agnis suis, et uniusquisque liber homo i blomam ferri*	—(but in both texts for South Petherton, *265*, 91 b)
Swell (*268*, 92)	*iacebat i virgata in Curi mansione regis . . . Et reddebat x solidos et viii denarios in firma regis*	—
Keinton Mandeville (*276b*, 92 b)	3 thegns *T.R.E.*	2 thegns *T.R.E.*
Alford (*277b*, 92 b)	*i de supradictis villanis reddit viii blumas ferri*	*de villanis viii blomas ferri*

DEVONSHIRE

Settlements

Boode (*126b*, 102b)	*i virgata terrae quae vocatur Boeurda*	*una virgata terrae* (*T.R.E.* in Braunton)
Kersford, Battishill, Combebow, Ebsworthy, Fernworthy and Way (*289*, 105b)	named places held with Bridestowe	place-names omitted
Warne, Burntown and Wringworthy (*318*, 108b)	named places held with Marytavy	place-names omitted
Shyttenbrook (*459*)	*Floherus habet i mansionem quae vocatur Sotrebroc quam tenuit Aluiet...Et reddidit gildum pro dimidia virgata quam possunt arare iiii boves et valet per annum ii solidos*	—

Assessment

Woodbury church (*96b*, 100b)	½ h 1 vg ½ fg	1 h 1 vg ½ fg
Sidbury (*118b*, 102)†	3 h	5 h
Hatherleigh near Okehampton (*178*, 103b), Walter	½ h ½ vg	3 vg
Burrington near Chulmleigh (*179*, 103b), manors added	2½ vg	½ h
Littleham near Exmouth (*184*, 104)	½ h in dem	1 vg in dem
Bolham in Clayhidon (*306*, 107b)	2½ h.	2 h
Dunkerswell (*338b*, 114)	1½ h 1 vg	1½ h
Bampton (*345b*, 111b)	1 h (details add to 3½ vg 1 fg)	1 h
Greenslinch (*469b*, 117)	2 vg (details add to 3 vg)	3 vg
Northwick (*481*, 118)	½ vg	1 vg

Manorial demesne expressed as carucates, not hides

North Tawton (*83*, 100)	3 carucates	—
Braunton (*83b*, 100)	1 carucate	—
Chittlehampton (*484*, 118)	5 carucates	—

Plough-lands

Exminster (*83*, 100), Battle Abbey	½	—
Axminster church (*84b*, 100b)	2	—

† See footnote on p. 414 below.

Plough-lands (*cont.*)

Sidbury (*118b*, 102)†	20	30
Coombe in Templeton (*133b*, 103)	1	—
Leigh near Meshaw (*179b*, 103b)	10	1
*Shobrooke (*215b*, 104b)	8	4
*Modbury (*217b*, 105b)	24	23
*Bolham in Clayhidon (*306*, 107b)	7	6
Liclemora (*319*, 109)	2½	2
Poulston (*323*, 109)	1½	1
Hooe (*334*, 110)	2½	2
Bampton (*345b*, 111b)	see pp. 239–40	
Hembury near Plymtree (*378*, 116b)	18	14
*Shapcombe (*378*, 116b)	2	3
*Dowland (*390*, 112b), Walter, 2nd entry	2	3
Stafford in Dolton (*456b*, 116)	1½	1
*Bulworthy (*487b*, 118)	2	3

Demesne teams

Sherford near Kingsbridge (*97b*, 100b)	1	—
Littleham near Exmouth (*184*, 104)	1	—
Bulkworthy (*211*, 104b)	7 ox	1
Feniton (*214b*, 105)	½	—
Newton Ferrers (*218b*, 105)	4 ox	½
Brixton in Broadwood Kelly (*292b*, 106)	1½	1
Dolton (*295*, 106b)	6	2
*Bolham in Clayhidon (*306*, 107b)	3	2
Culm Pyne (*306*, 107b)	2½	2
Creacombe near Witheridge (*310b*, 108)	½	—
Aller in Upottery (*313b*, 108b)	½	—
Tetcott (*319*, 108b)	1 t, 2 ox	1
Weston Peverel (*328*, 109b)	2½	2
Staddiscombe (*330b*, 109b)	½	1
Berry Pomeroy (*342*, 114b)	4	—

† This and other Sidbury discrepancies, as under *Assessment* and *Villagers' teams*, resulted from changes made in the copy of the Exeter Domesday used by the Exchequer clerk which were not added to the first draft of the Exeter Domesday.

Demesne teams (cont.)

Worlington near Witheridge (*379*, 116b)	½	—
Dodscott (*388b*, 112b)	1	—
Ash Thomas (*394b*, 112b)	1	—
Buckland-Tout-Saints (*396b*, 112)	1½	—
Pool (*396b*, 112)	1½	1
Marland (*407b*, 115b)	1	—
Sigford (*414b*, 115)	4 ox	½
Milton Damerel (*419*, 113)	2	1
Sutton in Halberton (*461*, 116), Aiulf	½	—
Burn in Silverton (*469b*, 117)	½	—
Ogwell (*470b*, 117)	2	—
Chittlehampton (*484*, 118)	3	5

Villagers' teams

Chillington (*97*, 100b)	36	—
Sidbury (*118b*, 102)†	18	25
Milford in Stowford (*121b*, 103)	3 ox	—
Thorne (*122*, 103)	1 ox	—
Hele in Ilfracombe (*128*, 102b)	2 ox	—
Spurway (*133*, 103)	2 ox	—
Colston (*133*, 103)	2 ox	—
Loxbeare (*134b*, 103)	1	—
*Peadhill (*134b*, 103)	½	—
Tavistock (*177*, 103b), Ermenald *et al.*	3 t, 9 ox, total for 5 holdings	4
Leigh near Meshaw (*179b*, 103b), Nigel	3	5
*Plymstock (*180*, 103b)	3	4
Yarcombe (*195b*, 104)	10	11
Shobrooke (*215b*, 104b)	4	8
Venn (*218b*, 105b)	1	—
Chichacott (*288*, 105b)	2 ox	—
Germansweek (*289*, 106)	1 t, 2 ox	1
Potheridge (*293b*, 106)	2	— (2 bordars in error for 2 teams)
Hele in Meeth (*294b*, 106b)	2	3
Mamhead (*297b*, 106b)	4	5
Sparkwell (*313*, 108)	1	—
Ashleigh (*317*, 108b)	3 t, 1 ox	3
Loventor (*320b*, 109)	*ii animalia in carrucam*	—
West Portlemouth (*322b*, 109)	*ii animalia in carrucam*	—

† See foot-note on p. 414 above.

Villagers' teams (cont.)

Ilton (*322b*, 109)	*vi animalia in carrucam*	1
Sorley (*323*, 109)	2 t *minus* 2 ox	2
Coleridge in Egg Buckland (*329*, 109b)	*i animal in carrucam*	—
Baccamoor (*331*, 110)	2	1
Fernhill (*332*, 110)	1 ox	—
Rapshays (*Oteri*, *340*, 114)	3 ox	½
Bampton (*345b*, 111b)	see pp. 239–40	
Dipford (*346*, 111b)	3	6
Densham (*367*, 111)	1	—
Dartington (*368b*, 111)	8½	8
Stowford near Lifton (*376*, 116b)	2 t, 3 ox	2
Dodscott (*388b*, 112b)	½	—
Ash Thomas (*394b*, 112b)	½	—
Whimple (*402b*, 110b)	1½	1
Combeinteignhead (*403*, 110b)	2	1
Kimber (*412*, 115)	1 t, 3 ox	1
*Weare Gifford (*412b*, 115)	5	5½
Goosewell (*417*, 111b)	2 ox	—
Bickham (*420b*, 113)	4 ox	½
Clyst in Cliston hd. (*457b*, 116b)	3	6
Cheldon (*459b*, 116)	2	—
Sutton in Halberton (*461*, 116), Aiulf	1	—
Galmpton in Churston Ferrers (*462*, 113b)	5 t, 4 ox	5½
Ogwell (*470b*, 117)	2	—
Staplehill (*472*, 117b)	3 ox	½
Twinyeo (*488b*, 118b)	3 ox	½

Demesne and villagers' teams

Wedfield (*211b*, 105)	1 dem t + 1 vill t, 2 vill ox	2
Rockbeare (*216*, 104b)	1 dem t + 1 vill ox	1
Hareston (*221b*, 105)	½ dem t + 4 vill ox	1
Appledore in Clannaborough (*295b*, 106b)	5 dem ox + 1 vill t	1½
Whitestone in Chittlehampton (*300b*, 107)	6 dem ox + 2 vill ox	1
Tilleslow (*317b*, 108b)	1 dem t + 3 vill ox	1½
Lupton (*320b*, 109)	1 dem t + 6 vill *animalia in carrucam*	2
South Allington (*324b*, 109)	1 dem t + 1 vill t, 2 vill ox	2
Butterford (*326b*, 109b), Torgis, 1st entry	6 dem ox + 1½ vill t	2

Demesne and villagers' teams (cont.)

Blachford (*327*, 109b)	6 dem ox + 2 vill t	3
Lower Credy (*340*, 114b)	1 dem t + 1 vill t, 7 vill ox	2
Coombe in Uplowman (*393b*, 112)	1 dem t + 2½ vill t	3

Population

Exminster (*83*, 100)	12 cot	—
South Molton (*83b*, 100)	12 sw	—
Shebbear (*94*, 101)	6 sw	—
Werrington (*98*, 101)	116 v	16 v
High Bickington (*110*, 101b)	2 sw	—
Halberton (*110b*, 101b)	5 sw who render 30 swine	—
Bury (*117*, 101b)	3 v	4 v
Hartleigh (*123b*, 102)	3 sw who render 11 swine	—
Buckland near North Molton (*129b*, 102b), Drogo, 1st entry	1 s	1 v, 2 s
Creely (*132b*, 103)	2 v	1 v
Hatherleigh near Okehampton (*178*, 103b), Walter	2 bo	—
Northam (*194*, 104)	1 sw	—
Culleigh (*210b*, 104b)	3 bo	4 bo
Buckland Brewer (*210b*, 104b)	3 sw	—
Dinnaton (*219*, 105b)	3 bo	4 bo
Okehampton (*288*, 105b)	31 v, 6 sw	21 v, —
Dunsland (*290b*, 106)	3 bo	4 bo
Highampton (*291*, 106)	1 sw	—
Inwardleigh (*291b*, 106)	1 sw	—
Potheridge (*293b*, 106)	5 bo	5 bo *cum* 2 bo (an error for 2 teams)
Chawleigh (*294b*, 106b)	3 sw who render 15 swine	—
Dolton (*295*, 106b)	7 s	—
Nymet Rowland (*295*, 106b)	2 sw who render 10 swine	—
Rashleigh (*295*, 106b)	2 sw who render 15 swine	—
Shapley in Chagford (*297b*, 106b), Robert	1 s	—
Holcombe Rogus (*299b*, 107)	6 bo	5 bo
Anstey (*300*, 107)	1 sw who renders 6 swine	—
Filleigh (*300b*, 107)	3 sw who render 15 swine	—
Lobb (*300b*, 107)	4 v	—
Chulmleigh (*309b*, 108)	5 sw who render 30 swine	—
Clawton (*318b*, 108b)	2 sw	—
Tetcott (*319*, 108b)	1 sw	—

Population (cont.)

Tetcott (*319*, 108b), ferling added	1 v, 1 sw	2 v
Henford (*319b*, 109)	8 bo	3 bo
Bridford (*319b*, 109)	1 sw who renders 10 swine	—
Brixham (*321*, 109)	5 cot	—
Charleton (*324*, 109)	12 s	11 s
Chivelstone (*325b*, 109b)	5 bo	3 bo
Membland (*327b*, 109b)	4 bo	3 bo
*Wollaton (*330*, 109b)	3 v	4 v
Bowlish (*336b*, 114)	2 s	2 bo
*Ogwell (*339*, 114), William, 1st entry	6 v, 6 bo	3 v, 3 bo
Stoodleigh in West Buckland (*345b*, 111b)	2 v	—
Ottery in Axminster hd. (*348*, 112)	1 sw who renders 10 swine	—
Dartington (*368b*, 111)	2 fishermen who render 80 salmon	—
Little Weare (*376b*, 116b)	2 bo	3 bo
Hembury near Plymtree (*378*, 116b)	6 s	—
Loosedon (*390*, 112b)	3 bo	2 bo
*Burlescombe (*391*, 112b)	5 bo (4 interlined)	4 bo
Withycombe Raleigh (*392b*, 112)	8 v	3 v
*Drayford (*393*, 112)	2 bo	3 bo
Pirzwell (*400*, 110b)	7 v	8 v
Lyn (*402*, 110b)	2 sw who render 20 swine	—
Whimple (*402b*, 110b)	3 bo	—
*Yarnscombe (*407*, 115)	3 v	8 v
Lamerton (*411*, 114b)	5 sw	—
Twigbear (*413*, 115)	3 s	4 s
Lobb (*414*, 115)	3 v	4 v
Stoke Rivers (*415*, 111)	3 bo	2 bo
Milton Damerel (*419*, 113)	11 bo, 7 bo	11 bo, 7 s
Westleigh (*420*, 113)	4 s	3 s
*Bramblecombe (*456b*, 116)	2 bo	3 bo
Ingsdon (*460*, 117)	4 bo	3 bo
Eveleigh (*469b*, 117b)	2 v	—
Rocombe (*470b*, 117)	2 v	—
*Huish in Tedburn St Mary (*471*, 117b)	4 bo	3 bo
Leigh near Colyton (*473*, 117b)	1 bo	—
*Bulworthy (*487b*, 118)	2 s	3 s
*Warbrightsleigh (*488*, 118b)	2 bo	3 bo

Rents of swineherds

Dolton (*295*, 106b)   2 sw who render 10 swine   2 sw

Values

Exminster (*83*, 100)	q.r. £8	—
Braunton (*83b*, 100)	q.r. £16 *ad pensum*	—
Hemyock (*84*, 100)	q.r. £6 *ad pensum*	—
East Budleigh (*84*, 100)	q.r. £10 *ad pensum*	—
Kingsteignton (*84*, 100)	q.r. £10 *ad pensum*	—
Axminster (*84b*, 100)	q.r. £26 *ad pondus et arsuram*	—
Kingskerswell (*85*, 100b)	q.r. £14 *ad pensum*	—
Diptford (*85b*, 100b)	q.r. 100s. *ad pensum*	—
West Alvington (*86*, 100b)	q.r. £7. 5s. 0d. *ad pensum*	—
Plympton St Mary (*86*, 100b)	q.r. £12. 10s. 0d. *ad pensum*	—
Yealmpton (*86b*, 100b)	q.r. £12. 10s. 0d. *ad pensum*	—
Walkhampton, Sutton in Plymouth, King's Tamerton (*87*, 100b)	*reddebant firmam hunius* (sic) *noctis cum suis appendiciis*	—
Bradstone (*93*, 101)	q.r. 60s. *ad pensum*	—
South Tawton (*93*, 100b)	q.r. £48 *ad pensum*	—
Black Torrington (*93b*, 101)	q.r. £18 *ad pensum*	—
Holsworthy (*93b*, 101)	q.r. £12 *ad pensum*	—
Hartland (*93b*, 100b)	q.r. £23 *ad pensum*	—
Tawstock (*94b*, 101)	q.r. £24 *ad pensum*	—
Witheridge (*96*, 100b)	q.r. £6	—
Moretonhampstead (*96*, 101)	q.r. £12 *ad pondus et arsuram*	—
Colaton Raleigh (*96b*, 101)	q.r. £8 *ad pensum*	—
Hempston (*96b*, 101)	q.r. 40s. *ad pensum*	—
Spitchwick (*97*, 101)	q.r. 60s. *ad pensum*	—
Whitford (*97*, 101)	q.r. £11 *ad pondus et combustionem*	—
Chillington (*97*, 100b)	q.r. £18	—
Langford in Ugborough (*97b*, 101)	q.r. £11 *ad pondus*	—
Kimworthy (*122*, 103)	q.r. 5s.	—
Worlington in Instow (*124b*, 102)	q.r. 25s.	*Olim* 30s.
Martinhoe (*125*, 102b)	q.r. 5s.	—
Buckland near North Molton (*129b*, 102b)	q.r. 3s.	—
Lower Credy (*132b*, 103)	q.r. 12s. 6d.	—
Coombe in Templeton (*133*, 103)	q.r. 10s.	—
Charton (*135*, 103b)	q.r. 10s. *de firma*	—

Values (cont.)

Thornbury near Holsworthy (*178b*, 103b)	*valet 60s. et q.r. valebat minus*	*Valet £3*
Leigh near Meshaw (*179b*, 103b), Robert	*q.r.* 12s.	—
Heathfield in Aveton Gifford (*182b*, 104)	*valet 40s. et q.r. 30s.*	—
Beer (*184*, 104)	*valet £4*	*Valet 60s.*
Bratton Fleming (*213*, 105), manors added	*valent 21s. 8d.*	*Valent 22s.*
Leigh in Silverton (*214*, 104b)	*q.r.* 5s.	—
Rockbeare (*216*, 104b)	*q.r.* 12d.	*Olim* 12s.
Leigh near Colyton (*217*, 104b)	*valet 20s.*	*valet 30s.*
Modbury (*221*, 105)	*q.r.* 2s.	—
Honeychurch (*292*, 106)	*q.r.* 15s.	—
Woolleigh (*294b*, 106b)	*valet 15s.*	*valet 20s.*
West Clyst (*309*, 107)	*q.r.* 15s.	*Olim* 20s.
Ford in Musbury (*313b*, 108)	*q.r.* 25s.	—
Raddon (*316*, 108b)	*valet 100s. (et 5 interlined)*	*valet 100s.*
Galmpton in South Huish (*322*, 109)	*valet 30s.*	*valet 50s.*
Tale (*338*, 114), Ralph, 1st entry	*q.r.* 10s.	—.
Ogwell (*339*, 114), William, 1st entry	*valet 10s.*	*valet 30s.*
Dunscombe (*340*, 114b)	*valet 20s. et q.r. tantundem*	*Olim et modo valet 30s.*
Lower Credy (*340b*, 114b)	*q.r.* 2s.	—
Dunstone (*342*, 114b)	*q.r.* 30d.	—
Luscombe (*368b*, 111)	*valet 15s.*	*valet 5s.*
Great Torrington (*376b*, 116b)	*valet £15*	*valet £20*
Villavin (*388*, 112b)	*q.r.* 10s.	*Olim* 100s.
Boehill (*395*, 112), Walter, 2nd entry	*reddit* 10s.	—
Waringstone (*Oteri*, *400b*, 110b)†	*q.r.* 12s.	—
Weare Gifford (*412b*, 115), count of Mortain	*q.r.* 5s.	—
Widey (*421b*, 113)	*q.r.* 30 (? s.)	—
Abbots Bickington (*456*, 117)	*reddit 20s. et q.r. tantundem*	—
Buckland in the Moor (*472b*, 117b)	*q.r.* 7s.	—

† The folios of the Exeter Domesday were misplaced in the binding of 1816, so that the entry begins on *400b* and continues on *403*.

Values (cont.)

Colscott (*481*, 118)	*valet* 10s.	*Valet* 5s.
St James's Church (*487*, 118b)	*valet* 40d.	—
Twinyeo (*488b*, 118b)	*valet* 30d.	*Valet* 20d.

Wood and underwood

Exminster (*83*, 100)	1 lg *in* l & b w	—
South Tawton (*93*, 100b)	2 lg × 2 f w	2 lg × 2 w
Holsworthy (*93b*, 101)	1 f w	1 ac w
North Molton (*94b*, 100b)	1 lg *in* l & *in* b u	1 lg *in* l & b w
King's Nympton (*98*, 101)	1 lg *in* l & b w	—
Tiverton (*98*, 100b)	see p. 258	
*Ide (*117b*, 101b)	4 ac u	3 ac u
Worlington near Witheridge (*134*, 103)	3 ac w	—
Loxbeare (*134b*, 103)	2 ac u	—
Burrington near Chulmleigh (*179*, 103b)	60 ac w & u	60 ac w
Leigh near Meshaw (*179b*, 103b), Nigel and Robert	39 ac w, total for 2 holdings	38 ac w
Down St Mary (*182*, 103b)	8 f u	7 f u
South Brent (*183b*, 104), Buckfast Abbey, 1st entry	5 ac w	4 ac w
Stockleigh in Meeth (*212*, 105)	2 ac w	—
Curscombe (*214b*, 105b)	2 ac u	2 ac w
Rockbeare (*216*, 104b)	4 ac w	3 ac w
Torridge (*221*, 105)	5 ac u	5 ac w
Germansweek (*289*, 106)	20 ac u	20 ac w
Dunterton (*289b*, 106)	160 ac w	60 ac w
Stockleigh in Meeth (*294*, 106)	2 ac w	2 ac u
Hele in Meeth (*294b*, 106b)	3 ac w	3 ac u
*Kingsford (*303*, 107)	4 ac w	4 (? or 8) ac w
Wolborough (*313*, 108)	30 ac u	30 ac w
Musbury (*313*, 108)	40 ac w	—
Clawton (*318b*, 108b)	1 lg *in* l & b w	1 lg w
Bridford (*319b*, 109)	1 lg × 1 f w & u	1 lg × 1 f w
West Portlemouth (*322b*, 109)	3 f × 1 f w	—
Leigh in Harberton (*324*, 109)	1 ac w	—
Butterford (*326b*, 109b), Turgis, 2nd entry	2 ac u	—
Coleridge in Egg Buckland (*329*, 109b)	30 ac w	30 ac u
Meavy (*329*, 109b)	½ lg u	½ ac u

Wood and underwood (cont.)

Ogwell (*339*, 114), William, 1st entry	12 ac u	—
Bampton (*345b*, 111b), manors added	2 ac w	—
Knowstone (*346b*, 111b), Rolf, 1st entry	5 f × 1 f w	—
Dunsford (*347*, 111b)	10 f w	10 ac w
Holne (*367b*, 111)	1 lg × 1 lg w	1 lg w
Fenacre (*391b*, 112b)	6 ac u	—
Murley (*393b*, 112)	9 ac w	8 ac w
Rifton (*410*, 115b)	5 ac w	—
Widworthy (*410*, 115b)	155 ac w	160 ac w
Ottery in Lamerton, Collacombe and Willestrew (*419*, 113)	4 f & 20 ac × 2 w	20 ac × 2 w
Thuborough (*419b*, 113)	2 ac u	2 ac w
Flete Damerel (*421*, 113)	10 ac u	10 ac w
*Widey (*421b*, 113)	½ lg × 3 f w	½ lg × 4 (? or 8) f w
Powderham (*457*, 111b)	1 lg w & 4 ac u	4 ac u
Chagford (*459*, 113b)	6 ac w	—
Ipplepen (*462*, 113b)	½ lg *in* l & b u	½ lg u
Brushford (*468*, 117)	6 ac w	—
Whitnole (*471b*, 117)	4 ac u	—
Staplehill (*472*, 117b)	6 ac w	—
Middlecott in Chagford (*483b*, 118b)	3 ac u	2 ac u
Bulworthy (*487b*, 118)	(damaged) u	1 ac w
Rousdon (*489*, 118)	10 ac u	—

Meadow

Modbury (*217b*, 105b)	1 (sic)	1 ac
Ludbrook (*219b*, 104b)	½ ac	2 ac
Pyworthy (*318b*, 108b)	1 lg *in* l & b	1 lg
Galmpton in South Huish (*322*, 109)	2 ac	—
Lambside (*327b*, 109b)	2 ac	3 ac
Georgeham (*408*, 115b)	1 ac	—
Chagford (*459*, 113b)	8 ac	—
Brushford (*468*, 117)	6 ac	—
Leigh near Colyton (*473*, 117b)	1½ ac	2 ac

Pasture

Wonford in Heavitree (*95b*, 100b)	½ lg × ½ lg	½ lg *in* l & b
Woodbury church (*96b*, 100b)	*communis pascua*	—
King's Nympton (*98*, 101)	see p. 265	

Pasture (cont.)

Rowley (*131*, 103)	30 ac	20 ac
Littleham near Exmouth (*184*, 104)	6 f *inter* l & b	6 f *in* l & b
Hele in Meeth (*294b*, 106b)	6 ac	—
Anstey (*300*, 107)	1 lg *in* l & *in* b	1 lg *in* l & b
*Prawle (*314b*, 108b)	63 ac	64 ac
Clawton (*318b*, 108b)	1 lg *in* l & b	1 lg
Pyworthy (*318b*, 108b)	1 lg *in* l & b	1 lg
Henford (*319b*, 109)	½ lg *in* l & b	½ lg
Meavy (*329b*, 109b), Turgis, 2nd entry	½ lg *in* l & b	½ lg
Wollaton (*330*, 109b)	2 f *in* l & b	2 f
Southbrook (*347*, 111b)	10 f *inter* l & b	10 f
Holne (*367b*, 111)	1 lg *inter* l & b	1 lg
Arlington (*371*, 115b)	½ lg *in* l & b	½ lg
Villavin (*388*, 112b)	½ lg *in* l & b	½ lg
Lupridge (*397*, 112b)	1 f *in* l & b	1 f
Lynton and Ilkerton (*402*, 110b)	2 lg × ½ lg	2 f × ½ lg
Georgeham (*408*, 115b)	20 ac	—
Wonford in Thornbury (*411b*, 114b)	2 f *in* l & b	2 f
Chagford (*459*, 113b)	4 ac	—
Mullacott (*469*, 117)	*alia terra iacet vastata ad pasturam*	—
Leigh near Meshaw (*475b*, 117b)	30 ac	25 ac

Fishery

Woodleigh (*421*, 113)	*i piscatoria*	—

Fishery and salt-pan

Buckland Monachorum (*417b*, 111b)	*i piscatio quae valet per annum x solidos et i salina*	*Ibi salina et piscaria reddens x solidos*

Salt-pan

Lobb (*414*, 115)	1	—

Waste

Buscombe (*127*, 102b)	*q.r. erat vastata*	—
Coombe in Templeton (*133b*, 103)	*est vacua*	—
Culbeer (*314b*, 108b)	*Ibi habet...nichilum*	—
after entry for Soar (*323*, 109)	*Hae ix praedictae mansiones sunt vastatae per Irlandinos homines*	—
Lower Credy (*340b*, 114b)	*alia terra est ita vastata quod non valet nisi ii solidos et q.r. tantundem*	*Valet ii solidos*

Waste (cont.)

Lupridge (*397*, 112b)	*q.r. erat vastata*	—
Badgworthy (*402b*, 110b)	*q.r. erat vastata*	—
Higher Molland (*409*, 115b)	*Alia medietas virgatae est*	—
	tota vastata	
Mullacott (*469*, 117)	*alia terra iacet vastata ad*	—
	pasturam	

Mills

Bovey Tracey (*135*, 102)	1 *reddens* 5s.	1 *reddens* 10s.
Honiton (*216b*, 104b)	1 *reddit* 7s. 6d.	1 *reddens* 6s. 6d.
Nymet Rowland (*295*, 106b)	1 *reddit* 20d.	—

Urban life

*Exeter (*120b*, 101b)	*xlviii domus...x ex illis*	*xlvii domus reddentes*
	reddunt x solidos et x	*x solidos et x denarios*
	denarios	

Miscellaneous

Otterton (*194b*, 104)	*ibi est mercatum in*	—
	dominica die	

CORNWALL

General

Winnianton manors (*99–100b*, *225b–8*; 120)	see pp. 309–10

Settlement

St Tudy (*204b*)	*Comes de Moritonio habet*	—
	i mansionem quae vocatur	
	Ecglostudic quae erat	
	Sancti Petrochi et reddebat	
	pro consuetudinem xxx	
	denarios in ecclesia Sancti	
	Petrochi	

Assessment

Trethake (*235*, 123)	½ h	—
Torwell and Tregarland (*235b*, 122)	1 h	—
Penpoll (*257b*, 122)	1 ac	—

Plough-lands

Treveniel (*244b*, 124)	2	1

Demesne teams

St Buryan (*207*, 121)	4 ox	½
Trescowe (*225*, 124)	1 ox	—
Pencarrow (*233*, 122b)	3 ox	—

Demesne teams (cont.)

Tremar (*234*, 124b)	1 ox	—
Trelowia (*234b*, 123)	4 ox	½
Dawna (*235*, 123)	4 ox	½
Killigorrick (*236b*, 122)	2 ox	—
Pengelly (*245*, 124b)	2 ox	—
Caradon (*245*, 123)	3 ox	—
Arrallas (*249b*, 123)	3 ox	½
Trenant in Fowey (*252b*, 124)	7 ox	1
Bissick (*254*, 124b)	2 ox	—

Villagers' teams

Tolcarne in North Hill (*181b*, 121b)	2 ox	—
Bossiney (*203b*, 121)	2 ox	—
St Neot (*207*, 121)	1 ox	—
Bodbrane (*229*, 124)	2 ox	—
Thorne (*233*, 123)	2 ox	—
Westcott (*239b*, 123b)	2 ox	—
Treslay (*240*, 123b)	3 ox	½
Dizzard (*243*, 124b)	1 ox	—
Hele (*243*, 124b)	3 ox	½
Trevallack (*245*, 124b)	2 ox	—
Caradon (*245*, 123)	3 ox	—
Brannel (*249*, 121b)	6	20
Trevellion (*252*, 124)	2 ox	—
Ashton (*258b*, 122)	3 ox	½

Demesne and villagers' teams

St Stephens by Launceston (*206b*, 120b)	3 dem t + 6 vill t	3
Trenance in St Keverne (*224*, 124b)	7 dem ox + 1 vill t	2
Trewince (*224*, 124b)	½ dem t + 1 vill t, 2 vill ox	2
Crawle (*224b*, 123)	7 dem ox + 6 vill ox	2
Manely (*229b*, 124)	6 dem ox + 2 vill t	3
Lanwarnick (*231*, 124b)	2 dem ox + 2 vill ox	½
South Boduel (*231*, 125)	2 dem ox + 2 vill ox	½
Draynes (*231*, 124b)	1 dem ox + 4 vill ox	½
Havet (*234*, 124)	3 dem ox + 3 vill ox	1
Trewolland (*234b*, 123)	½ dem t + 2 vill ox	1
Trethake (*235*, 123)	1 dem t + 3 vill ox	1½
Tresparrett (*238*, 123b)	1 dem t + 3 vill ox	1½
St Juliot (*238b*, 122b)	2 dem ox + 2 vill ox	½
Hornacott (*239b*, 123b)	1 dem t + 3 vill ox	1½
Alvacott (*239b*, 123b)	1 dem t + 3 vill ox	1½
Lamellen (*240b*, 123b)	1 dem t + 3 vill ox	1½
Worthyvale (*241*, 123)	2½ dem t + 4 vill t	6

Demesne and villagers' teams (cont.)

Trenuth (*241*, 123b)	1½ dem t + 1½ vill t	2½
Lancarffe (*241b*, 123b)	1 dem t + 7 vill ox	2
Lesnewth (*242*, 124b)	½ dem t + 3 vill ox	1
Treluggan in Landrake (*244b*, 124)	½ dem t + 3 vill ox	1
Treveniel (*244b*, 124)	6 dem ox + 2 vill ox	1
Goodern (*248*, 122b)	1 dem t + 3 vill ox	1½
Burthy (*248b*, 122b)	1 dem t + 4 vill ox	1½
Lanescot (*248b*, 122b)	1 dem t + 6 vill ox	2
Lantyan (*252*, 124)	4 dem ox + 1 vill t	1½
Pentewan (*253b*, 124b)	½ dem t + 1 vill t, 2 vill ox	2
Perranuthnoe (*255*, 124b)	5 dem ox + 1 vill t	1½
Tremoane (*257*, 122)	4 dem ox + 2 vill ox	1
Landulph (*260*, 122b)	6 dem ox + ½ vill t	1
Trefreock (*263*, 125)	1 dem t + 2 vill ox	1
Tregrill (*264*, 124)	2 dem ox + 2 vill t	2

Population

Milton (*232*, 123)	13 v	14 v
*Trenance in St Austell (*245b*, 125)	6 (? or 1) bo	1 bo
*Week St Mary (*259*, 122b)	3 (? or 6) v	6 v
*Delabole (*261*, 125)	6 (? or 3) bo	3 bo

Values

Rinsey (*99*, 120)	q.r. 40s.	—
Rinsey (*100*, 120)	q.r. 12s.	—
Truthwall in Ludgvan (*208b*, 120b)	q.r. 20s.	—
Trevelyan (*229b*, 122)	valet 3s. et q.r. tantundem	Olim et modo valet 2s.
Botelet (*230b*, 124)	valet 20s.	valet 50s.
Pengold (*238b*, 125)	reddit 10s. et q.r. tantundem	Olim et modo valet 5s.
Rosebenault (*241*, 123b)	q.r. 20s.	—
Treluggan in Landrake (*244b*, 124)	q.r. 20s.	Olim 10s.
Caradon (*245*, 123)	valet 5s.	valet 10s.
Treverres (*254*, 124b)	valet 3s.	valet 5s.
Rinsey (*262b*, 124b)	q.r. 7s.	—

Wood and underwood

Liskeard (*228*, 121b)	300 ac w	400 ac w
Thurlibeer (*233b*, 124)	8 ac w	8 ac u
Froxton (*334b*, 125)	40 ac w	40 ac u

Pasture

St Stephens by Launceston (*206b*, 120b)	3 lg × 2 lg	3 lg
Trecan (*229*, 122)	5 ac	10 ac
Otterham (*259b*, 122b)	1 lg *in* l & *in* b	1 lg l & b

Miscellaneous

Coswarth (*111b*, 120b)	*De hac mansione habebat Santus Petrochus pro consuetudinem xxx denarios et i bovem et vii oves...sed modo ablata est inde*	*De hoc manerio habebat Santus Petrochus T.R.E. pro consuetudinem xxx denarios aut i bovem*
Methleigh (*199*, 120b)	*annuale forum*	*forum*

MATERIAL NOT APPEARING
IN EXETER TEXT BUT INCLUDED IN
EXCHEQUER VERSION

DORSET

Settlement

Seaborough (*154 bis*, 87b *bis*)	*Seveberga...Seveberga*	*Seveberge...alia Seveberge*

Plough-lands

Long Bredy (*37b*, 78)	—	*Terra est ix carucis*†
Nettlecombe (*38*, 78)	—	*Terra est* [ ]

Meadow

Winterborne, Eastern Group (*47b*, 82)	*ii agri*	*ii acrae prati*

Pasture

Spettisbury (*47b*, 82)	*in alio loco* 2½ f × 1½ f *pascuae*	*alio loco super aquam pastura* 2½ f × 1½ f

Urban life

Shaftesbury (*11*, 75)	—	*Ibi habet abbatissa cl et unum burgenses et xx mansuras vacuas et i hortum. Valent lxv solidos*

† The *ix* is probably postscriptal, and obtained by adding together the 3 + 5 + 1 teams of the Exeter text, which has no plough-land formula.

SOMERSET

Values

Woolmersdon (*371 b*, 97), 1½ virgates added	*dimidia virgata valet per annum v solidos et virgata tantundem*	*una virgata terrae et dimidia . . . Valet et valuit x solidos*

Mint

Taunton (*174*, 87b)	—	*de moneta l solidos*

Miscellaneous

Charlton in Shepton Mallet (*167b*, 90b)	*Cerletona . . . una mansio*	*in Cerletone et alibi*

DEVONSHIRE

Settlement

Dinnaton (*85b*, 100b)	*Dunitona . . . Dunitona*	*Dunitone et altera Dunitone*

Plough-lands

Warcombe (*129*, 102b)	—	*Terra est ii carucis†*

Demesne teams

Up Exe (*132*, 103)	—	*In dominio est i caruca‡*

CORNWALL

Assessment

St Enoder (*203*, 121)	*In ea est i hida terrae*	*Ibi est una hida quae nunquam geldavit*
Crantock (*206*, 121)	*In ea sunt iii hidae ii agri minus*	*Ibi sunt iii hidae ii acras minus quae nunquam geldaverunt*

Teams

St Keverne (*205b*, 121)	—	*Ibi est i caruca§*

† Quite possibly arrived at by adding together the one demesne team and the one villagers' team in the Exeter text.

‡ The Exeter text gives only a villagers' team, which was added postscriptally to Exchequer version: the clerk may accordingly have inscribed a demesne team in error, and omitted to delete it.

§ The Exeter text records 8 *animalia* and 30 *oves* as the livestock of the holding. Could the 8 *animalia* refer to the team of the Exchequer text? Reginald Lennard assumed this to be so—'Domesday Plough-teams: the South-Western Evidence', *Eng. Hist. Rev.* LX (1945), p. 218.

APPENDIX III

SUMMARY OF THE DOMESDAY BOOK FOR
THE SOUTH-WESTERN COUNTIES

County	Assessment	Plough-lands	Plough-teams
Wiltshire	Hides, virgates and occasional acres 5-hide unit in about 30% of places One carucate at one place	Normally *Terra est n carucis*	Understocking in 37% of entries Excess teams in 6% of entries
Dorset	Hides, virgates and occasional acres 5-hide unit in about 20% of places Some non-gelding carucates	E.D. Normally *hanc terram possunt arare n carrucae* D.B. Normally *Terra est n carucis*	Understocking in 46% of entries Excess teams in 8% of entries
Somerset	Hides, virgates, ferlings and occasional acres 5-hide unit in about 20% of places Some non-gelding carucates	As for Dorset	Understocking in 51% of entries Excess teams in 14% of entries
Devon	Hides, virgates, ferlings and (once) acres 5-hide unit rare Carucates at 4 places	As for Dorset	Understocking in 71% of entries Excess teams in 6% of entries
Cornwall	Hides, virgates, ferlings and acres Hide comprised only 12 acres 5-hide unit rare Distinction invariably made between assessment and gelding hides	As for Dorset	Understocking in 94% of entries Excess teams in only one entry

Population	Values	No. of Place-names
Main groups: villeins, bordars, serfs, coscets, and, far below, cottars and coliberts 87 swineherds 54 unnamed thegns for 1066	Normally for 2 dates *Firma noctis* on 6 manors, commuted on 2 by 1086 *Firma noctis* unmentioned on another 2 but apparently commuted by 1086	335
Main groups: bordars, villeins, serfs, and, far below, cottars and coscets 56 salt-workers 3 *ancillae* 340 unnamed thegns for 1066	Normally for 2 dates *Firma noctis* on 5 groups of 21 manors and at 3 boroughs, all uncommuted in 1086	319
Main groups: villeins, bordars, serfs, and, far below, cottars and coliberts 84 swineherds 307 unnamed thegns for 1066	Normally for 2 dates *Firma noctis* on 5 groups of 12 manors, all commuted by 1086	611
Main groups: villeins, bordars, serfs As many as 370 swineherds and 61 salt-workers One *ancilla* 141 unnamed thegns for 1066	Normally for 2 dates *Firma noctis* on one group of 3 places, commuted by 1086	983
Main groups: bordars, villeins, serfs 40 *cervisarii* 17 unnamed thegns for 1066	Normally for 2 dates Decrease in value on three quarters of holdings	330

County	Wood	Meadow	Pasture	Marsh
Wiltshire	Normally in linear dimensions but frequently in acres Underwood rare	Normally in acres Amounts usually under 30 acres Occasional linear dimensions	For 75% of places Some linear dimensions, some acres	None
Dorset	Normally in linear dimensions, but frequently in acres Some underwood	As for Wiltshire	As for Wiltshire	None
Somerset	Normally in linear dimensions, but frequently in acres Frequent underwood	Normally in acres Amounts usually under 30 acres Once in linear dimensions	For 50% of places Normally in acres but frequently in linear dimensions Common pasture for 2 places	*Morae* for 11 places. Normally in acres or linear dimensions
Devon	Normally in acres but frequently in linear dimensions Frequent underwood	As for Wiltshire	For 75% of places Normally in acres but frequently in linear dimensions Common pasture for 12 places	None
Cornwall	Normally in acres but frequently in linear dimensions Some underwood	Rare. Entered for only 42 places Amounts usually 1 or 2 acres	For nearly 90% of places Normally in acres but frequently in linear dimensions	None

Fisheries	Salt-pans	Waste	Mills	Churches
None	None	None But waste houses at Malmesbury	For 197 places Usually with money renders Frequent fractions	For 29 places
For 3 places Fishermen at 2 places; eel render from mills at a third place	For 5 places Salt-workers at 3 of these Some money renders	In 1086 for 2 places Also waste houses at Bridport, Dorchester, Shaftesbury, Wareham	For 166 places Usually with money renders Occasional fractions	For 12 places
For 10 places Some money renders; one eel render	None	In 1086 for 2 places Also waste houses at Bath. At an earlier date for 7 places	For 250 places Usually with money renders Occasional fractions	For 17 places
For 16 places Some money renders; one salmon render	For 28 places Salt-workers at 6 of these Some money renders; two renders in kind	In 1086 for 7 places. Also waste houses at Barnstaple, Exeter, Lydford At an earlier date for 5 places; 9 other places, wasted by Irishmen, had recovered or partially recovered by 1086	For 80 places Usually with money renders Only one fraction	For 9 places
None	For one place	In 1086 for 2 places	For only 5 places, all with money renders	None

County	Miscellaneous	Boroughs
Wiltshire	Forests for Chute, Downton, Grovely, Laverstock, Melchet, Milford Vineyards for Bradford on Avon, Lacock, Tollard Royal, Wilcot Market for Bradford on Avon Mint for Malmesbury *Ferraria* for Fifield Bavant E.D. entries for only one place, and for 2 others in Somerset folios	Bedwyn Bradford on Avon Calne Cricklade Malmesbury Marlborough Salisbury Tilshead Warminster Wilton
Dorset	Forest for Horton (*foresta de Winburne*) Vineyards for Durweston, Wootton Moneyers in 1066 for Bridport, Dorchester, Shaftesbury, Wareham Castle for Corfe Castle (in Kingston manor) *Bruaria* for Boveridge E.D. entries for 12 fiefs (140 places) and for the boroughs	Bridport Dorchester Shaftesbury Wareham Wimborne Minster
Somerset	Park at Donyatt Vineyards for Glastonbury, North Curry, Meare, Panborough, and for Muchelney *et al.* Markets for Crewkerne, Frome, Ilchester, Ilminster, Milborne Port, Milverton, Taunton Mints for Bath, Taunton Castles for Dunster, Montacute	Axbridge Bath Bruton Frome Ilchester Langport Milborne Port Milverton Taunton
Devon	*Parcus bestiarum* for Winkleigh Markets for Okehampton, Otterton Castle for Okehampton *Morae* for Molland and Sherford in Brixton (presumably upland moors) E.D. lacks 46 entries for 45 places and one anonymous holding	Barnstaple Exeter Lydford Okehampton Totnes
Cornwall	Markets for Bodmin, Liskeard, St Germans, St Stephens by Launceston, Trematon Annual fair at Methleigh Castles at Launceston, Trematon	Bodmin

SOUTH-WESTERN COUNTIES
Summary of Rural Population in 1086

This summary includes the apparently rural element
in the boroughs—see the respective county summaries.

A. Total Figures

	Villeins	Bordars	Cottars	Serfs	Coscets	Coli-berts	Others	Total
Wiltshire	3,418	2,766	276	1,550	1,385	232	108	9,735
Dorset	2,569	3,032	196	1,243	182	33	89	7,344
Somerset	5,239	4,743	390	2,106	56	208	116	12,858
Devonshire	8,519	4,876	36	3,323	70	32	452	17,308
Cornwall	1,702	2,421	—	1,148	—	49	40	5,360
Total	21,447	17,838	898	9,370	1,693	554	805	52,605

B. Percentages

	Villeins	Bordars	Cottars	Serfs	Coscets	Coli-berts	Others
Wiltshire	35·1	28·4	2·9	15·9	14·2	2·4	1·1
Dorset	34·9	41·3	2·7	16·9	2·5	0·5	1·2
Somerset	40·8	36·9	3·0	16·4	0·4	1·6	0·9
Devonshire	49·2	28·2	0·2	19·2	0·4	0·2	2·6
Cornwall	31·8	45·1	—	21·4	—	0·9	0·8
Total	40·8	33·9	1·7	17·8	3·2	1·1	1·5

SOUTH-WESTERN COUNTIES
General Summary

For the various doubts associated with individual figures, see the text. The assessment,
plough-lands, plough-teams and rural population of the boroughs are included in
these totals, but it must be noted that the information given for the boroughs is often
very fragmentary. The assessment is for 1066, and includes non-gelding hides.

	Settle-ments	Assess-ment	Plough-lands	Plough-teams	Rural pop.	Boroughs
Wiltshire	335	3,893 h	3,360	2,926	9,735	10
Dorset	319	2,351 h	2,275	1,842	7,344	5
Somerset	611	2,902 h	4,775	3,886	12,858	9
Devonshire	983	1,159 h	7,932	5,759	17,308	5
Cornwall	330	428 h	2,563	1,218	5,360	1
Total	2,578	10,733 h	20,905	15,631	52,605	30

h = hides

There were also carucates and non-gelding carucates for all counties except Cornwall.

EXTENSION AND TRANSLATION OF FRONTISPIECE

(Folio *438* of the *Liber Exoniensis*)

EXTENSION

TERRA WILLELMI DE OU IN SUMERSETA

Willelmus habet i mansionem quae vocatur Watelega quam tenuit Alestanus die qua rex Edwardus fuit vivus et mortuus. Et reddidit Gildum pro i hida. Hanc potest arare i carruca. Ibi habet Willelmus ii villanos et vi quadragenarias nemoris in longitudine et iiii in latitudine, et valet per annum x solidos et quando Willelmus recepit valebat tantundem.

Willelmus habet i mansionem quae vocatur Hantona quam tenuit Alestannus die qua rex Edwardus fuit vivus et mortuus et reddidit gildum pro xiii hidis. Has possunt arare xii carrucae. Inde habet Willelmus v hidas et iiii carrucas in dominio et villani viii hidas et x carrucas. Ibi habet Willelmus xvi villanos et xxiiii bordarios et v servos et xxxvi animalia et xliiii porcos et cc oves x minus et ii molendinos qui reddunt vii solidos et vi denarios et i leugam nemoris in longitudine et dimidiam in latitudine et lx agros prati, et valet xv libras et quando Willelmus recepit valebat xii libras.

Willelmus habet i mansionem quae vocatur Geveltona quam tenuit Alestanus de Boscoma die qua rex Edwardus fuit vivus et mortuus. Et reddidit Gildum pro viii hidis. Has possunt arare viii carrucae. Modo tenet Radulfus Blouuet de Willelmo. De his habet Radulfus iiii hidas in dominio et iii carrucas et villani habent iiii hidas et v carrucas. Ibi habet Radulfus vi villanos et iiii bordarios et totidem servos et ii roncinos et ii equas indomitas et xii animalia et xvi porcos et c oves et ii molendinos qui reddunt per annum xxx solidos et lxxxx agros prati et xl agros pascuae et valet per annum viiii libras et quando Willelmus recepit valebat tantundem. Huic mansioni sunt additae ii hidae terrae quas tenuerunt v tagni pariter...

TRANSLATION

LAND OF WILLIAM OF EU IN SOMERSET

William has one manor which is called Whatley [in Winsham], which Ælfstan held on the day on which King Edward was alive and dead. And it rendered geld for one hide. One team can plough this. There William has 2 villeins and 6 furlongs of wood in length and 4 in breadth, and it is worth 10*s.* a year, and as much when William received it.

William has one manor which is called Hinton St George, which Ælfstan held on the day on which King Edward was alive and dead, and it rendered geld for 13 hides. Twelve teams can plough these. Thence William has 5 hides and 4 teams in demesne and the villeins 8 hides and 10 teams. There William has 16 villeins and 24 bordars and 5 serfs and 36 'animals' and 44 swine and 200 sheep less 10 and 2 mills which render 7s. 6d. and one league of wood in length and a half in breadth and 60 acres of meadow, and it is worth £15, and when William received it, it was worth £12.

William has one manor which is called Yeovilton, which Ælfstan of Boscumbe [in Wiltshire] held on the day on which King Edward was alive and dead. And it rendered geld for 8 hides. Eight teams can plough these. Now Ralf Blouet holds it of William. Of these Ralf has 4 hides in demesne and 3 teams and the villeins have 4 hides and 5 teams. There Ralf has 6 villeins and 4 bordars and as many serfs and 2 rounceys and 2 unbroken mares and 12 'animals' and 16 swine and 100 sheep and 2 mills which render 30s. a year and 90 acres of meadow and 40 acres of pasture, and it is worth £9 a year and when William received it, it was worth as much. To this manor have been added two hides of land which five thegns held in parage...

INDEX

Abbas Combe, *see* Combe, Abbas and Temple
Abbots Bickington, 420
Abbotsbury, 106, 116, 125, 126
Abbotskerswell, 229, 257, 266, 384
Abbots Wootton, *see* Wootton, Abbots and Fitzpaine
Abdick Hundred, 143
Acres
    geld, 14–15, 17, 80–1, 84, 149, 150, 155, 236, 237, 239, 306, 310, 348, 349, 430
    linear, 34, 43, 102, 108, 176, 258, 266, 376, 384
    of marsh, 179, 185–6, 354, 432
    of meadow, 39, 41, 43, 104, 105, 179, 180, 262, 263, 328–9, 354, 380, 381, 432
    of pasture, 42, 43, 44, 45, 106, 108, 109, 181, 183–4, 264, 266, 267, 269, 330, 331, 332, 354, 382, 384, 385, 432
    of underwood, 36, 102, 177, 258 n., 259, 260, 325, 326, 372, 376
    of vineyard, 60, 125, 209
    of wood, 33, 34, 35, 99, 101, 103, 173, 174, 175–6, 256, 258, 325–6, 327, 354, 372, 376, 377, 432
Adber, 71, 94, 95 n., 100, 140, 374, 398
Addeston, *see* Winterbourne (on R. Till, Wilts.)
Adsborough, 171, 186
Afflington, 80, 107, 398
Affpuddle, 67, 73, 74, 95, 124, 397
Afton, 228
Albretesberga Hundred, 76
Alcombe, 182 n.
Aldbourne, 4 n., 26, 30, 40, 42, 50, 56, 64
Alderbury, 15, 49
Alderbury Hundred, 10, 37
Alderstone, 10
Alderton, 19, 22, 26 n., 45, 46, 52, 53
Alder-wood, 36, 258, 376
Aldwick, 143
Alexander, J. J., 295, 346
Alford, 163, 211, 412
Alfoxton, 140
Alhampton, 140, 174, 375, 402, 409

All Cannings, 40
Allen, R., 73, 76
Aller (in South Molton, Devon), 241 n.
Aller (in Upottery, Devon), 414
Aller (in Carhampton, Somerset), 142, 168, 408
Aller (in Sampford Brett, Somerset), 143, 400
Aller (near Somerton, Somerset), 143
Allerford, 210, 212
Al(l)eriga Hundred, 224, 278
All Hallows Farm, 73 n., 76, 97, 122
Allington (Dorset), 94
Allington (near All Cannings, Wilts.), 15 n., 26
Almer, 78
Almsworthy, 146, 185, 189, 190, 360, 411
Alston, 274
Altarnun, 304
Alton Pancras, 83
Alvacott, 313, 425
Alverton, 307, 321, 323, 345
Alvestone, 11 n.
Alwinestona, 232 n., 277
Amesbury, 4 n., 15 n., 30, 31, 33, 37, 48, 50, 64, 375
Ancillae, 90, 95, 247, 251, 431
Andersey, 151, 405
Anderson, *see* Winterborne (eastern group, Dorset)
Anderson, O. S., 224 n.
Angli, 22, 26, 28, 29, 92, 163, 352, 370
'Animals', 60, 123, 205, 206, 208, 286, 288, 337
Animals, 'otiose', 206
Anonymous holdings, 6, 8 n., 45, 47, 72, 81 n., 87 n., 91 n., 93, 101, 112, 113, 114, 135, 138 n., 139, 141, 168, 176, 184, 190, 192, 223, 228, 251, 273, 276, 277, 299, 341
Anstey, East and West, 256, 265, 417, 423
Ansty, 43
Antony, 312, 328, 352
Appendages of manors, 4, 31, 64, 70, 97, 120, 122, 138, 140, 200, 201, 203, 254, 356